# Carotenoids and Retinal Disease

# Carotenoids and Retinal Disease

Edited by
## John T. Landrum • John M. Nolan

CRC Press
Taylor & Francis Group
Boca Raton London New York

CRC Press is an imprint of the
Taylor & Francis Group, an **informa** business

MATLAB® is a trademark of The MathWorks, Inc. and is used with permission. The MathWorks does not warrant the accuracy of the text or exercises in this book. This book's use or discussion of MAT-LAB® software or related products does not constitute endorsement or sponsorship by The MathWorks of a particular pedagogical approach or particular use of the MATLAB® software.

Cover microphotograph of the macula, courtesy of Dr. Max Snodderly.

CRC Press
Taylor & Francis Group
6000 Broken Sound Parkway NW, Suite 300
Boca Raton, FL 33487-2742

© 2014 by Taylor & Francis Group, LLC
CRC Press is an imprint of Taylor & Francis Group, an Informa business

No claim to original U.S. Government works

Printed on acid-free paper
Version Date: 20130904

International Standard Book Number-13: 978-1-4665-0204-8 (Hardback)

This book contains information obtained from authentic and highly regarded sources. Reasonable efforts have been made to publish reliable data and information, but the author and publisher cannot assume responsibility for the validity of all materials or the consequences of their use. The authors and publishers have attempted to trace the copyright holders of all material reproduced in this publication and apologize to copyright holders if permission to publish in this form has not been obtained. If any copyright material has not been acknowledged please write and let us know so we may rectify in any future reprint.

Except as permitted under U.S. Copyright Law, no part of this book may be reprinted, reproduced, transmitted, or utilized in any form by any electronic, mechanical, or other means, now known or hereafter invented, including photocopying, microfilming, and recording, or in any information storage or retrieval system, without written permission from the publishers.

For permission to photocopy or use material electronically from this work, please access www.copyright.com (http://www.copyright.com/) or contact the Copyright Clearance Center, Inc. (CCC), 222 Rosewood Drive, Danvers, MA 01923, 978-750-8400. CCC is a not-for-profit organization that provides licenses and registration for a variety of users. For organizations that have been granted a photocopy license by the CCC, a separate system of payment has been arranged.

**Trademark Notice:** Product or corporate names may be trademarks or registered trademarks, and are used only for identification and explanation without intent to infringe.

**Visit the Taylor & Francis Web site at**
http://www.taylorandfrancis.com

**and the CRC Press Web site at**
http://www.crcpress.com

# *Dedication*

This volume is dedicated to Dr. Alan Howard, Wilkins Fellow of Downing College, Cambridge University, and Honorary Professor at University of Ulster. His career at Cambridge University spans more than 60 years. He has held positions in the Department of Investigative Medicine and Pathology as well as the Dunn Nutritional Laboratory, before founding his own laboratory at Papworth Hospital, Cambridge, until it closed in 1999. He has published over 250 original articles in medical literature and authored and/or edited 8 major books on coronary heart disease and obesity. He was cofounder of the *International Journal of Obesity* and was involved in the foundation of the International Congress on Obesity and the European Atherosclerosis Society. Dr. Howard's inspiration and sponsorship were critically important to the establishment of the Macular Carotenoids Conferences that led the participants to collaborate on the contributions found in this work. The Macular Carotenoids Conference held in July 2011 was the first in this ongoing series of conferences to be held at Downing College, Cambridge University. This conference was conceived by Dr. Howard and principally sponsored by the Howard Foundation, a charitable trust that supports biomedical research at the University of Cambridge. Founded by Dr. Howard in 1982, the Howard Foundation provides support and encouragement for research into nutritional sciences, particularly emphasizing efforts to understand the nutritional factors that influence the genesis and etiology of cardiovascular disease, obesity, and antioxidants including carotenoids and their functional role in the reduction of risk for AMD. The Howard Foundation also maintains a strong supportive relationship with Downing College at Cambridge University. In 1970, Dr. Howard invented the internationally acclaimed Cambridge Diet, a scientifically designed, very low calorie diet, as an effort to address obesity and consequent cardiovascular heart disease. In 1984, he established a highly successful business venture,

Cambridge Nutrition Ltd, to internationally market the Cambridge Diet (currently "Cambridge Weight Plan®"), which has continuously supported the work of the Howard Foundation. In 2005, Cambridge Nutrition Ltd was acquired by investors providing resources that the Howard Foundation committed to construction of the Howard Theatre (2009), which is now the home site of the Macular Carotenoids Conferences, the Conference on the Science and Economics of Climate Change, and the upcoming "Blue Skies" meeting on prostate cancer, among a long list of other scholarly and artistic events. The Howard Foundation has also funded the recently renovated car park and new Howard Gate, complementing two other buildings at Downing College: the Howard Building (1987) and Howard Lodge (1994).

Dr. Howard's career is unique in that he has successfully bridged the divide between academic scholarship and research and entrepreneurship, making the benefits of his scientific accomplishments available to the public. Trained as a nutritionist in Cambridge at the Medical Research Council's Dunn Nutritional Laboratory in the 1950s, his lifelong research interests have been in the field of nutrition, especially in nutritional relationships associated with coronary heart disease and the treatment of obesity. He has served the research community devoting his expertise to the organization of international and national meetings, an accomplishment for which he has earned widespread renown. Between 1960 and 1995, he was a founding member and first secretary of the European Atherosclerosis Discussion Group; organized the First International Symposium on Atherosclerosis in Athens; a founding member and secretary of the Association for the Study of Obesity (United Kingdom); a joint organizer of the First International Congress on Obesity in London; founded and edited the *International Journal of Obesity*; organized satellite meetings on very low calorie diets in Ischia, Cambridge, and Kyoto; founded and organized the First International Symposium on Clinical Nutrition in London; and acted as secretary and then chairman of the Food Education Society (United Kingdom).

Following a remarkably productive research career in the Departments of Pathology and Medicine, University of Cambridge, Dr. Howard established a private laboratory at nearby Papworth Hospital where he continued to conduct research in cardiovascular heart disease, especially antioxidants. A major project completed by Dr. Howard at the Papworth Hospital was a comparison of the dietary habits of people from Toulouse, France, where the incidence of cardiovascular heart disease was reported to be notably low, particularly when compared to that observed for people from Belfast in Northern Ireland, where cardiovascular heart disease is very high. He collaborated in this work with Cambridge biochemist Dr. David Thurnham (later Howard Professor at the University of Ulster) to apply modern analytical methods to the analysis of vitamin E and carotenoids among the populations of Toulouse and Belfast. Both cities, although similarly highly industrialized, have populations whose food preferences differ significantly. Toulouse is renowned for its *Le Cordon Bleu*, a high animal fat cuisine, consumed with large amounts of fresh fruit and vegetables, and locally produced red wine. This contrasts with the average diet among residents of Belfast, which also includes high levels of animal fats, but is characterized by a limited and low intake of fruit and vegetables, and the principal alcoholic beverage is beer. The chief finding of their collaboration was that the average concentration of the carotenoid lutein in plasma of Toulouse residents was twice as high as that of Belfast residents (Howard et al. 1996). Dr. Howard concluded that the so-called "French Paradox" was due to the high consumption of fruit and vegetables (especially spinach with its high content of lutein) and red wine, which contains potent antioxidant polyphenols.

Pursuing this research interest, Dr. Howard attended the 1995 Gordon Conference on Carotenoids at Ventura in California, where he met Drs. Richard Bone and John Landrum and became aware of their work on the macular carotenoids, lutein, zeaxanthin, and *meso*-zeaxanthin, and macular degeneration. Dr. Howard initiated a collaboration to explore the potential benefits of *meso*-zeaxanthin for the putative prevention and treatment of AMD with Professors Bone and Landrum following a visit to Florida International University in Miami. This collaboration culminated in the successful commercialization of supplements containing *meso*-zeaxanthin produced from marigold petals by Industrial Orgánica SA of Monterrey, Mexico. Dr. Howard has since expanded his support of clinical research through development of collaboration with Professor Stephen Beatty, an ophthalmologist, and Professor John Nolan, a nutrition scientist, and the Macular Pigment Research Group at Waterford Institute of Technology in Waterford, Ireland. Dr. Howard's efforts and support of research continue to further the understanding of the functional nature of carotenoids in the human eye, especially that of *meso*-zeaxanthin in protecting the retina from AMD. In July 2013, Dr. Howard hosted the Second Macular Carotenoids Conference, which convened at the Howard Theatre in Downing College, bringing together the international macular carotenoids community.

At the 2009 commemoration of the 800th anniversary of the founding of Cambridge University celebrated at Buckingham Palace, Dr. Howard was recognized by the Chancellor of Cambridge University, the Duke of Edinburgh, for his dedication and tenacity and honored by the award of the Chancellor's 800th Anniversary Medal for Outstanding Philanthropy.

## REFERENCE

Howard, AN; Williams, NR; Palmer, CR; Cambou, JP; Evans, AE; Foote, JW; Marques-Vidal, P; McCrum, EE; Rudavets, JB; Nigdikar, SV; Radjput-Williams, J; Thurnham, DL. Do hydroxy-carotenoids prevent coronorary heart disease? A comparison between Belfast and Toulouse. *Int J Vitam Res*. 1996; 66(2); 113–18.

# Contents

Foreword ......................................................................................................................xi
Acknowledgments ........................................................................................................xv
Editors .......................................................................................................................xvii
Contributors ...............................................................................................................xxi

**Chapter 1** Macular Pigment: From Discovery to Function ...................................1

John T. Landrum, Richard A. Bone, Martha Neuringer, and Yisi Cao

**Chapter 2** Risk Factors for Age-Related Macular Degeneration and
Their Relationship with the Macular Carotenoids ............................23

Tos T. J. M. Berendschot

**Chapter 3** Epidemiology and Aetiopathogenesis of Age-Related Macular
Degeneration ......................................................................................41

Sobha Sivaprasad and Phil Hykin

**Chapter 4** Relationships of Lutein and Zeaxanthin to Age-Related
Macular Degeneration: Epidemiological Evidence ...........................63

Julie A. Mares

**Chapter 5** Clinical Trials Investigating the Macular Carotenoids .....................75

Sarah Sabour-Pickett, John M. Nolan, and Stephen Beatty

**Chapter 6** The Promise of Molecular Genetics for Investigating the
Influence of Macular Xanthophyllys on Advanced Age-Related
Macular Degeneration .......................................................................93

John Paul SanGiovanni and Martha Neuringer

**Chapter 7** A Review of Recent Data on the Bioavailability of Lutein and
Zeaxanthin .......................................................................................129

Mareike Beck and Wolfgang Schalch

**Chapter 8** Multiple Influences of Xanthophylls within the Visual System ....... 147

Billy R. Hammond, Jr. and James G. Elliott

**Chapter 9** Transport and Retinal Capture of the Macular Carotenoids............ 171

*Binxing Li and Paul S. Bernstein*

**Chapter 10** Measurement and Interpretation of Macular Carotenoids in Human Serum ................................................................................. 187

*David I. Thurnham, Katherine A. Meagher, Eithne Connolly, and John M. Nolan*

**Chapter 11** Xanthophyll–Membrane Interactions: Implications for Age-Related Macular Degeneration ................................................. 203

*Witold K. Subczynski, Anna Wisniewska-Becker, and Justyna Widomska*

**Chapter 12** Light Distribution on the Retina: Implications for Age-Related Macular Degeneration ..................................................................... 223

*Richard A. Bone, Jorge C. Gibert, and Anirbaan Mukherjee*

**Index** .................................................................................................................. 235

# Foreword

This volume had its origins among the organizers and contributors at the first Macular Carotenoids Conference held at Downing College, Cambridge University, in July 2011. That conference provided an open forum for many eminent international investigators in this field. Contributions from this international conference have been updated and "fleshed out." As a result, *Carotenoids and Retinal Disease* presents an up-to-date, thorough, and accessible volume devoted to the chemistry, pathobiology, visual science, and medical and public health import of the macular carotenoids.

What purposes are served by a reference on carotenoids and retinal disease? Pioneering studies beginning in the 1980s showed macular pigment is entirely composed of lutein, zeaxanthin, and the isomer *meso*-zeaxanthin. The broader health significance of these "retinal carotenoids" became evident from seminal observations of a reproducible, inverse relationship between macular pigment density and advance of age-related macular degeneration (AMD), the most common cause of acquired blindness for much of the world. Moreover, macular pigment is essential to maintain visual performance including dark adaptation, contrast, and recovery from photostress, obvious advantages during human evolution. Unlike provitamin A carotenoids (α-carotene, β-carotene, α-cryptoxanthin, and β-cryptoxanthin), the macular pigments possess distinctive vitamin A–*independent* visual functions. Lutein and zeaxanthin are transported across the retinal pigment epithelium to accumulate in neuroretina cells including photoreceptor outer segments. In the neuroretina, they assume a cooperative functionality with colocalized omega-3 long-chain polyunsaturated fatty acids. Photoreceptors, the body's most rapidly replenishing cell type, are exposed to harsh conditions of oxidative and actinic stress. This book advances our understanding about how macular carotenoids prevent otherwise inevitable damage and vision loss. This centrality for eye health, as well as the deleterious effects of a deficiency state, supports the standing of macular carotenoids as vitamins. To date, this health link is unique among carotenoids.

Initial epidemiological and case cohort studies of macular carotenoid intakes and protection from retinal disease prevalence/severity have led to dietary supplementation trials. Additionally, dietary modification is increasingly promoted and supplementation is becoming common in several countries. These dietary supplements may have significant public health benefits.

This book spans the breadth of numerous disciplines from the lab bench to population-based investigation. Black-and-white and color figures and photographs enliven the text and enforce concepts of retina topography, visual processing, genetic and cell signal pathways, disease etiopathogenesis, and epidemiology. Chapter 1 traces eighteenth-century observations of a yellowish region (the *macula lutea*) through increasingly refined retinal characterization in the nineteenth and twentieth centuries to the functional attribution of the macular carotenoids during recent decades. This chapter also introduces essentials of comparative macular anatomy.

Chapter 5 elegantly presents the importance of macular pigment from the perspective of vision science. The complex visual interactions of regional (macular) distribution of lutein and zeaxanthin, photoreceptor density, and light exposure in AMD are discussed in Chapter 12.

Understanding the highly selective ocular accumulation of lutein and zeaxanthin—from among the dozens of dietary carotenoids—requires insights derived from chemistry (Chapters 1, 7, 8, 11, and 12), visual science (Chapters 1, 8, and 9), and bioavailability studies (Chapter 7). Their transport and metabolism are intertwined with lipid biology and they appear to be trapped in selective sites by the recently discovered lutein- and xanthophyll-binding proteins (Chapter 9). Chapter 11 explains the high membrane solubility of these dipolar, terminally dihydroxylated molecules, and the distinctive perpendicular orientation in retinal membrane bilayers that maximizes light quenching and free radical scavenging activities. Technical advances in xanthophyll processing and insights derived from xanthophyll-free animal models are reviewed in Chapter 7. Chapter 10 explains contemporary separation and isolation methods for lutein, zeaxanthin, and *meso*-zeaxanthin by high-performance liquid chromatography.

Key sections of this volume provide a comprehensive treatment of the nutritional, lifestyle, and genetic risks for AMD. Chapters 2 and 3 trace AMD risk factors, epidemiology, pathogenesis, and classifications. Chapter 4 recounts recent longitudinal studies and clinical trials that revealed the relationship between dietary lutein and zeaxanthin in advanced AMD. Treatment of at-risk and affected individuals is potentially crucial to avoid visual loss (Chapters 4 and 5). Whether xanthophyll supplementation decreases risk of vision loss in AMD remains uncertain. However, clinical trials show supplements can improve visual performance and macular pigment optical density. Results from a large multicenter controlled, randomized trial, the Age-Related Eye Disease Study 2, may answer more fully unresolved questions.

Major AMD risk factors are poor nutrition, smoking, lifestyle, and genetics. Hypothesis-driven gene studies and genome-wide association studies are advancing this field rapidly. Chapter 6 recounts the current status of genetics of macular pigment and AMD and numerous chapters reference genetic information. Gene polymorphisms and deleterious gene mutations have been discovered in many steps in macular xanthophyll absorption, transport, retinal uptake, accumulation, and association with processes implicated in AMD pathogenesis. Chapter 6 also describes genetic and bioinformatics strategies essential to this research. Data emerging from large-scale genotyping projects undoubtedly will enhance approaches to macular carotenoids in retinal disease.

New advances in the relevant science and health context assuredly will become more prominent in coming years. Intriguing evidence suggests lutein and zeaxanthin play a wider role in suppressing severity of other common neovascular retinopathies, namely, diabetic retinopathy and retinopathy of prematurity. The macular pigments may have a role, currently anecdotal, as adjunct therapies in several heritable retinal diseases. Finally, the neuroretina is an extension of the brain and investigations show lutein and zeaxanthin also accumulate in certain brain regions. Preliminary studies show cognitive improvements after xanthophyll supplementation, plausibly through some similar mechanisms.

In summary, this book is a valuable reference for all those who are investigators or treat people at risk of vision loss. Investigators will recognize areas in which research and application are most needed. Clinicians from several disciplines will be instructed and stimulated to further interdisciplinary collaborations on prevention and treatment of retinal disease.

**Lewis P. Rubin**
*Texas Tech University Health Sciences Center at El Paso*

MATLAB® is a registered trademark of The MathWorks, Inc. For product information, please contact:

The MathWorks, Inc.
3 Apple Hill Drive
Natick, MA 01760-2098 USA
Tel: 508 647 7000
Fax: 508-647-7001
E-mail: info@mathworks.com
Web: www.mathworks.com

# Acknowledgments

John T. Landrum and John M. Nolan are greatly appreciative of the assistance that Sarah O'Regan has provided throughout this project. Her organizational contribution and coordination have been essential to their successful collaboration. John T. Landrum especially wishes to thank Eileen Landrum for her extraordinary patience, encouragement, and tolerance with his commitment to this project. John M. Nolan would like to thank Jane Nolan for her continued support and encouragement during this project.

# Editors

**John T. Landrum, PhD**, is a professor in the Department of Chemistry and Biochemistry at Florida International University (FIU) where in addition to his role as a faculty member he serves as a director at the Office of Pre-Health Professions Advising for the College of Arts and Sciences. He joined the faculty at the FIU in August 1980.

Dr. Landrum earned BS and MS degrees in chemistry (*cum laude*) from California State University, Long Beach, in 1975 and 1978, respectively, and completed his MS thesis ("The cooperative binding of oxygen by hemocyanin"). In 1980, he completed his PhD in chemistry at the University of Southern California (USC). He was recognized by the USC for his graduate research in 1978 and was awarded the USC Graduate Research Award for Outstanding Research. His PhD dissertation ("Synthetic models toward cytochrome c oxidase") focused on the study of small molecular models to investigate structural and magnetic properties of porphyrin complexes to provide fundamental insight into possible structures of the two copper, two iron active sites of the terminal electron acceptor of the electron transport chain.

FIU was newly established when Dr. Landrum joined the faculty there and he has taught courses at all academic levels within the Department of Chemistry and Biochemistry. His accomplishments in teaching were honored with an Excellence in Teaching Award in 1991. He was instrumental in establishing a Master of Science degree program in chemistry at FIU and served as the first graduate program director (1987–1992). Dr. Landrum served as an associate dean of the FIU graduate school between 2006 and 2007. In 2008, he was invited to assume his current position as a director of the College of Arts and Sciences' Office of Pre-Health Professions Advising. Upon arrival at the FIU, he established an active research program involving undergraduates in studies of porphyrin metal complexes as models for the biological function of transition metals in natural systems. His interest in carotenoids and their functions in biological systems was triggered by a collaboration with Dr. Richard A. Bone (Department of Physics, FIU), which began in the early 1980s and led to the first definitive characterization of the human macular pigment.

His research efforts over the past 30 years have been primarily devoted to understanding the nature of the carotenoids present in the human macula, including their identity, distribution, transport, and metabolism. Over this period, he and his collaborators have shown that the macular pigment is composed of the carotenoids lutein, zeaxanthin, and *meso*-zeaxanthin. He has been able to demonstrate that these carotenoids have a protective function within the retina where they reduce the risk of age-related macular degeneration, the leading cause of vision loss among adults, and that dietary supplements of these carotenoids can increase pigmentation. His current research efforts are focused on understanding the mechanisms of biological recognition of individual carotenoids, their absorption and transport, and their role in the developing human eye.

In 2004, Dr. Landrum's contributions in the field of chemistry were recognized by FIU with an award for Excellence in Research. Since becoming a faculty member at FIU, he has been awarded numerous grants in support of his research efforts. He has directed the research of 16 graduate and over 100 undergraduate students and has authored or coauthored 66 articles and chapters in peer reviewed journals and books. In addition to the current volume, he is the editor of *Carotenoids: Physical, Chemical, and Biological Functions and Properties*. He has frequently been an invited speaker and has presented his research to audiences at more than 40 major international conferences and symposia since the early 1990s. He is the president-elect of the International Carotenoid Society. In 2004, he served as a vice-chairman for the Gordon Research Conference on Carotenoids and in 2007 he served as a chairman for this prestigious conference. He served as a chairman for the Macula and Nutrition Group (2000–2004); as a chairman (2008) and steering committee member of the Carotenoid Interactive Research Group (2002–2008); as a council member and treasurer for the International Carotenoid Society (2005–present); and as an associate editor for *Archives of Biochemistry and Biophysics*. He has also served as an editor or coeditor for several special editions on the current progress in the field of carotenoid research for the journal *Archives of Biochemistry and Biophysics*. Dr. Landrum is a member of the American Chemical Society; the Association for Research in Vision and Ophthalmology; the International Carotenoid Society (founding member); the Carotenoid Interactive Research Group; the International Research Society, Sigma Xi; and the Macula and Nutrition Group (founding member).

**John M. Nolan, PhD**, is a Fulbright Scholar, Howard and European Research Council (ERC) Fellow, adjunct professor of Trinity College Dublin, principal investigator of the Macular Pigment Research Group, Waterford Institute of Technology, Ireland.

Professor Nolan's research career began in 2002 at the Waterford Institute of Technology, Ireland, with his PhD entitled "Determinants of macular pigment in healthy subjects." This work focused on the study of a nutritional pigment at the back of the eye, referred to as macular pigment (MP), and its relationship with risk factors for age-related macular degeneration (AMD), the leading cause of age-related blindness in the developed world. This was the largest ever cross-sectional study of its type, and one of the most important findings to emanate from this research was that healthy middle-aged offspring of patients with AMD have a significant lack of MP. This novel observation suggested that a lack of this dietary pigment at the back of the eye may contribute to the increased risk of AMD among the sons and daughters of sufferers of this disease. The findings from his work initiated important study of AMD risk factors and their link with MP in the normal population.

In 2005, Professor Nolan was awarded his PhD and in the same year was awarded an Irish–American Fulbright scholarship for designing a study to further this research initiative. This new study was an initial investigation into the relationship between MP and foveal (retinal) architecture and identified, for the first time, that a distinctive feature of foveal architecture, namely, foveal width, determines, in part, a person's MP level.

In 2012, Professor Nolan was named a Howard Fellow by the Howard Foundation, Cambridge, United Kingdom, in recognition of his research into nutrition and AMD.

This fellowship supports his continued research in the Macular Pigment Research Group (MPRG) and provides a strong collaboration with Dr. Alan Howard and the Howard Foundation. Professor Nolan is also chairman of the International Macular Carotenoids Conference (www.macularcarotenoids.org), held at the Howard Foundation, Downing College, Cambridge University, United Kingdom.

In 2013, Professor Nolan was named an adjunct professor of Trinity College Dublin in recognition of his role in directing the AMD component of The Irish Longitudinal Study on Ageing (TILDA). TILDA is the most detailed study on aging ever undertaken in Ireland and is investigating health, lifestyle, and financial situations of up to 10,000 people in Ireland as they grow older, including generating unique 25-year longitudinal scientific data on MP and AMD on this selected sample.

Professor Nolan is now the principal investigator of the MPRG (www.mprg.ie), a research group that he established in 2006 with his colleague, Professor Stephen Beatty, a consultant ophthalmologist. The MPRG leads world-class research initiatives in the role of eye nutrition for vision and prevention of blindness and is now the largest group worldwide studying macular carotenoids. The group adopts a multidisciplinary research approach to vision science, by incorporating expertise in biochemistry, ophthalmology, optometry, nutrition, genetics, and statistics. Professor Nolan has secured over €5 million in research funding to support ongoing studies and has successfully supervised 10 students to MSc, PhD, and MD level qualifications to date.

A significant and current research interest at the MPRG is the study of *meso*-zeaxanthin, the central macular carotenoid. It is because of the unique potential of this particular carotenoid that the Howard Foundation and the European Research Council are actively supporting studies within the group. Indeed, one of the most important findings of Professor Nolan's research career to date was that subjects who have atypical (and undesirable) central dips in their MP spatial profile can, uniquely, normalize these dips following enrichment with this central carotenoid, *meso*-zeaxanthin.

# Contributors

**Stephen Beatty**
Macular Pigment Research Group
Department of Chemical and Life
 Sciences
Waterford Institute of Technology
Waterford, Ireland

**Mareike Beck**
DSM Nutritional Products Ltd.
Human Nutrition and Health
Kaiseraugst, Switzerland

**Tos T. J. M. Berendschot**
University Eye Clinic Maastricht
Maastricht, the Netherlands

**Paul S. Bernstein**
Department of Ophthalmology and
 Visual Sciences
Moran Eye Center
University of Utah School of Medicine
Salt Lake City, Utah

**Richard A. Bone**
Department of Physics
Florida International University
Miami, Florida

**Yisi Cao**
Department of Chemistry and
 Biochemistry
Florida International University
Miami, Florida

**Eithne Connolly**
Macular Pigment Research Group
Department of Chemical and Life
 Sciences
Waterford Institute of Technology
Waterford, Ireland

**James G. Elliott**
DSM Nutritional Products, Inc.
Parsippany, New Jersey

**Jorge C. Gibert**
Department of Physics
Florida International University
Miami, Florida

**Billy R. Hammond Jr.**
Vision Sciences Laboratory
University of Georgia
Athens, Georgia

**Phil Hykin**
Medical Retina Department
Moorfields Eye Hospital
London, United Kingdom

**John T. Landrum**
Department of Chemistry and
 Biochemistry
Florida International University
Miami, Florida

**Binxing Li**
Department of Ophthalmology and
 Visual Sciences
Moran Eye Center
University of Utah School of
 Medicine
Salt Lake City, Utah

**Julie A. Mares**
Department of Ophthalmology and
 Visual Sciences
University of Wisconsin–Madison
Madison, Wisconsin

**Katherine A. Meagher**
Macular Pigment Research Group
Department of Chemical and Life Sciences
Waterford Institute of Technology
Waterford, Ireland

**Anirbaan Mukherjee**
Department of Physics
Florida International University
Miami, Florida

**Martha Neuringer**
Division of Neuroscience
Oregon National Primate Research Center
Oregon Health & Science University
Portland, Oregon

**John M. Nolan**
Macular Pigment Research Group
Department of Chemical and Life Sciences
Waterford Institute of Technology
Waterford, Ireland

**Sarah Sabour-Pickett**
Macular Pigment Research Group
Department of Chemical and Life Sciences
Waterford Institute of Technology
Waterford, Ireland

**John Paul SanGiovanni**
National Eye Institute
Clinical Trials Branch
Bethesda, Maryland

**Wolfgang Schalch**
DSM Nutritional Products Ltd.
Human Nutrition and Health
Kaiseraugst, Switzerland

**Sobha Sivaprasad**
Medical Retina Department
Moorfields Eye Hospital
London, United Kingdom

**Witold K. Subczynski**
Department of Biophysics
Medical College of Wisconsin
Milwaukee, Wisconsin

**David I. Thurnham**
Northern Ireland Centre for Food & Health (NICHE)
School of Biomedical Sciences
University of Ulster
Coleraine, United Kingdom

**Justyna Widomska**
Department of Biophysics
Medical University
Lublin, Poland

**Anna Wisniewska-Becker**
Department of Biophysics
Jagiellonian University
Krakow, Poland

# 1 Macular Pigment
## From Discovery to Function

John T. Landrum, Richard A. Bone,
Martha Neuringer, and Yisi Cao

### CONTENTS

1.1 Current Perspective on Macular Pigment ........................................................ 1
1.2 Early Reports of the Macular Pigment ........................................................... 2
1.3 Role of the Development of Ophthalmoscopic Tools in Characterizing
    Macular Pigment ........................................................................................... 4
1.4 Entoptical Phenomena, Color Perception, and the Macular Pigment ............... 6
1.5 Modern Measurements of Macular Pigment ................................................... 8
    1.5.1 Early to Mid-Twentieth Century: 1900–1963 ........................................ 8
    1.5.2 The Late-Twentieth Century through the Present: 1970–2013 ............... 9
1.6 Macular Pigment in the Developing Eye ....................................................... 12
1.7 Functional Role of Xanthophylls in the Human Retina ................................. 13
Acknowledgments .................................................................................................. 16
References .............................................................................................................. 16

## 1.1 CURRENT PERSPECTIVE ON MACULAR PIGMENT

An awareness of the presence of the "*macula lutea*" dates to a time during the last two decades of the eighteenth century, something more than two centuries ago (Nussbaum et al. 1981; Whitehead et al. 2006). Perhaps, it should be a surprise that fundamental facts needed to describe the macular pigments awaited discovery until the end of the twentieth century (Landrum and Bone 2001). A firm determination of the identities of the macular carotenoids, lutein and zeaxanthin, was established and confirmed during the mid-1980s providing the firm standing essential for further detailed research (Bone et al. 1985, 1988; Handelman et al. 1988). Since that time, the level of interest and enthusiasm that researchers have shown in the macular pigment has grown dramatically. The burgeoning enthusiasm underlying current research into the functional role of the macular pigments is also paralleled by an increasing interest that is now being devoted to the role of other non-vitamin A carotenoids. Bone et al. (1993) demonstrated that the rarely encountered carotenoid, meso-zeaxanthin, comprises approximately 30% of the total macular pigment within the central fovea. The ability of carotenoids to act as antioxidants within the body is now widely recognized, and it is accepted that the retina is subject to high oxidative loads (Landrum 2013). The Age-Related Eye Disease Study I (AREDS I) demonstrated that antioxidant consumption significantly retards the progression of early stages of age-related macular degeneration (AMD) and the development of wet neovascular AMD (SanGiovanni et al. 2007). The possibility that the macular pigment has a role in the preservation of central vision among the increasing population of seniors is ample motivation for the intense scrutiny to which the macular carotenoids are now subject. In the United States, the Age-Related Eye Disease Study II (AREDS II), in which lutein and zeaxanthin are specific components, is approaching conclusion and continued research on the transport, localization, and actions of the carotenoids into the macula is certain to remain an important and productive line of investigation for the foreseeable future (Chew et al. 2012). The potentially broad importance of the non-provitamin A carotenoids in maintaining optimal human health also beckons researchers who are investigating the actions and role of carotenoids and their metabolites in cell functions.

## 1.2 EARLY REPORTS OF THE MACULAR PIGMENT

In the absence of specialized equipment, a careful and detailed anatomical dissection of the eye is required to observe the macula and the macular pigment. A structure measuring approximately 5 mm in diameter, the macula, is typified by a thinning of the retinal layers that produces the fovea or retinal pit (Figure 1.1).

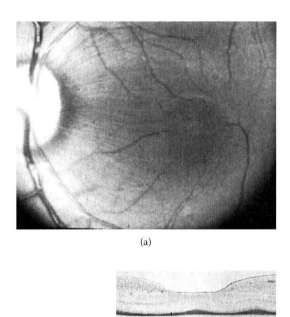

(a)

(b)

**FIGURE 1.1** (See color insert.) Microscopic cross sections of macula aligned with a fundus photograph of the macula. (a) The macaque fundus and (b) microscopic cross section of macaque fovea in natural light. (Photos courtesy of M. Neuringer and D. M. Snodderly. With permission.)

Everard Home (1798), in his secondhand report to the Royal Society of Samuel Thomas von Soemmerring's observations (*vide infra*) and augmented by his own, described the *macula lutea* as being three lines (1 line = 1/12 in., ~2 mm) in diameter. The investigations of von Soemmerring were communicated to Home by a Geneva surgeon, Maunoir. Home confirmed the report with his own firsthand observations. When viewed anteriorly with a modern ophthalmoscope, the fovea stands out and is bracketed above and below by the vascular arcade of the superior and inferior venous arches that radiate from the optic nerve head. The first report of the yellow pigmentation in the central human retina was made by Francesco Buzzi in 1782 and was followed by the independent and more detailed descriptions of von Soemmerring in 1795 (Buzzi 1782; von Soemmerring 1799, 1801). von Soemmerring was noted for the remarkable quality of his anatomical illustrations and described his observations of the macula in the following words, "In the center of the retina is a very conspicuous round hole with a golden yellow margin around which blood vessels form a beautiful wreath" (Söhn 1801; von Soemmerring 1801). By contrast to Soemmerring's belief in a central hole, Buzzi recognized that the central macula was a thin, but intact, section of the retina (Home 1798). According to Home, von Soemmerring may have confused the perception of the "blind spot" with this "hole" and suggested that it might regulate the retina's response to bright light by expanding and contracting. Such speculation provides a sense of the enormous strides that have been made in our expanding knowledge since that period. In Home's report to the Royal Society, he noted that Michaelis, an anatomist in Paris, failed to find either the central yellow spot or a "hole" in any animal other than humans. This comment reflects the extent to which researchers of the period were aware of and interested in the *macula lutea* and the possibility that it was an important fundamental characteristic of the human retina. Home described the *macula lutea* as being more readily visible in the eye two days after death, initiating a belief, that persisted well into the twentieth century, that this coloration must develop posthumously. As late as 1951, Hartridge remained a strong advocate of this perspective (Hartridge 1951). Home, a careful student of the literature, noted the works of Buzzi as well as those of Michaelis and Reil (Home 1798). Michaelis' work, as reported by Home, followed from that of von Soemmerring and extended his observations to the immature eye noting that the yellow spot is not visibly detectable in fetuses of seven to eight months or in newborn infants. Modern analytical methods have provided new insight on the pigmentation in young eyes (Bone et al. 1985, 1988). Both von Soemmerring and Michaelis observed a pale yellow spot in children, as young as 1 year (Home 1798). These physicians also noted that the pigmentation seemed to diminish with age and was lacking in diseased eyes. The extent to which macular pigment varies with age continues to be a point about which there is some controversy (Bone et al. 1988; Berendschot and van Norren 2004; Bernstein et al. 2004; Nolan et al. 2010). These eighteenth-century reports are strikingly accurate observations, and they were made at a time when our knowledge of the underlying science of physiology was in its infancy, limited by the absence of suitable tools and technology to readily enable reliable and repeatable experimentation and measurement. Such limitations led to reports of the type ascribed to Fermin in 1796. According to Freidenwald (1902), Fermin reported observing a luminosity associated with the retina, and in the first quarter of the nineteenth century some researchers even suggested that the retina was able to illuminate objects enabling individuals to read in the dark! Some confusion about the macular pigment seems to have arisen from the study of the nonhuman eye and the tendency of at least a few early researchers to generalize their observations of eyes from different species. In his efforts to reproduce accounts of the yellow spot in humans and to understand its absence among other animals, Home became one of the first to observe the *macula lutea* in nonhuman primates. He recounted his careful dissection of the eye of a "monkey" and his confirmation of the yellow macular pigment. Research using primates has been extensively utilized during the past 30 years, often providing opportunities to undertake experiments that would be impractical or even impossible to conduct in humans (Wolin and Massopust 1967; Malinow et al. 1980; Snodderly et al. 1984).

## 1.3 ROLE OF THE DEVELOPMENT OF OPHTHALMOSCOPIC TOOLS IN CHARACTERIZING MACULAR PIGMENT

During the early part of the nineteenth century, a number of reports expanded on the techniques useful to observe the posterior of the living eye. As early as 1824, Purkinje described a method to illuminate the living human retina by use of a concave mirror (Albert and Miller 1973; Weale 1994). In 1846, William Cumming published his technique and observations and, in 1847, Charles Babbage, having read Cumming's work, produced a simple ophthalmoscope. This accomplishment was belatedly described by Wharton-Jones in 1854 three years after von Helmholtz's description of his own invention, and von Helmholtz is now generally regarded as the inventor (Cumming 1846; von Helmholtz 1851, 1916; Lyons 1940; Wharton-Jones 1954). Babbage was the Lucasian Professor of Mathematics at Cambridge from 1828 to 1839 and is most well-known for his contributions to information science and the invention of a computing machine that enabled the production of accurate logarithm tables (Warrick 2007). von Helmholtz described his ophthalmoscope in an 1851 monograph while only 29 (von Helmholtz 1851). Albrecht von Graefe, also a young man at this time and soon to be the force behind the establishment of *Archiv für Ophthalmologie* in 1855, became an immediate devotee of von Helmholtz's new ophthalmoscope, reportedly shouting, "Helmholtz has unveiled a new world to us" (McMullen 1917). von Helmholtz reported on his observations of the macular pigment in his *Handbuch for Physiologischen Optik* published in German in two volumes in 1855 and 1866 and later translated to English in 1924 (von Helmholtz 1924). In von Helmholtz's original monograph on the ophthalmoscope, he recounts his first observations of the retina and accurately describes observing the yellow spot in the avascular foveal region (von Helmholtz 1951).

In 1866, Schultz described his observations of the macula emphasizing its yellow-orange color, a quality he attributed to the internal layers of the retina; he unsuccessfully attempted to extract the pigment, albeit with water (Schultz 1866). Schultz may have been the first to ascribe a function to the *macula lutea* with his suggestion that it could reduce chromatic aberration (see Chapter 8). Further descriptions of the macular pigment in postmortem eyes were described by Schwalbe and Schmidt-Rimpler in 1874. They correctly identified the reason that von Soemmerring, Buzzi, Home, and contemporaries only observed the *macula lutea* consistently in dissected eyes after some extended period postmortem. They reasoned that in freshly enucleated eyes the retina remained intimately associated with the underlying retinal pigment epithelium (RPE) and vascular choroid but that after a period it separates making the yellow carotenoids more easily detectable. Significant amounts of the light entering the eye are reflected from the sclera in the living eye as well as the freshly postmortem eye. Much of the light entering the eye passes through the entire depth of the retina and is reflected from the sclera after penetrating to the choroid. The observed reflected hue is dominated by the strong absorbance of hemoglobin resulting in the dominance of typical red reflection of the fundus and the "red-eye" artifact, which are the bane of modern photographers. In the postmortem retina, the photoreceptors have separated from the underlying RPE and the newly generated interface reflects forward a much larger percentage of the light, which passes through neural retina. This interface also serves to reflect backward, away from the observer, reddish light scattered forward from the sclera after it has passed through the choroid. Under these conditions, the absorbance of the yellow carotenoids within the inner retina contrasts more sharply with the surrounding, and essentially unpigmented, perifovea (Figure 1.2). Quantitative approaches to explaining the transmittance and reflectance of light within and through the eye have been a fundamental component to many advances during the latter part of the twentieth century (Berendschot et al. 2003). Models variously treat the retina in the living eye as composed of multiple absorbing and reflecting layers in which the macular pigment, the photopigments, hemoglobin, and the highly reflective "white" sclera account for the total fundus reflectance. One of the earliest approaches to modeling the reflectance of light during fundus observations was reported by Brindley and Wilmer (1952).

Descriptions of the bright yellow pigment in living eyes appear several times in the early twentieth century (Gullstrand 1906a,b; Chevalleraeu and Polyak 1907; Dimmer 1907; Hoeve 1912; Lottrup-Andersen 1913). Each reported that under favorable conditions, illumination with red-free light or in unique cases, where arterial occlusion resulted in diminished choroidal perfusion with hemoglobin, the *macula lutea* was readily visible. In a series of papers, Vogt reported a number of observations he made in red-free light, which provided sound support for the presence of macular pigment within the living retina (Vogt 1921a,b, 1937). As mentioned earlier, some controversy would remain concerning the presence of the macular pigment within the living retina until at least 1951 when Hartridge argued that this coloration arose from the presence of the blue-absorbing shortwave photoreceptors (Hartridge 1951). Wald later showed this to be incorrect, demonstrating that the shortwave photoreceptors are largely absent within the central fovea (Wald 1967).

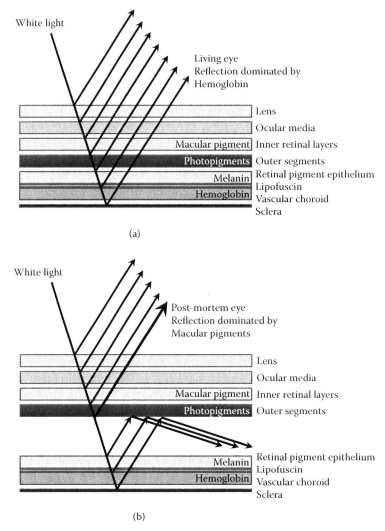

**FIGURE 1.2** A schematic diagram of the layered structure of the retina illustrating how light passing through the (a) living or (b) postmortem retina would be reflected and absorbed at various points accounting for an increased visibility of the *macula lutea* in the postmortem eye.

## 1.4 ENTOPTICAL PHENOMENA, COLOR PERCEPTION, AND THE MACULAR PIGMENT

Entoptical detection of the *macula lutea* and its effects on color matching has been repeatedly described by researchers. Haidinger's classic work on the detection of polarized blue light and the observation of Haidinger's "brushes or tufts" was published in 1844 (Haidinger 1844). Haidinger's brushes are readily observed when a subject's gaze is directed on a uniform and brightly illuminated source of white or blue light. An LED computer screen is a very effective source. Interposing a polarizing filter oriented with the plane of polarization from the screen will make observation of the brushes easiest. When a white background is used, the brushes are distinctly yellow. Using a deep blue background will produce a dark brush that can be somewhat difficult to distinguish against the dark background. Individuals with low levels of macular pigment find observing Haidinger's brushes exceedingly difficult or impossible, a fact that led to some confusion until quantitative *in vivo* assessments of macular pigment demonstrated that there is a wide intersubject variability and some subjects possess virtually no pigmentation. The descriptions of Haidinger and James Clerk Maxwell of entoptical effects are the most well-known reports of the nineteenth century (Maxwell 1855, 1856a,b). Maxwell's interest in light and color perception led him to investigate and carefully describe many entoptical phenomena including "Maxwell's spot," which was later commented on by von Helmholtz (1924). The variability of results in color-matching experiments noted by many investigators during the latter half of the nineteenth century and Maxwell's spot were consistently recognized as originating from the absorption of blue light by the macular pigment (Maxwell 1855; Maxwell 1856a; Hering 1893). A plausible explanation of Haidinger's polarized brushes was not addressed until the mid-1980s (Bone 1980; Bone and Landrum 1983, 1984). Bone and Landrum showed that xanthophyll molecules exhibit a high dichroic ratio and naturally orient when incorporated within membranes. The radial symmetry of the nerve axons of the fovea provides the organizing structure necessary for the observation of the "brushes" (Figure 1.3). In 1878, Ewald reported the yellow spot could be easily observed on first awakening if a subject fixates on a white background (Ewald 1878). Ewald correctly ascribed this observation to the *macula lutea*. Haab (1879) reported a similar observation a year later. Early in the next century, Kolmer (1936), studying the postmortem retina of freshly enucleated eyes, showed that the central yellow pigment was readily

**FIGURE 1.3 (See color insert.)** The macular carotenoids are organized by their preference to align perpendicularly to the membranes of the axons that are themselves radially displayed within the fovea. Polarized light is absorbed only by those molecules that are aligned generally parallel with the plane of polarization resulting in attenuation of the beam in the zone depicted with the hourglass shape producing the Haidinger's brush entoptical effect.

observable when the neural retina was removed, separated from the RPE, and placed on a glass slide. He noted observing the pigment only within the inner layers of the macula and was unsuccessful in his attempts to identify it.

## 1.5 MODERN MEASUREMENTS OF MACULAR PIGMENT

### 1.5.1 Early to Mid-Twentieth Century: 1900–1963

With the ophthalmoscope universally available to researchers and the advent of intense and easily controlled light sources during the first half of the twentieth century, many observers described detecting the macular pigment directly in the living eye, particularly when using red-free light sources. In 1922, Holm described observing the yellow pigment in a freshly enucleated eye in which the retinal layers still maintained their integrity (Holm 1922). Nevertheless, a small cadre persisted in the belief that the yellow color was associated with the sclera or some other structure of the eye (Segal 1950). Gullstrand was among these and argued that the *macula lutea* was a postmortem artifact and that it originated in the sclera, leaking into the retina after death (Gullstrand 1906a,b). Some confusion about the macular pigment may have originated from experiments on nonprimate eyes and the failure of some early researchers to appreciate the large differences between species (Krause 1934). Krause's popular text *The Biochemistry of the Eye* included data supporting the presence of xanthophyll in the bovine sclera but not retina, an inclusion that almost certainly led many students to discount the significance of the *macula lutea* in humans (Krause 1934). However, in 1933, Walls and Judd reported observing the yellow macular pigment and in their review, put forward the belief that the macular pigment must function identically to the oil droplets of birds (Walls and Judd 1933). They were among the first researchers of the twentieth century to ascribe a visually significant function to the *macula lutea*.

Visual psychophysical techniques depend on an observer's direct perception to measure an optical stimulus (Howells et al. 2011). In vivo macular pigment levels can be measured with high accuracy and over the entire visual spectrum utilizing a number of psychophysical approaches. These include heterochromatic flicker photometry (HFP), dichroic spectrometry, color matching, threshold sensitivity, and minimum motion spectrometry (Bone and Sparrock 1971; Stabell and Stabell 1980; Pease and Adams 1983; Pease et al. 1987; Werner et al. 1987; Bone et al. 1992; Moreland et al. 1998; Wooten et al. 1999; Beatty et al. 2000a; Bone and Landrum 2004; van der Veen et al. 2009). Ruddock (1963, 1972) investigated color matching and in 1963 demonstrated convincingly that the absorbance of blue light by the macular pigment strongly influences color matches made by observers when fixating on stimuli directed on the fovea but not when stimuli were directed into the parafovea, where it is now known that macular pigment concentrations are ~100× less than within the central fovea. Each of these psychophysical techniques differs to some extent in the underlying assumptions that influence the quality and magnitude of the measurement, but each depends directly on the absorbance of light by the macular pigment within the inner layers of the retina of the fovea as compared to that in the peripheral retina. Herbert Ives described the method of heterochromatic photometry in a seminal paper published in 1912 (Ives 1912; Buckley and Darrow 1956). (Some literature reports attribute this work to his father Frederick Ives, a highly successful businessman whose inventions greatly influenced the development of color printing and photography [http://www.nasonline.org/publications/biographical-memoirs/memoir-pdfs/ives-herbert.pdf].)

### 1.5.2 THE LATE-TWENTIETH CENTURY THROUGH THE PRESENT: 1970–2013

Bone applied HFP to measure macular pigment optical densities in 49 individuals as a function of wavelength in 1971 demonstrating the spectral properties of the macular pigment were consistent with that of the macular extract spectrum published by Wald in 1945 (Bone and Sparrock 1971). Bone's data convincingly showed that a large variability exists in the amount of pigmentation between subjects, evidence that, in the light of future work, indicates the influence that diet as well as genetic and environmental factors have on pigmentation. Subsequent investigators have pursued many refinements in the flicker technique and have contributed to the development of a fundamental theory and practices for measurement of macular pigment. Measurement of macular pigment by HFP can be now reliably compared and interpreted between research groups (Werner and Wooten 1979; Werner et al. 1987; Wooten et al. 1999; Beatty et al. 2000; Mellerio et al. 2002; Bone and Landrum 2004). HFP is now beginning to be adopted as a standard of good practice during clinical eye examinations (Jeffrey B. Morris, pers. comm.).

Activity in research, measurement, and characterization of the macular carotenoids accelerated in the early 1980s. Malinow et al. (1980) followed the early suggestion of Wald et al. (1941) that the macular pigment was of dietary origin. Malinow and coworkers demonstrated that the *macula lutea* does not develop in macaques whose diet is depleted in sources of xanthophylls. One outcome of this work was evidence that the macular pigment is involved in retinal health. Retinas of macaques on xanthophyll-depleted diets showed characteristics similar to those seen in the aging human eye. Soon thereafter, in their exceptional study, Snodderly et al. (1984, 1992) demonstrated conclusively, using microspectrophotometry, that the *macula lutea* is concentrated principally within the inner layers of the retina and that the optical density of the pigment could be readily imaged through the use of blue light. Bone et al. (1985) pioneered the use of modern high-performance liquid chromatography methods to analyze human retina extracts and discovered that the macular pigment was composed exclusively of the two isomeric carotenoids, lutein and zeaxanthin. These three reports combined to ensure the recognition that the human and primate retinas clearly contain a yellow pigment and that this pigment must be actively transported to its very specific location. It would be impossible to explain the absence of the many other carotenoids found in serum and other tissues without postulating a regulated mechanism responsible for the localization of macular carotenoids. In an extension of our initial study of the human macular pigment, we analyzed the distribution of the pigment in the eyes of over 100 of subjects, 10 from each decade of life (Bone et al. 1988). These data revealed that zeaxanthin is uniquely concentrated within the central fovea (Figure 1.4), where its concentration exceeds that of lutein, a surprising observation given that

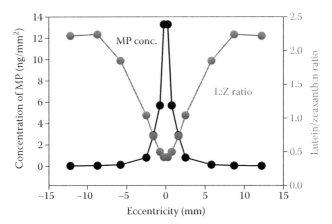

**FIGURE 1.4** The macular pigment concentration varies from a low near zero in the peripheral retina to a maximum of exceeding 10 ng/mm$^2$ in the central fovea (left axis), the ratio of L to Z varies across the fovea having a value near 2.2 in the perifovea to a minimum in the central fovea of approximately 0.2.

lutein is more abundant in both the human diet and serum (Bone et al. 2000; Beatty et al. 2004). A complexity of this observation is that the lutein : zeaxanthin ratio increases across the fovea from the central foveal value of approximately 0.2 to a value of approximately 2.2 in the peripheral retina (Bone et al. 1988). This lutein/zeaxanthin distribution overlays a region of the retina where the large changes in the rod : cone ratio also occurs (Østerberg 1935). This coincidence inspired the suggestion that zeaxanthin appeared to be preferentially associated with cone cells and therefore accumulated at the highest concentrations within the central fovea where the cone density is at its highest (Østerberg 1935; Curcio et al. 1987). A further illustration of the sophistication and specificity of xanthophyll metabolism and regulation within the retina was uncovered with the discovery by Bone and Landrum that approximately 50% of the zeaxanthin within the retina is present in the form of an uncommon stereoisomer of zeaxanthin, meso-zeaxanthin (Bone et al. 1993; Landrum et al. 1999). The origin of this isomer was shown to be the biochemical conversion of lutein to form meso-zeaxanthin within the retina (Johnson et al. 2005). The entire biochemical mechanism needed to explain this asymmetric physical distribution of the xanthophylls remains to be elucidated as of this writing. Data appear to be consistent with a hypothesis that the process by which lutein is converted to meso-zeaxanthin involves oxidation and reduction. We and others have observed the presence of small quantities of oxolutein in the retina, a metabolite resulting from the facile oxidation of the hydroxyl group present on the lutein epsilon ring (Khachik et al. 1997; Landrum et al. 2001, 2002). An analogous pathway has been demonstrated for canthaxanthin, which was consumed by individuals in large doses as an oral tanning agent (Gupta et al. 1985). Goralczyk et al. (1997) described the accumulation of canthaxanthin and its metabolites in the fovea and perifovea. They observed that significant amounts of reduction products (isozeaxanthin and 4-hydroxy-4'-keto-β,β-carotene) of the canthaxanthin keto groups are present in the perifovea (Goralczyk et al. 1997).

It is plausible that oxolutein might be efficiently reduced by this pathway to form meso-zeaxanthin. Similar pathways are known in other animal species (Maoka et al. 1986). Traces of S,S-zeaxanthin have also been detected within the retina and its presence would be consistent with this hypothesized reductive pathway (Bone et al. 1993). The recent isolation by Bernstein et al. (see Chapter 9) of two distinct human xanthophyll-binding proteins, GSTP 1 and the StAR D3, which specifically bind lutein and zeaxanthin, move us a step closer to answering the mechanism involved in the accumulation of L and Z within the retina (Bhosale et al. 2004; Li et al. 2011). Antibody staining has provided evidence that these two proteins are, like the xanthophylls, not uniformly distributed within the retinal layers (Li et al. 2011). They may be functioning as transporters, binding proteins, or both. Multiple factors that influence macular pigment levels in healthy individuals have been addressed by numerous researchers and include dietary intake, retinal thickness, body mass index, smoking, and oxidative stress (Beatty et al. 2000; Liew et al. 2006; Mares et al. 2006; Waldstein et al. 2012; Zheng et al. 2013). There have also emerged a number of genetic factors that are associated with macular pigment levels (Hammond et al. 1995; Borel et al. 1998; Cardinault et al. 2003; Liew et al. 2006; Zerbib et al. 2009; Reboul and Borel 2011).

Efforts to measure the visual pigments within the retina *in vivo* motivated the development of various ophthalmoscope designs, many of which have the ability to determine both spectral and spatial characteristics of light reflected in fundus observations (Elsner et al. 1990; Gellermann et al. 1998; Chen et al. 2001). An excellent review on the research progress in the use of fundus reflectance in vision research was published in 2003 by Berendschot et al. Differences in the foveal and peripheral reflectance measurements arise from structural and biochemical differences. After bleaching the retinal photopigments, the persistence of a difference between these regions is principally due to the *macula lutea* and has been repeatedly confirmed by research groups (Elsner et al. 1998; Berendschot et al. 2003; van de Kraats et al. 2006). Sophisticated models have been developed to account for the reflectance measurements and treat the eye as a structure possessing a number of structural layers each producing light scatter as well as containing light-absorbing species within them (Berendschot et al. 2003). Curve fitting in modern methods incorporates and corrects for the presence of the lens optical density and scattering, the imperfections of the optic media, absorbance by the macular pigment, and the visual pigments such as hemoglobin, lipofuscin, and melanin (Berendschot et al. 2003). Utilizing the autofluorescence of the underlying lipofuscin, measurements of the retinal distribution of the macular pigment can now be rapidly achieved although numerous assumptions enter into the computer model that must be used to deconvolute the spectral measurement (Delori 2004; Delori et al. 2011). Resonance Raman spectroscopy has also been adapted to the ophthalmoscope for imaging the distribution and quantity of macular pigment *in vivo* by Gellerman et al. (1998).

Imaging techniques suited to the measurement of macular pigment emerged and matured in the 1990s. Reflectance, autofluorescence, and *in vivo* resonance Raman spectroscopy have provided measurements of the spatial profile of the carotenoid distribution within the retina (Gellermann et al. 1998; Chen et al. 2001; Trieschmann et al. 2003, 2006, 2008). The distributions of the carotenoids observed in the macula

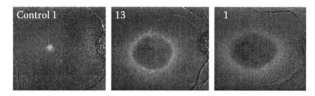

**FIGURE 1.5** The macular pigment distribution is uniquely distributed as a ring in the retinas of individuals having macular telangiectasia. (Adapted from Charbel Issa, P., et al., *Exp Eye Res*, 89(1), 25–31, 2009.)

can be classified essentially into three types: (1) narrow normal distributions with a sharp central peak with a width at half height of approximately 2.6° eccentricity and centered at the center of the fovea, (2) a broader distribution also centered on the fovea but exhibiting "shoulders" extending into the perifoveal region as far as about 7° eccentricity, and (3) a broader Gaussian shape. Chen suggested that these are distribution changes correlated with age and that the narrow type distribution is typical of younger subjects and the distribution width increases with age. A distribution with shoulders was first demonstrated in the macaque retina by Snodderly et al. (1984). Chen et al. (2001) reported that although the distributions increase in width with age, no difference in the peak, optical density was seen in different age groups. This would appear to conflict with other techniques since it would imply that an increase in the total carotenoid content of the retina must occur with increasing age. Resonance Raman spatial profiles also show macular pigment to be distributed in narrow and wide profiles. Gellerman reported more unusual distributions in which a surrounding ring of carotenoid pigmentation is observed. In some cases, this ring appears to be fragmented (Gellermann et al. 1998). An unusual macular pigment distribution is observed in the eyes of individuals diagnosed with the disease MacTel, macular telangiectasia (Charbel Issa et al. 2009). Xanthophylls within the central fovea are depleted, but a surrounding ring of pigmentation is observed at approximately 6° eccentricity, Figure 1.5 (Charbel Issa et al. 2009). Kirby et al. (2010) also described unusual distributions of the macular pigment for older patients as well as for those who were smokers. They observed a central dip in the measured macular pigment optical densities at 0.5° eccentricity among patients of increased age or among smokers compared to younger and nonsmokers.

## 1.6 MACULAR PIGMENT IN THE DEVELOPING EYE

A remaining void in our understanding of the macular pigment is the natural history of the deposition of the xanthophylls during the maturation of the retina and the fovea in the young eye. As early as the report by Home in his presentation to the Royal Society in 1798, evidence was strong that no visible yellow pigment is found in the retinas of fetuses and newborn infants but a faint color was detectable in young children (Home 1798). The macaque studies of Malinow et al. (1980) produced depletion of the macular pigment through deprivation of xanthophylls in the diet and provided unambiguous evidence that the macular pigment is derived from dietary components that cannot be biosynthesized by primates. In our 1988 paper on the distribution of the macular pigment, we investigated the manner in which the quantity of xanthophylls changed with age and analyzed eyes from each decade of life. Aside from the first decade, it was found that the macular pigment is largely constant throughout life. In fact, in the postmortem analysis of eyes from 10 donors ranging from age 0 to 9, the total pigment level appeared to be consistent with adult levels, with the exception of those below the age of about 1 year. The most striking observation made from these analyses was that the proportions of lutein and zeaxanthin in the retina were not typical of adult retinas until approximately the age of 3 years. For all eyes analyzed from donors at ages less than 2 years, the amount of lutein exceeded that of zeaxanthin and the lutein : zeaxanthin ratio averaged $1.44 \pm 0.16$ compared with a value of $0.77 \pm 0.2$ obtained from the average of all adult eyes in the same research study (Bone et al. 1988). Clearly, these data indicate that young eyes in which the fovea is not fully developed differ in their ability to accumulate and metabolize lutein and zeaxanthin compared to those in which the macula has fully matured. In recent years, analyzing retinas obtained from young macaques, we were able to measure the total xanthophyll level as well as the amounts of the individual carotenoids, lutein, zeaxanthin, and meso-zeaxanthin. Meso-zeaxanthin is either absent or below the level of detection in eyes prior to birth. Levels of meso-zeaxanthin increase soon after birth and levels increase proportional to age during maturation of the eye. The retina of macaques matures at a rate roughly four times that of humans (Boothe et al. 1985; Provis et al. 2005). The neonatal lutein to zeaxanthin ratio, like that in humans, is greater than 1 (~1.5), and the ratio drops to values less than 1 after approximately 6 months of age. These data are entirely consistent with those of humans for whom the foveal development age at approximately 2 years matches with that of macaques at approximately 6 months due to their fourfold more rapid development. Analysis of meso-zeaxanthin in the human retina during development has not, as yet, been reported but is expected to follow the pattern we have observed in macaques.

## 1.7 FUNCTIONAL ROLE OF XANTHOPHYLLS IN THE HUMAN RETINA

The earliest evidence of a functional role for the *macula lutea* was recognized from its ability to influence the ability of individuals to perform color matches. Schultz (1866) postulated that the light-filtering ability of the yellow macular pigment can improve visual function by reducing the effects of chromatic aberration in the eye. The refractive index of the components of the eye vary with wavelength with the consequence that the eye cannot simultaneously produce a perfect focus for all wavelengths of light and images projected onto the retina may have "fuzzy," polychromatic boundaries that may diminish perceptual acuity. Filtering out a large proportion of blue light, it was reasoned, would produce more distinct images by sharpening boundaries and improve visual acuity under conditions of high levels of blue light. Hammond and Elliot discuss these hypotheses fully in Chapter 8.

Epidemiological studies of diet and serum in the early 1990s indicated that xanthophylls are associated with significant reductions in the risk for AMD in a U.S. study population (Seddon et al. 1994). Kirschfeld argued as early as 1982 that the *macula lutea*'s ability to filter blue light might function to preserve the critically important cell layers of the retina from excessive levels of high-energy "blue" light capable of generating reactive oxygen species (ROS) and photooxidative damage (Kirschfeld 1982). This hypothesis has largely increased in popularity as our knowledge of the macular pigment and its relationship to disease processes has improved. In a case study of postmortem eyes, we found that the odds ratios for AMD were lower among those individuals who had higher macular pigment levels (Bone et al. 2001). The presence of macular pigment, even at modest levels, appears to be associated with a reduction in the risk for AMD. The protection conferred by the macular pigment may manifest itself by multiple mechanisms. The foremost among these is the ability of carotenoids to absorb "blue" light that reaches the inner retina and to attenuate its intensity prior to entering the sensitive photoreceptors and RPE. Protection of the eye against acute light damage by the macular pigment has been inadvertently tested and reported in a number of instances. Intense light used during surgical procedures has been noted to produce lesions in the retina. Macula sparing has been associated with the "footprint" of the *macula lutea* (Bernstein and Ginsberg 1964; Weiter et al. 1988). In controlled exposure to laser light, lesions are reduced due to the presence of xanthophylls in the central retina as compared to those produced in the unpigmented peripheral retina (Barker et al. 2011).

Krinsky (1971) elegantly described the mechanism by which carotenoids are essential to survival in photic aerobic environments and protect plants from the ROS generated during the photosynthetic steps. Carotenoids, including the xanthophylls, are efficient at quenching triplet excited-state photosensitizers and singlet oxygen, which are generated by energetic light (Cantrell and Truscott 2004; Edge and Truscott 2010). Carotenoids also effectively interrupt the free radical chain mechanism by which ROS react with and damage membrane components, proteins, and other essential biological structures. Although it is evident from Snodderly's now classic cross-sectional microphotograph of the macula that xanthophylls are concentrated at the highest levels within the inner retinal levels or Henle fibers, one can readily see that pigmentation extends into the photoreceptor outer layers (Snodderly et al. 1984). Rapp et al. (2000) directly demonstrated by analysis of rod outer segments that the xanthophylls are present in these critical structures. The retina is an aerobic tissue and the photoreceptors function in an environment of relatively high oxygen tension in the presence of significant levels of high-energy (blue) light capable of photoexcitation of the naturally occurring photosensitizers of oxygen and generating singlet oxygen. The presence of xanthophylls within this actinic environment provides the possibility to ameliorate the risk of damage posed to the retina, analogous to that in plants, by the presence of light and oxygen. Barker et al. (2011) recently reported clear evidence that macular pigment protects against acute exposure of the retina to shortwave laser light within the central retina in contrast to the peripheral retina (Figure 1.6). Acute exposure to intense light produces a lesion in the retina which develops over a significant period, approximately 24 hours. The biochemistry of the cell death process associated with high intensity light exposure

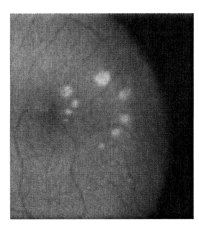

**FIGURE 1.6** **(See color insert.)** Exposure of the macaque retina to blue light produces lesions that are dependent on the total energy of the exposure. Two separate sets of five exposures were made in arcs: (a) the outer arc produced larger lesions at each exposure than the identical energy exposures associated with the (b) inner arc where the macular pigment attenuates the intensity of the damaging light reaching the photoreceptors and retinal pigment epithelium. The lowest energy exposure within the central yellow pigmented region failed to produce a detectable lesion. (From Barker, F.M. et al., *Invest Ophthalmol Vis Sci.*, 52: 3934, 2011. With permission.)

is mediated by numerous cellular events, many of which are triggered by ROS that induce production of cytokines, mitogens, and other cellular signals (Landrum 2013). The ability of the macular pigment to intercept ROS reduces their ability to trigger these pathways, which lead to inflammation, apoptosis, and cell death. This hypothesis is consistent with the observations of Barker et al. but has not yet been demonstrated to be significant for chronic, low-level exposure to shortwave light.

In recent years, there has developed a strong body of evidence that carotenoids can provide benefits to cells not only through their intact presence but also through the presence of their cleavage products. Similar to vitamin A, they appear to be able to exert an influence and participate in intracellular signaling (Mein and Wang 2010; Landrum 2013). Carotenoid cleavage enzymes produce apocarotenal *in vivo*, which have been associated with up- and down-regulation of gene transcription (Lobo et al. 2010, 2012). The high level of metabolic activity associated with visual function in the outer segments of the photoreceptors may be influenced by these metabolites that would be well positioned to participate as regulators in fine-tuning respiratory pathways in the retina.

The evolutionary advantage that is conferred on humans and nonhuman primates by the macular pigment seems unlikely to be associated with a reduction in the risk of AMD. Evolutionary success is generally ascribed to a trait that enhances the survival and reproduction of future generations. Macular pigment is a feature found in humans as well as in the more short-lived primates. Improved survival rates are unlikely to be associated with the ability of the macular carotenoids to protect against a loss of vision that occurs uniquely among humans at ages of more than 60 years. Early humans were as short-lived as are most primates, as such it is unlikely that they would be able to benefit from a trait that fails to assert its benefit early in life. Because the *macula lutea* is a feature of the eye that evolved in a common ancestor of modern primates and humans, it clearly provides a biological advantage that must have been important to that ancestral species as well as modern humans and primates. Although the ability of the *macula lutea* to protect the retina from actinic stress and damage is certainly real, its ability to contribute as survival quality among the younger and reproductively significant members of the species has not been established. It is notable that the young eye possesses a lens and optical media, which have a remarkable clarity and transmittance to blue light. This suggests that among young individuals, the retina is at a much greater risk of damage from photooxidative stress than that in older individuals. It may be that the importance of protecting the young retina is the principal function of the *macula lutea*. Hypotheses for the function(s) of the macular pigment have been proposed, including a faster recovery from glare and an ability to distinguish objects seen against a background possessing large amounts of scattered blue light (see Chapter 8).

Many higher animals are efficient at mobilizing and utilizing the xanthophylls (Weedon 1971). Knowing that carotenoids are ancient biological molecules and that their de novo synthesis is completely lacking among higher animals, it appears probable that the proteins and pathways that are essential for their metabolism may be shared among many diverse species. As described earlier in this chapter, the binding protein for lutein found in the human eye is a member of the large family of steroidal acute regulatory domain proteins and is ubiquitously found in animals. The protein responsible for lutein binding in the silk worm appears to have evolved from the same ancestral protein as that found in humans (Li et al. 2011). Similar comparisons exist among the class of carotenoid cleavage enzymes (Kiefer et al. 2001; von Lintig and Vogt 2004). An increased understanding of the biochemistry and the nature of the metabolic pathways by which they utilize carotenoids will likely continue to enrich our understanding of the actions of carotenoids and provide further evidence to their potential benefits (Tabunoki et al. 2002; Landrum et al. 2010).

## ACKNOWLEDGMENTS

This work was supported by a grant from Wyeth Nutrition and Four Leaf Japan Co. Ltd. Martha Neuringer wishes to acknowledge support from National Institutes of Health grant P51OD011092.

## REFERENCES

Albert, D. M. and W. H. Miller (1973). "Jan Purkinje and the ophthalmoscope." *Am J Ophthalmol* **76**(4): 494–499.

Barker, F. M., 2nd, D. M. Snodderly, et al. (2011). "Nutritional manipulation of primate retinas, V: Effects of lutein, zeaxanthin, and n-3 fatty acids on retinal sensitivity to blue-light-induced damage." *Invest Ophthalmol Vis Sci* **52**(7): 3934–3942.

Beatty, S., H. H. Koh, et al. (2000a). "Macular pigment optical density measurement: A novel compact instrument." *Ophthalmic Physiol Opt* **20**(2): 105–111.

Beatty, S., H. H. Koh, et al. (2000b). "The role of oxidative stress in the pathogenesis of age-related macular degeneration." *Surv Ophthalmol* **45**: 115–134.

Beatty, S., J. Nolan, et al. (2004). "Macular pigment optical density and its relationship with serum and dietary levels of lutein and zeaxanthin." *Arch Biochem Biophys* **430**(1): 70–76.
Berendschot, T. T. and D. van Norren. (2004). "Objective determination of the macular pigment optical density using fundus reflectance spectroscopy." *Arch Biochem Biophys* **430**: 149–155.
Berendschot, T. T., P. J. DeLint, et al. (2003). "Fundus reflectance—Historical and present ideas." *Prog Retin Eye Res* **22**(2): 171–200.
Bernstein, H. N. and J. Ginsberg (1964). "The pathology of chloroquine retinopathy." *Arch Ophthalmol* **71**: 238–245.
Bernstein, P., D. Zhao, et al. (2004). "Resonance Raman measurement of macular carotenoids in the living human eye." *Arch Biochem Biophys* **430**: 163–169.
Bhosale, P., A. J. Larson, et al. (2004). "Identification and characterization of a Pi isoform of glutathione S-transferase (GSTP1) as a zeaxanthin-binding protein in the macula of the human eye." *J Biol Chem* **279**(47): 49447–49454.
Bone, R. A (1980). "The role of the macular pigment in the dtection of polarized light." *Vision Res* **20**: 213–220.
Bone, R. A. and J. M. B. Sparrock (1971). "Comparison of macular pigment densities in human eyes." *Vision Res* **11**: 1057–1064.
Bone, R. A. and J. T. Landrum (1983). "Dichroism of lutein: A possible basis for Haidinger's brushes." *Appl Opt* **22**: 775–776.
Bone, R. A. and J. T. Landrum (1984). "Macular pigment in henle fiber membranes: A model for Haidinger's brushes." *Vision Res* **24**: 103–108.
Bone, R. and J. T. Landrum (2004). "Heterochromatic flicker photometry." *Arch Biochem Biophys* **430**: 137–142.
Bone, R. A., J. T. Landrum, et al. (1985). "Preliminary identification of the human macular pigment." *Vision Res* **25**: 1531–1535.
Bone, R. A., J. T. Landrum, et al. (1988). "Analysis of the macular pigment by HPLC: Retinal distribution and age study." *Invest Ophthalmol Vis Sci* **29**: 843–849.
Bone, R. A., J. T. Landrum, et al. (1992). "Optical density spectra of the macular pigment in vivo and in vitro." *Vision Res* **32**: 105–110.
Bone, R. A., J. T. Landrum, et al. (1993). "Stereochemistry of the human macular carotenoids." *Invest Ophthalmol Vis Sci* **34**: 2033–2040.
Bone, R. A., J. T. Landrum, et al. (2000). "Lutein and zeaxanthin in eyes, serum and diet of human subjects." *Exp Eye Res* **71**: 239–245.
Bone, R. A., J. T. Landrum, et al. (2001). "Macular pigment in donor eyes with and without AMD: A case-control study." *Invest Ophthalmol Vis Sci* **42**: 235–240.
Boothe, R. G., V. Dobson, et al. (1985). "Postnatal development of vision in human and non-human primates." *Annu Rev Neurosci* **8**: 495–545.
Borel, P., P. Grolier, et al. (1998). "Low and high responders to pharmacological doses of beta-carotene: Proportion in the population, mechanisms involved and consequences on beta-carotene metabolism." *J Lipid Res* **39**(11): 2250–2260.
Brindley, G. and E. Willmer (1952). "The reflexion of light from the macular and peripheral fundus oculi in man." *J Physiol* **116**: 350–356.
Buckley, O. E. and K. K. Darrow (1956). *Herbert Eugene Ives 1882–1953*. Washington, DC, National Academy of Sciences.
Buzzi, F. (1782). "Nuove sperienze fatte sull' occhio umano." *Opuscoli Scetti Sulle Scienze e Sulle Arti* **5**: 87.
Cantrell, A. and T. G. Truscott (2004). Carotenoids and radicals interactions with other nutrients. In *Carotenoids in Health and Disease*. Edited by N. I. Krinsky, S. T. Mayne, and H. Sies. New York, Marcel Dekker: 31–52.

Cardinault, N., J. M. Gorrand, et al. (2003). "Short-term supplementation with lutein affects biomarkers of lutein status similarly in young and elderly subjects." *Exp Gerontol* **38**(5): 573–582.
Charbel Issa, P., R. L. van der Veen, et al. (2009). "Quantification of reduced macular pigment optical density in the central retina in macular telangiectasia type 2." *Exp Eye Res* **89**(1): 25–31.
Chen, S.-J., Y. C. Chang, et al. (2001). "The spatial distribution of macular pigment in humans." *Curr Eye Res* **23**: 422–434.
Chevalleraeu, A. and A. Polyak (1907). "De la coloration jaune de la macula." *Ann Oculist* **138**: 241.
Chew, E. Y., T. Clemons, et al. (2012). "The Age-Related Eye Disease Study 2 (AREDS2): Study design and baseline characteristics (AREDS2 report number 1)." *Ophthalmology* **119**(11): 2282–2289.
Cumming, W. (1846). "On a luminous appearance of the human eye, and its application to the dectection of disease of the retina and posterior part of the eye." *Med-Chir Trans* **11**: 283–296.
Curcio, C. A., K. R. Sloan, Jr., et al. (1987). "Distribution of cones in human and monkey retina: individual variability and radial asymmetry." *Science* **236**(4801): 579–582.
Delori, F. (2004). "Autofluorescence method to measure macular pigment optical densities fluorometry and autofluorescence imaging." *Arch Biochem Biophys* **430**: 156–162.
Delori, F., J. P. Greenberg, et al. (2011). "Quantitative measurements of autofluorescence with the scanning laser ophthalmoscope." *Invest Ophthalmol Vis Sci* **52**(13): 9379–9390.
Dimmer, F. (1907). "Die *Macula lutea* der menschlichen Netzhaut und die durch die bedingten entotischen Erscheinungen." *Albrecht von Graefes Arch Ophthalmol* **65**: 486.
Edge, R. and T. G. Truscott (2010). Properties of carotenoid radicals and excited states and their potential role in biological systems. In *Carotenoids: Physical, Chemical, and Biological Functions and Properties*. Edited by J. T. Landrum. Boca Raton, FL, CRC Press: 283–307.
Elsner, A. E., S. A. Burns, et al. (1990). Quantitative reflectometry with the SLO. In *Scanning laser ophthalmoloscopy and tomography*. Edited by J. E. Naseman and R. O. W. Burk. Munich, Germany, Quintessenz-Verlag: 109–121.
Elsner, A. E., S. A. Burns, et al. (1998). "Foveal cone photopigment distribution: Small alterations associated with macular pigment distribution." *Invest Ophthalmol Vis Sci* **39**(12): 2394–2404.
Ewald, A. (1878). "Ueber die entoptische Wahrnehnung der *Macula lutea* und des Sehpurpurs." *Unters Physiol Inst Univ Heidelberg* **2**: 241.
Friedenwald, H. (1902). "The history of the invention and of the development of the ophthalmoscope." *J Am Med Assoc* **38**: 549–552.
Gellermann, W., M. Yoshida, et al. (1998). "Raman detection of pigments in the human retina." *Proc Opt Biopsy II* **3250**: 8–17.
Goralczyk, R., S. Buser, et al. (1997). "Occurrence of birefringent retinal inclusions in cynomolgus monkeys after high doses of canthaxanthin." *Invest Ophthalmol Vis Sci* **38**: 741–752.
Gullstrand, A. (1906a). "Die Farbe der Macula centralis retinae." *Albrecht von Graefes Arch Ophthalmol* **62**: 1.
Gullstrand, A. (1906b). "Zur Maculafrage." *Albrecht von Graefes Arch Ophthalmol* **66**: 141.
Gupta, A. K., H. F. Haberman, et al. (1985). "Canthaxanthin." *Int J Dermatol* **24**(8): 528–532.
Haab, O. (1879). "Die Farbe der *Macula lutea* und die entoptishe Wahrenmung der Sehpurpurs." *Ann Oculist* **84**: 179–188.
Haidinger, W. (1844). "Uber das direkte Erkennen des polarisirten Lichts und der Lage der Polarisationsebene." In *Annalen der Physik und Chemie*. Edited by J. C. Poggendorff. Berlin, Germany, Johann Ambrosius Barth, Volume **3**: 29–39.

Hammond, B. R., Jr., K. Fuld, et al. (1995). "Macular pigment density in monozygotic twins." *Invest Ophthalmol Vis Sci* **36**: 2531–2541.

Handelman, G. J., E. A. Dratz, et al. (1988). "Carotenoids in the human macula and whole retina." *Invest Ophthalmol Vis Sci* **29**: 850–855.

Hartridge, H. (1951). "Macular pigment." *Nature* **167**(4237): 76–77.

Hering, E. (1893). "Uber den Einflub der *Macula lutea* auf spectrale Farbengleichungen." *Pfluger Arch Ges Physiol* **54**: 227–312.

Hoeve, J. (1912). "Die Farbe der *Macula lutea*." *Albrecht von Graefes Arch Ophthalmol* **80**: 132.

Holm, E. (1922). "Das gelbe Maculapigment und seine optishe Bedeutung." *Albrecht von Graefes Arch Ophthalmol* **108**: 1–85.

Home, E. (1798). "An account of the orifice in the retina of the human eye, discovered by Professor Soemmering. To which are added, proofs of this appearance being extended to the eyes of other animals." *Philos Trans Roy Soc London Pt 2* **88**: 332–345.

Howells, O., F. Eperjesi, et al. (2011). "Measuring macular pigment optical density in vivo: A review of techniques." *Graefes Arch Clin Exp Ophthalmol* **249**(3): 315–347.

Ives, H. E. (1912). "Studies in the photometry of lights of different colors: I. Spectral luminosity curves obtained by the equality of brightness photometer and the flicker photometer under similar conditions." *Philos Mag* **6**(24): 149–188.

Johnson, E. J., M. Neuringer, et al. (2005). "Nutritional manipulation of primate retinas. III. Effects of lutein or zeaxanthin supplementation on adipose tissue and retina of xanthophyll-free monkeys." *Invest Ophthalmol Vis Sci* **46**(2): 692–702.

Khachik, F., P. S. Bernstein, et al. (1997). "Identification of lutein and zeaxanthin oxidation products in human and monkey retinas." *Invest Ophthalmol Vis Sci* **38**(9): 1802–1811.

Kiefer, C., S. Hessel, et al. (2001). "Identification and characterization of a mammalian enzyme catalyzing the asymmetric oxidative cleavage of provitamin A." *J Biol Chem* **276**(17): 14110–14116.

Kirby, M. L., S. Beatty, et al. (2010). "A central dip in the macular pigment spatial profile is associated with age and smoking." *Invest Ophthalmol Vis Sci* **51**(12): 6722–6728.

Kirschfeld, K. (1982). "Carotenoid pigments: Their possible role in protecting against photo-oxidation in eyes and photoreceptor cell." *Proc R Soc Lond B* **216**: 71–85.

Kolmer, W. (1936). "Die Netzhaut (retina)." In *Handbuch der Mikroskopischen Anatomie des Menschen*. Edited by W. Mollendorff. Berlin, Julius Springer: 295.

Krause, A. C. (1934). *The Biochemistry of the Eye*. Baltimore, MD, Johns Hopkins Press.

Krinsky, N. I. (1971). "Function." In *Carotenoids*. Edited by O. Isler. Basel, Birkhauser-Verlag: 669–716.

Landrum, J. T. (2013). Reactive oxygen and nitrogen species in biological systems: Reactions and regulation by carotenoids. In *Nutrition and Health: Carotenoids and Human Health*. Edited by S. A. Tanumihardjo. New York, Springer Humana Press: 57–101.

Landrum, J. T., D. Callejas, et al. (2010). Specific accumulation of lutein within the epidermis of butterfly larvae. In *Carotenoids: Physical, Chemical, and Biological Functions and Properties*. Edited by J. T. Landrum. Boca Raton, FL, CRC Press: 525–535.

Landrum, J. T. and R. A. Bone (2001). "Lutein, zeaxanthin, and the macular pigment." *Arch Biochem Biophys* **385**(1): 28–40.

Landrum, J. T., R. A. Bone, et al. (1999). Analysis of zeaxanthin distribution within individual human retinas. In *Methods in Enzymology*. Edited by L. Packer. San Diego, CA, Academic Press. **299**: 457–467.

Landrum, J. T., R. A. Bone, et al. (2001). "Minor carotenoids in the human retina." *IOVS* **42**: ARVO abstract.

Landrum, J. T., R. A. Bone, et al. (2002). A model for carotenoid metabolism in the human retina. 13th International Carotenoid Symposium, Honolulu, Hawaii, January 6–11.

Li, B., P. Vachali, et al. (2011). "Identification of StARD3 as a lutein-binding protein in the macula of the primate retina." *Biochemistry* **50**(13): 2541–2549.

Liew, S. H., C. E. Gilbert, et al. (2006). "Central retinal thickness is positively correlated with macular pigment optical density." *Exp Eye Res* **82**(5): 915–920.

Lobo, G. P., J. Amengual, et al. (2012). "Mammalian carotenoid-oxygenases: Key players for carotenoid function and homeostasis." *Biochim Biophys Acta* **1821**(1): 78–87.

Lobo, G. P., S. Hessel, et al. (2010). "ISX is a retinoic acid-sensitive gatekeeper that controls intestinal beta,beta-carotene absorption and vitamin A production." *FASEB J* **24**(6): 1656–1666.

Lottrup-Andersen, C. (1913). "Ein Fall von akuter Ischame der Retina mit sehr Deutlichem Hervortreten der gelben Farbe der *Macula lutea*." *Klin Monatsbl Augenheitkd* **57**: 740–742.

Lyons, H. G. (1940). "Charles Babbage and the ophthalmoscope." *Notes Rec R Soc Lond* **3**: 146–148.

Malinow, M. R., L. Feeney-Burns, et al. (1980). "Diet-related macular anomalies in monkeys." *Invest Ophthalmol Vis Sci* **19**: 857–863.

Maoka, T., A. Arai, et al. (1986). "The first isolation of enantiomeric and meso-zeaxanthin in nature." *Comp Biochem Physiol* **83B**: 121–124.

Mares, J. A., T. L. LaRowe, et al. (2006). "Predictors of optical density of lutein and zeaxanthin in retinas of older women in the Carotenoids in Age-Related Eye Disease Study, an ancillary study of the Women's Health Initiative." *Am J Clin Nutr* **84**(5): 1107–1122.

Maxwell, J. C. (1855). "Experiments on colour, as perceived by the eye, with remarks on colour-blindness." *Trans Roy Soc Edinb* **21**(11): 275–297.

Maxwell, J. C. (1856a). "On the theory of compound colors with reference to mixtures of blue and yellow light." *Edinb J* **4**: 335–337.

Maxwell, J. C. (1856b). "On the unequal sensibility of the foramen centrale to light of different colours." *Rep Brit Assoc* **2**: 12.

McMullen, W. (1917). "The evolution of the ophthalmoscope: Communications" *Br J Ophthalmol* **1**: 593–599.

Mein, J. R. and X.-D. Wang (2010). Oxidative metabolites of lycopene and their biological functions. In *Carotenoids: Physical, Chemical, and Biological Functions and Properties*. Edited by J. T. Landrum. Boca Raton, FL, CRC Press: 417–435.

Mellerio, J., S. Ahmadi-Lari, et al. (2002). "A portable instrument for measuring macular pigment with central fixation." *Curr Eye Res* **25**(1): 37–47.

Moreland, J. D., A. G. Robson, et al. (1998). "Macular pigment and the colour-specificity of visual evoked potentials." *Vision Res* **38**(21): 3241–3245.

Nolan, J. M., R. Kenny, et al. (2010). "Macular pigment optical density in an ageing Irish population: The Irish Longitudinal Study on Ageing." *Ophthalmic Res* **44**(2): 131–139.

Nussbaum, J. J., R. C. Pruett, et al. (1981). "Historic perspectives macular yellow pigment: The first 200 years." *Retina* **1**: 296–310.

Østerberg, G. (1935). "Topography of the layer of rods and cones in the human retina." *Acta Ophthalmol* **6**: 1.

Pease, P. L. and A. J. Adams (1983). "Macular pigment difference spectrum from sensitivity measures of a single cone mechanism." *Am J Optom Physiol Opt* **60**(8): 667–672.

Pease, P. L., A. J. Adams, et al. (1987). "Optical density of human macular pigment." *Vision Res* **27**(5): 705–710.

Provis, J. M., P. L. Penfold, et al. (2005). "Anatomy and development of the macula: Specialisation and the vulnerability to macular degeneration." *Clin Exp Optom* **88**(5): 269–281.

Rapp, L. M., S. S. Seema, et al. (2000). "Lutein and zeaxanthin concentrations in rod outer segment membranes from perifoveal and peripheral human retina." *Invest Ophthalmol Vis Sci* **41**: 1200–1209.

Reboul, E. and P. Borel (2011). "Proteins involved in uptake, intracellular transport and basolateral secretion of fat-soluble vitamins and carotenoids by mammalian enterocytes." *Prog Lipid Res* **50**(4): 388–402.

Ruddock, K. H. (1963). "Evidence for macular pigmentation from colour matching data." *Vision Res.* **3**: 417.

Ruddock, K. H. (1972). *Light Transmission through the Ocular Media and Macular Pigment and Its Significance for Psychophysical Investigation.* Berlin, Germany: Springer-Verlag.

SanGiovanni, J. P., E. Y. Chew, et al. (2007). "The relationship of dietary carotenoid and vitamin A, E, and C intake with age-related macular degeneration in a case-control study: AREDS Report No. 22." *Arch Ophthalmol* **125**(9): 1225–1232.

Schmidt-Rimpler, H. (1874). "Die farbe der *macula lutea* im auge der menschen." *Centralbl Für dem Medic Wissenschaft*, XII, 900–902.

Schultz, M. (1866). "Sur la lache jaune de la retinae et son influence sur la vue normale et sure les anomalies de perception des coleurs (traduit par Leber)." *Journ d'anat et de physiol de ch Robin*: 440–446.

Schwalbe, G. (1874). Microscopische anatomie des Schnerven der Netzhaut und des Glaskorpers. In *Handbuch der gesammten Augenheilkunde*. Edited by A. Graefe and T. Saemisch. Leipzig, Germany, Engelmann, Volume **1**: 430.

Seddon, J. M., U. A. Ajani, et al. (1994). "Dietary carotenoids, vitamins A, C, and E and advanced age-related macular degeneration." *JAMA* **272**: 1413–1420.

Segal, J. (1950). "Localization du pigment maculaire de la retine." *CR Soc Biol (Paris)* **144**: 1630.

Snodderly, D. M., J. D. Auron, et al. (1984). "The macular pigment. II. Spatial distribution in primate retinas." *Invest Ophthalmol Vis Sci* **25**: 674–685.

Snodderly, D. M., R. S. Weinhaus, et al. (1992). "Neural-vascular relationships in central retina of macaque monkeys (Macaca fascicularis)." *J Neurosci* **12**(4): 1169–1193.

Stabell, U. and B. Stabell (1980). "Variation in density of macular pigmentation and in short-wave cone sensitivity with eccentricity." *J Opt Soc Am* **70**(6): 706–711.

Tabunoki, H., H. Sugiyama, et al. (2002). "Isolation, characterization, and cDNA sequence of a carotenoid binding protein from the silk gland of *Bombyx mori* larvae." *J Biol Chem* **277**: 32133–32140.

Trieschmann, M., B. Heimes, et al. (2006). "Macular pigment optical density measurement in autofluorescence imaging: Comparison of one- and two-wavelength methods." *Graefes Arch Clin Exp Ophthalmol* **244**(12): 1565–1574.

Trieschmann, M., F. J. van Kuijk, et al. (2008). "Macular pigment in the human retina: Histological evaluation of localization and distribution." *Eye (Lond)* **22**(1): 132–137.

Trieschmann, M., G. Spital, et al. (2003). "Macular pigment: Quantitative analysis on autofluorescence images." *Graefes Arch Clin Exp Ophthalmol* **241**(12): 1006–1012.

van de Kraats, J., T. T. Berendschot, et al. (2006). "Fast assessment of the central macular pigment density with natural pupil using the macular pigment reflectometer." *J Biomed Opt* **11**(6): 064031.

van der Veen, R. L., T. T. Berendschot, et al. (2009). "A new desktop instrument for measuring macular pigment optical density based on a novel technique for setting flicker thresholds." *Ophthalmic Physiol Opt* **29**(2): 127–137.

Vogt, A. (1921a). "Ophthalmoskopische Untersuchungen der *Macula lutea* im rotfreien Licht." *Klin. Monatsbl Augenheitkd* **66**: 321.

Vogt, A. (1921b). "Zur Farbe der *Macula lutea*." *Klin. Monatsbl Augenheitkd* **60**: 449.

Vogt, A. (1937). Untershuchungen des Auges im rotfreien Licht. *Aberhalden Hand d Biol Arbeitsmethod* **Abt. 5 Teil (6)1**: 365.

von Helmholtz, H. (1851). *Beschreibung eines Augen-Spiegels* [*A Description of an Ophthalmoscope*]. Berlin, Germany: A. Förstner'sche Verlagsbuchlandlung.

von Helmholtz, H. (1916). *The Description of an Ophthalmoscope*. Chicago, IL: Cleveland Press. (This is the English translation by Thomas H. Shastid of the 1851 publication by Helmholtz.)

von Helmholtz, H. (1924). *Helmholtz's Treatise on Physiological Optics*. Menasha, WI: George Banta Publishing Company.

von Helmholtz, H. (1951). "Description of an ophthalmoscope for examining the retina of the living eye." *AMA Arch Ophthalmol* **46**: 565–583.
von Lintig, J. and K. Vogt (2004). "Vitamin A formation in animals: Molecular identification and functional characterization of carotene cleaving enzymes." *J Nutr* **134**(1): 251S–256S.
von Soemmerring, S. T. (1799). "De Foramina centrali limbo luteo cincto retinae humanae." *Comment Soc Reg Sci Goetting* **13**: 3.
von Soemmerring, S. T. (1801). *Abbildungen des menschliche Auges*. Frankfurt, Germany, Varentrapp and Wenner, http://www.zvab.com/buch-suchen/titel/abbildungen-des-menschlichen-auges/autor/soemmerring.
Wald, G. (1967). "Blue-blindness in the normal fovea." *J Opt Soc Am* **57**: 1289.
Wald, G., W. R. Carroll, et al. (1941). "The human excretion of carotenoids and vitamin A." *Science* **94**(2430): 95–96.
Waldstein, S. M., D. Hickey, et al. (2012). "Two-wavelength fundus autofluorescence and macular pigment optical density imaging in diabetic macular oedema." *Eye (Lond)* **26**(8): 1078–1085.
Walls, G. L. and H. D. Judd (1933). "The intraocular colour filters of vertebraes." *Br J Ophthalmol* **17**: 641–705.
Warrick, P. (2007). *Charles Babbage and the Countess*. Bloomington, IN: Authorhouse.
Weale, R. (1994). "On the invention of the ophthalmoscope." *Doc Ophthalmol* **86**(2): 163–166.
Weedon, B. C. L. (1971). Occurrence. In *Carotenoids*. Edited by O. Isler. Basel, Switzerland, Birkhäuser Verlag: 29–59.
Weiter, J. J., F. C. Delori, et al. (1988). "Central sparing in annular macular degeneration." *Am J Ophthalmol* **106**: 286–292.
Werner, J. S. and B. R. Wooten (1979). "Opponent chromatic response functions for an average observer." *Percept Psychophys* **25**(5): 371–374.
Werner, J. S., S. K. Donnelly, et al. (1987). "Aging and the human macular pigment density. Appended with translations from the work of Max Schultz and Ewald Hering." *Vision Res* **27**: 257–268.
Wharton-Jones, T. (1954). "Report on the ophthalmoscope." *Br Foreign Med-Chirl Rev* **24**: 425–432.
Whitehead, A. J., J. A. Mares, et al. (2006). "Macular pigment: A review of current knowledge." *Arch Ophthalmol* **124**(7): 1038–1045.
Wolin, L. R. and L. C. Massopust, Jr. (1967). "Characteristics of the ocular fundus in primates." *J Anat* **101**(Pt 4): 693–699.
Wooten, B. R., B. R. Hammond, Jr., et al. (1999). "A practical method for measuring macular pigment optical density." *Invest Ophthalmol Vis Sci* **40**: 2481–2489.
Zerbib, J., J. M. Seddon, et al. (2009). "rs5888 variant of SCARB1 gene is a possible susceptibility factor for age-related macular degeneration." *PLoS One* **4**(10): e7341.
Zheng, W., Z. Zhang, et al. (2013). "Macular pigment optical density and its relationship with refractive status and foveal thickness in Chinese school-aged children." *Curr Eye Res* **38**(1): 168–173.

# 2 Risk Factors for Age-Related Macular Degeneration and Their Relationship with the Macular Carotenoids

*Tos T. J. M. Berendschot*

## CONTENTS

2.1 Introduction ............................................................................................. 23
2.2 Age ........................................................................................................... 24
2.3 Smoking ................................................................................................... 28
2.4 Diet ........................................................................................................... 29
2.5 Sex ............................................................................................................ 30
2.6 Body Mass Index ..................................................................................... 30
2.7 (Sun) Light ............................................................................................... 30
2.8 Genetics .................................................................................................... 31
2.9 Conclusions .............................................................................................. 31
References ......................................................................................................... 32

## 2.1 INTRODUCTION

There is a growing body of scientific evidence and a biologically plausible rationale that the macular carotenoids (lutein, zeaxanthin, and meso-zeaxanthin, collectively known as macular pigment [MP] [Bone et al. 1988; Davies and Morland 2004]) exert a protective effect in the macula. Firstly, MP is a short-wavelength (blue) light filter, absorbing light between 390 and 540 nm (Vos 1972; Bone et al. 1992; DeMarco et al. 1992; Sharpe et al. 1998), thereby decreasing chances for photochemical light damage (Landrum et al. 1997b). In addition, the macular carotenoids are capable of scavenging free radicals and singlet oxygen (Khachik et al. 1997; Bohm et al. 2012; Stahl and Sies 2012). Finally, it has been shown that lutein suppresses inflammation (Izumi-Nagai et al. 2007; Berendschot et al. 2012; Kijlstra et al. 2012). A paper by Seddon et al. (1994), which reported that increasing the consumption of foods rich in certain carotenoids, in particular dark green, leafy vegetables (a rich source of MP), may decrease the risk of developing late age-related macular degeneration

(AMD), generated interest and study into the relationship between diet, MP, and AMD (Mares-Perlman 1999; Beatty et al. 2001; Bone et al. 2001; Berendschot et al. 2002; Mares-Perlman et al. 2002; Snellen et al. 2002; Gale et al. 2003; Mozaffarieh et al. 2003; Cho et al. 2004; Richer et al. 2004; Delcourt et al. 2006; Chong et al. 2007; Nolan et al. 2007b; Richer et al. 2007; Loane et al. 2008; O'Connell et al. 2008; Beatty et al. 2009; Carpentier et al. 2009; Tsika et al. 2009).

To date, several risk factors for AMD have been identified (Age-Related Eye Disease Study Research Group 2000; Evans 2001; Hyman and Neborsky 2002; Miyazaki et al. 2003; Wang et al. 2003; Klein et al. 2004; Clemons et al. 2005; O'Connell et al. 2008). This chapter discusses the relationship between known and putative risk factors for AMD with MP. For instance, it is possible that cigarette smoking (an established risk factor for AMD) reduces MP levels and that the apparent increased risk for AMD in smokers could well result from the lack of MP (given the lower retinal antioxidant and light-filtering properties of such an individual). Of course, if there is no relationship between cigarette smoking and MP levels, then smoking must increase the risk for AMD through another pathway.

In the following sections, some of the more important risk factors for AMD and their association with MP are discussed.

## 2.2 AGE

The most established and accepted risk factor for AMD is aging itself (van Leeuwen et al. 2003; Evans et al. 2005; Seddon et al. 2011). Its association with MP has been studied using all the different measurement techniques currently available. With the exception of Raman spectroscopy, most studies report none or only modest correlations between MP and age (see Table 2.1).

Fifteen studies that used heterochromatic flicker photometry (HFP) found no age effect (Werner et al. 1987; Ciulla et al. 2001; Delori et al. 2001; Mellerio et al. 2002; Bernstein et al. 2004; Ciulla and Hammond 2004; Liew et al. 2005; Iannaccone et al. 2007; Richer et al. 2007; Connolly et al. 2011; Raman et al. 2011; Tsika et al. 2011; Raman et al. 2012; Yu et al. 2012; Zheng et al. 2012). One of these included even subjects with cataracts and AMD (Ciulla and Hammond 2004). Eight studies found a small decline in macular pigment optical density (MPOD) with age using HFP (Hammond and Caruso-Avery 2000; Beatty et al. 2001; Berendschot and van Norren 2005; Neelam et al. 2005; Nolan et al. 2004, 2007b, 2008; Yu et al. 2012). After correction for body fat, in the study of Nolan et al. (2004), the significance of the relationship persisted for males only. Raman et al. (2011) found an increase from 20 to 39 years and a decline thereafter with age. One study found a decline in MPOD with age (van der Veen et al. 2009). Nolan et al. assessed MPOD every month for 24 consecutive months in four healthy subjects aged between 23 and 51 years. They concluded that fluctuations in serum concentrations of lutein and zeaxanthin, in the absence of dietary modification or supplementation, are associated with stable MP optical density (Nolan et al. 2006).

Using color matching as another psychophysical technique, Davies and Morland (2002) found no age effect.

## TABLE 2.1
### Studies that Reported on the Relationship between Age and MPOD

| Principal Author | Technique | Test Stimulus | Parafoveal Stimuli (Eccentricity) | Sample Number | Age Range (years) | Age Effect | Correlation Coefficient | Significance |
|---|---|---|---|---|---|---|---|---|
| Beatty et al. (2001) | HFP | 0.95° | 6° | 46 | 21–81 | Decline | −0.48 | $p < .05$ |
| Berendschot and van Norren (2005) | HFP | 1° | 5° | 53 | 19–76 | Decline | −0.42 | $p < .01$ |
| Bernstein et al. (2004) | HFP | 1.5° | 8° | 40 | 18–61 | Decline | −0.279 | $p = ns$ |
| Ciulla et al. (2001) | HFP | 1° | 4° | 280 | 18–50 | None | — | $p = ns$ |
| Ciulla and Hammond (2004) | HFP | 1° | 4° | 390 | 18–88 | None | 0.04 | $p = ns$ |
| Connolly et al. (2011) | HFP | 1° | 7° | 44 | 18–61 | None | — | $p = ns$ |
| Delori et al. (2001) | HFP | 0.8° | 5.5° | 30 | 15–80 | None | — | $p = ns$ |
| Hammond and Caruso-Avery (2000) | HFP | 1° | 4° | 217 | 18–90 | Decline | −0.14 | $p < .05$ |
| Iannaccone et al. (2007) | HFP | 1° | 7° | 222 | 69–86 | Decline | −0.026 | $p = ns$ |
| Liew et al. (2005) | HFP | 1° | 5° | 150 | 18–50 | None | — | $p = ns$ |
| Mellerio et al. (2002) | HFP | 1° | 5° | 124 | 18–84 | None | −0.06 | $p = ns$ |
| Neelam et al. (2005) | HFP | 1° | 5° | 125 | 20–60 | Decline | −0.181 | $p < 0.05$ |
| Nolan et al. (2004) | HFP | 1° | 5° | 100 | 22–60 | Decline | −0.359 | $p < 0.01$ |
| Nolan et al. (2007a) | HFP | 1° | 5° | 800 | 20–60 | Decline | −0.286 | $p < 0.01$ |
| Nolan et al. (2008) | HFP | 0.5° | 7° | 59 | 19–57 | Decline | −0.252 | $p < 0.05$ |
| Raman et al. (2011) | HFP | 1° | 7° | 161 | 20–60+ | Decline | — | |
| Raman et al. (2012) | HFP | 0.5° | 7° | 62 | 50+ | None | — | $p = ns$ |

*(Continued)*

## TABLE 2.1 (Continued)
## Studies that Reported on the Relationship between Age and MPOD

| Principal Author | Technique | Test Stimulus | Parafoveal Stimuli (Eccentricity) | Sample Number | Age Range (years) | Age Effect | Correlation Coefficient | Significance |
|---|---|---|---|---|---|---|---|---|
| Richer et al. (2007) | HFP | 1° | 7° | 90 | — | None | — | — |
| Tsika et al. (2011) | HFP | 0.5° | 8° | 102 | 55–88 | None | — | $p = $ ns |
| van der Veen et al. (2009) | HFP | 0.5° | 8° | 26 | 22–64 | Increase | 0.083 | $p < 0.1$ |
| Werner et al. (1987) | HFP | 1° | 5° | 50 | 10–90 | Decline | −0.21 | $p = $ ns |
| Yu et al. (2012) | HFP | 0.25° | 7° | 281 | 17–85 | Decline | −0.17 | $p = 0.014$ |
| Yu et al., 2012) | HFP | 0.5° to 1.75° | 7° | 281 | 17–85 | None | Positive, negative | $p = $ ns |
| Zheng et al. (2012) | HFP | 1° | 7° | 94 | 6–12 | Increase | 0.02 | $p = $ ns |
| Davies and Morland (2002) | CM | 1° | 5° | 68 | 25–72 | None | — | $p = $ ns |
| Berendschot and van Norren (2004) | REF | | | 138 | 18–76 | None | 0.058 | $p = $ ns |
| Berendschot and van Norren (2005) | REF | | | 134 | 18–70 | None | 0.05 | $p = $ ns |
| Berendschot et al. (2002) | REF | | | 435 | 55–91 | Increase | 0.15 | $p = .002$ |
| Broekmans (2002) | REF | | | 376 | 18–75 | None | 0.035 | $p = $ ns |
| Chen et al. (2001) | REF | | | 54 | 20–84 | None | — | — |
| Delori et al. (2001) | REF | | | 159 | 15–80 | Increase | — | — |
| Kilbride et al. (1989) | REF | | | 7 | — | Decline | — | — |
| Wüstemeyer et al. (2003) | REF | | | 109 | 16–76 | Decline | −0.44 | $p < .01$ |

# Risk Factors for Age-Related Macular Degeneration

| Study | Method | N | Age | Effect | Value | p |
|---|---|---|---|---|---|---|
| Zagers and van Norren (2003), Zagers (2004) | REF | 38 | 18–64 | None | 0.058 | — |
| Berendschot and van Norren (2005) | AF | 53 | 18–70 | Decline | −0.08 | $p = ns$ |
| Delori et al. (2001) | AF | 159 | 15–80 | Increase | — | — |
| Dietzel et al. (2011) | AF | 369 | 60–80 | Increase | 0.14 | $p < .01$ |
| Hogg et al. (2012) | AF | 86 | 55–76 | None | — | $p = ns$ |
| Jahn et al. (2005) | AF | 146 | 50–88 | Negative | −0.11 | $p = ns$ |
| Liew et al. (2005) | AF | 150 | 18–50 | Increase | 0.17 | $p < 0.05$ |
| Sasamoto et al. (2011) | AF | 43 | 65 stdev = 9.1 | None | — | $p = ns$ |
| Tanito et al. (2012) | AF | 22 | 28–58 | None | — | $p = ns$ |
| Trieschmann et al. (2006) | AF | 108 | 51–87 | Decline | — | $p = ns$ |
| Wüstemeyer et al. (2003) | AF | 109 | 16–76 | None | 0.03 | $p = ns$ |
| Bernstein et al. (2004) | RS | 40 | 18–61 | Decline | −0.55 | $p < .01$ |
| Gellermann et al. (2002a) | RS | 140 | 21–84 | Decline | −0.664 | $p < .01$ |
| Neelam et al. (2005) | RS | 125 | 20–60 | Decline | −0.433 | $p < .01$ |
| Tanito et al. (2012) | RS | 22 | 28–58 | None | — | $p = ns$ |
| Bone et al. (1988) | HPLC | 87 | 3–95 | None | — | $p = ns$ |
| Bone et al. (2001) | HPLC | 56 | 58–98 | Increase | 0.34 | $p = .01$ |
| Handelman et al. (1988) | HPLC | 16 | 1 week to 81 years | None | — | $p = ns$ |

AF, autofluorescence; CM, color matching; HFP, heterochromatic flicker photometry; HPLC, high performance liquid chromatography; REF, reflectance; RS, Raman Spectroscopy.

Using fundus reflectance, Delori et al. (2001) found a 17% higher MPOD for older subjects (aged 65–80 years) than for younger subjects, aged 15–30 years. Analyzing the MP spatial distribution from reflectance maps, three studies found no change in peak optical density with age (Kilbride et al. 1989; Chen et al. 2001; Wüstemeyer et al. 2003). Zagers et al.'s data using the directional reflectance method also showed no change with age (Zagers and van Norren 2003; Zagers 2004). Using reflectance spectroscopy (Berendschot and van Norren 2004), three studies showed no age effect (Broekmans et al. 2002; Berendschot and van Norren 2004, 2005), whereas a small increase with age was observed in a fourth study (Berendschot et al. 2002).

Using autofluorescence, eight studies (Delori et al. 2001; Wüstemeyer et al. 2003; Berendschot and van Norren 2005; Jahn et al. 2005; Trieschmann et al. 2006; Sasamoto et al. 2011; Hogg et al. 2012; Tanito et al. 2012) found no relationship with age and two found a small increase (Liew et al. 2005; Dietzel et al. 2011) with age.

High-performance liquid chromatography has been employed to study age effects in donor eyes. Both Bone et al. (1988) and Handelman et al. (1988) found no age effect. Bone et al. (2001) in another study comparing donor eyes with and without AMD found a slight increase with age.

Studies using Raman spectroscopy initially all showed a strong decrease with age (Gellermann et al. 2002; Bernstein et al. 2004; Gellermann and Bernstein 2004). This seemed strange in view of the results from all other methods. A number of possible causes like aberrations, scatter, and fixation were looked at (Wooten and Hammond 2003a,b; Bernstein and Gellermann 2003a,b). The most important parameter seems to be pupil size. Excluding eyes with a pupil diameter smaller than 7 mm causes the age-related effect to decline to nonsignificance (Neelam et al. 2005).

Taken together, the data is inconclusive. Most studies found no age effect; a few found either a slight decrease or a slight increase. This brings us to conclude that the relationship between age and MPOD, if any, is small, which suggests that age is an independent risk factor for AMD (Gellermann et al. 2002; Bernstein et al. 2004; Gellermann and Bernstein 2004).

## 2.3 SMOKING

Cigarette smoking is the most important, modifiable risk factor for AMD (Age-Related Eye Disease Study Research Group 2000; Klaver et al. 1997; van Leeuwen et al. 2003; Tomany et al. 2004; Seddon et al. 2006; Chakravarthy et al. 2007; Klein et al. 2010). Most studies that have investigated the relationship between MP and cigarette smoking have reported an inverse association between these variables (Hammond et al. 1996b; Hammond and Caruso-Avery 2000; Beatty et al. 2001; Curran-Celentano et al. 2001; Broekmans et al. 2002; Mellerio et al. 2002; Nolan et al. 2007a,b; Dietzel et al. 2011). Moreover, studies have shown that there is a dose–response relationship between cigarette smoking and MP (Nolan et al. 2007b). Of interest, the increased risk of AMD in cigarette smokers may, at least in part, be caused by the relative lack of MP in smokers when compared to nonsmokers. Note that both the increased risk of AMD and the decrease in MP may have common underlying causes. For example, possible explanations to account for a relative lack of MP among cigarette smokers include a poor diet (with consequentially reduced levels of antioxidants in diet and serum [Dallongeville et al. 1998; Nolan et al. 2007b]) and/or increased overall oxidant load associated with tobacco use (with a consequential excessive depletion of L and/or Z and/or MZ at the macula) (Beatty et al. 2000). Furthermore, the prooxidant effect of smoking causes a decrease in serum carotenoids concentration (Hammond et al. 1997; Berendschot et al. 2000; Dietzel et al. 2011), which in turn lowers the MPOD (see Section 2.4) (Seddon et al. 1994).

## 2.4 DIET

Seddon et al. (1994) were the first to suggest an inverse association between intake of green vegetables and the risk of AMD. They proposed a key role for lutein, one of the major carotenoids of MP. Following this suggestion, many papers have been published that have studied the relationship between dietary intake and serum concentrations of L and Z, and MP and also the relationship between these carotenoids and their relationship with AMD (Mares-Perlman 1999; Beatty et al. 2001; Bone et al. 2001; Berendschot et al. 2002; Mares-Perlman et al. 2002; Snellen et al. 2002; Gale et al. 2003; Mozaffarieh et al. 2003; Cho et al. 2004; Richer et al. 2004; Delcourt et al. 2006; Chong et al. 2007; Nolan et al. 2007b; Richer et al. 2007; Loane et al. 2008; O'Connell et al. 2008; Beatty et al. 2009; Carpentier et al. 2009; Tsika et al. 2009). Humans are unable to synthesize carotenoids. Serum levels of L and Z are therefore dependent on dietary intake (Landrum et al. 1999; Johnson et al. 2000, 2005; Broekmans et al. 2002; Loane et al. 2009). Lutein serum concentrations and the amount of MP are correlated, particularly for males (Hammond et al. 1997; Chung et al. 2004; Burke et al. 2005; Wenzel et al. 2006). Furthermore, it has been shown that MP can be increased by dietary modification (Landrum et al. 1997a; Berendschot et al. 2000; Bone et al. 2003; Neuringer et al. 2004; Johnson et al. 2005, 2008; Köpcke et al. 2005; Snodderly et al. 2005; Bhosale et al. 2007; Bone et al. 2007; Richer et al. 2007; Schalch et al. 2007; Rougier et al. 2008; Tanito et al. 2009; Zeimer et al. 2009) or by supplements (Koh et al. 2004; Trieschmann et al. 2007) in healthy subjects as well as in subjects with a diseased macula (Bone et al. 1993, 1997). Meso-zeaxanthin, which has not yet been studied extensively in foods, is believed to be formed in the retina by conversion from lutein; however, detailed analysis of foods for MZ is merited (Izumi-Nagai et al. 2007). Thus, as with smoking, a possible explanation of the observed inverse association between a diet low in lutein and zeaxanthin and AMD could again be due, in part, to the fact that low intake of these carotenoids and MP may be correlated. Note that lutein has also an anti-inflammatory effect, which may act systemically (Sommerburg et al. 1998; Granado et al. 2006; Thurnham 2007; Maiani et al. 2008).

Foods that are yellow or green in color, such as corn and egg yolk, as well as green vegetables like spinach, broccoli, and kale, may contain a high concentration of lutein. Orange peppers, egg yolk, and sweet corn are good sources of zeaxanthin (Sommerburg et al. 1998; Granado et al. 2006; Thurnham 2007; Maiani et al. 2008). In general, fruit and vegetables contain five times more lutein than zeaxanthin. A typical Western diet contains 1.3–3 mg/day of lutein and zeaxanthin combined (Hammond et al. 1997). This results in serum levels that are far below those that may be achieved following supplementation with a carotenoid formulation.

## 2.5 SEX

Females have a higher life expectancy than males, and it is possible that their higher life expectancy explains the increased prevalence of AMD in females. However, when longevity is taken into account, gender differences in AMD risk generally disappear (Hammond et al. 1996a; Hammond and Caruso-Avery 2000; Broekmans et al. 2002; Mellerio et al. 2002; Nolan et al. 2007b; Raman et al. 2011). Nevertheless, in line with the apparent higher risk of AMD for females, a number of studies (Beatty et al. 2001; Ciulla et al. 2001; Curran-Celentano et al. 2001; Tang et al. 2004; Burke et al. 2005; Iannaccone et al. 2007), although not all (Greene et al. 2006; Waters et al. 2007; Kirby et al. 2009; Loane et al. 2010), reported that females have a lower MP than males. Several mechanisms have been proposed to explain this observation. First, high-density lipoprotein (HDL) is important for the transport of L and Z in serum (Schaefer et al. 1983; Wenzel et al. 2007; Nolan et al. 2007b). As such, differences in lipid concentrations may influence serum concentrations of the macular carotenoids, with possible effects on MP (Hammond et al. 1996a). Secondly, steroid hormones may affect lutein metabolism (Hammond et al. 2002; Nolan et al. 2004). Finally, females generally have a higher percentage body fat, a factor known to influence serum carotenoid concentrations (Clemons et al. 2005; Moeini et al. 2005).

## 2.6 BODY MASS INDEX

Body mass index (BMI) has been found to be associated with progression of AMD (Hammond et al. 2002; Nolan et al. 2007b). Both Hammond et al. and Nolan et al. (2004) found an inverse relationship between BMI and MP. In line with this, in another study by Nolan et al., it was observed that a significant inverse relationship exists between the percentage of body fat and MP (Roodenburg et al. 2000; Brown et al. 2004; Wenzel et al. 2006; Johnson et al. 2008). Again, this modifiable risk factor could act through a decrease in MP, but for efficient uptake of the retinal carotenoids, a certain amount of fat in the diet is recommended (Winkler et al. 1999; Beatty et al. 2000; Hollyfield 2010).

## 2.7 (SUN) LIGHT

Oxidative damage is thought to play an important role in the development of AMD (Sparrow et al. 2000; Sparrow and Boulton 2005). This can occur because of light interacting with photosensitizers in the presence of oxygen, which produces singlet oxygen (Vos 1972; Bone et al. 1992; DeMarco et al. 1992; Sharpe et al. 1998). MP absorbs blue light (Thomson et al. 2002; Landrum and Bone 2004; Kim et al. 2006; Nolan et al. 2007b; Barker et al. 2011; Lien and Hammond 2011) and thus may decrease photochemical light damage, resulting from singlet oxygen-associated oxidation (Barker et al. 2011). Barker et al. (2011) have shown that the fovea is less sensitive to blue light–induced damage than the parafovea. They also proved that foveal protection was absent in xanthophyll-free animals but was restored by supplementation (Fletcher et al. 2008). Comparing data on sunlight exposure of AMD and control subjects, Fletcher et al. found a significant association between blue light exposure and neovascular AMD (Mellerio et al. 2002). It is possible, as discussed with the other factors presented earlier, that the risk that sunlight presents for AMD is due to a lack of MP caused by light exposure (Wenzel et al. 2003).

However, MP seems unaffected by light and oxidation throughout the day (Jahn et al. 2006; Nolan et al. 2006) and over a longer 1-year period (Nolan et al. 2009; Obana et al. 2011). By contrast, differences in MP have been observed between subjects with clear and yellow-tinted intraocular lenses (IOLs), indicating that excessive blue light exposure is associated inversely with MPOD, since clear IOLs transmit higher intensities of blue light than yellow-tinted IOLs (Chamberlain et al. 2006; Despriet et al. 2006; Fagerness et al. 2009; Anderson et al. 2010; Hecker and Edwards 2010; Khandhadia et al. 2012).

## 2.8 GENETICS

A recent and very exciting field in the study of AMD involves investigation of single-nucleotide polymorphism (SNP) to the prevalence of pathology (Liew et al. 2005). In particular, variants in several genes encoding complement pathway proteins have been associated with AMD. Genetic factors must play an important role in determining the MPOD. In the healthy eye, heritability estimates of 0.67 and 0.85 have been estimated (Nolan et al. 2007b). Furthermore, subjects with a confirmed family history of age-related maculopathy (ARM) showed significantly lower MPOD than subjects with no known family history of disease (Zerbib et al. 2009). A possible candidate for this observation is the *SCARB1* gene (Reynolds et al. 2010). It is involved in the metabolism of lutein and expressed in the retinal pigment epithelium. It interacts with *APOE*, a gene that may be involved in AMD. Another pathway of action may be through the *LIPC* gene (Greene et al. 2006; Waters et al. 2007; Kirby et al. 2009; Loane et al., 2010). The HDL-raising allele of the *LIPC* gene (T) was associated with a reduced risk of AMD. Higher total cholesterol and low-density lipoprotein levels were associated with increased risk, whereas higher HDL levels tended to be associated with reduced risk of AMD, in line with the observation that HDL is involved in the transport of lutein (Borel et al. 2011). Yet another study suggests that genetic variants in *BCMO1* and *CD36* genes can modulate blood and retina concentrations of lutein (Zhang et al. 2002; Rougier et al. 2011) and thus the MPOD. Recently, Meyers et al. (2012) investigated genetic determinants of the density of lutein and zeaxanthin in the macula of women from the Carotenoids in Age-Related Eye Disease Study. They found six genes to be associated with MPOD: five SNPs in three genes related to HDL levels or cholesterol transport (*ABCA1*, *ABCG5*, and *LIPC*); three SNPs in *SCARB1*, which encodes a plasma membrane scavenger lipoprotein receptor, which has been related to carotenoid uptake; two SNPs in *RPE65*, which encodes a retinoid-binding protein in the retinoid visual cycle; and four SNPs from *BCMO1*, which encodes a provitamin A carotenoid cleavage enzyme. Finally, diseases related to AMD, like Stargardt's disease (Aleman et al. 2007) and ABCA4-associated retinal degenerations, also show a reduction in MPOD.

## 2.9 CONCLUSIONS

This chapter discussed the relationship between known and putative risk factors for AMD with MP. Aging, the most established and accepted risk factor for AMD, seems to be an independent risk factor for AMD. However, other factors like gender, smoking, diet, and body mass influence the MPOD and therefore their association with AMD may at least in part be due to a difference in MPOD. Furthermore, the risk that sunlight presents for AMD may be due to a lack of MP, although MP itself seems unaffected by light and oxidation throughout the day. Finally, several genes that play a role in AMD are also associated with MP density.

# REFERENCES

Age-Related Eye Disease Study Research Group. (2000): Risk factors associated with age-related macular degeneration. A case-control study in the age-related eye disease study: age-related eye disease study report number 3. Age-Related Eye Disease Study Research Group. *Ophthalmology* 107: 2224–2232.

Aleman TS, Cideciyan AV, Windsor EAM, Schwartz SB, Swider M, Chico JD, Sumaroka A, et al. (2007): Macular pigment and lutein supplementation in ABCA4-associated retinal degenerations. *Invest Ophthalmol Vis Sci* 48: 1319–1329.

Anderson DH, Radeke MJ, Gallo NB, Chapin EA, Johnson PT, Curletti CR, Hancox LS, et al. (2010): The pivotal role of the complement system in aging and age-related macular degeneration: hypothesis re-visited. *Prog Retin Eye Res* 29: 95–112.

Barker FM, Snodderly DM, Johnson EJ, Schalch W, Koepcke W, Gerss J, Neuringer M. (2011): Nutritional manipulation of primate retinas, V: effects of lutein, zeaxanthin, and n-3 fatty acids on retinal sensitivity to blue-light-induced damage. *Invest Ophthalmol Vis Sci* 52: 3934–3942.

Beatty S, Koh H, Phil M, Henson D, Boulton M. (2000): The role of oxidative stress in the pathogenesis of age-related macular degeneration. *Surv Ophthalmol* 45: 115–134.

Beatty S, Murray IJ, Henson DB, Carden D, Koh H, Boulton ME. (2001): Macular pigment and risk for age-related macular degeneration in subjects from a Northern European population. *Invest Ophthalmol Vis Sci* 42: 439–446.

Beatty S, Stevenson M, Nolan JM, Woodside J, The CARMA Study Group, Chakravarthy U. (2009): Longitudinal relationships between macular pigment and serum lutein in patients enrolled in the CARMA clinical trial (carotenoids and co-antioxidants in age-related maculopathy). *Invest Ophthalmol Vis Sci* 50: 1719.

Berendschot TTJM, Goldbohm RA, Klöpping WA, van de Kraats J, van Norel J, van Norren D. (2000): Influence of lutein supplementation on macular pigment, assessed with two objective techniques. *Invest Ophthalmol Vis Sci* 41: 3322–3326.

Berendschot TTJM, Tian Y, van der Veen RLP, Makridaki M, Murray IJ, Kijlstra A. (2012): Lutein decreases complement factor D in age-related macular degeneration. *Acta Ophthalmol* 90. doi: 10.1111/j.1755-3768.2012.3663.x.

Berendschot TTJM, van Norren D. (2004): Objective determination of the macular pigment optical density using fundus reflectance spectroscopy. *Arch Biochem Biophys* 430: 149–155.

Berendschot TTJM, van Norren D. (2005): On the age dependency of the macular pigment optical density. *Exp Eye Res* 81: 602–609.

Berendschot TTJM, Willemse-Assink JJM, Bastiaanse M, de Jong PTVM, van Norren D. (2002): Macular pigment and melanin in age-related maculopathy in a general population. *Invest Ophthalmol Vis Sci* 43: 1928–1932.

Bernstein PS, Gellermann W. (2003a): Author response: assessment of the Raman method of measuring human macular pigment. *Invest Ophthalmol Vis Sci* [serial online] Available at http://www.iovs.org/cgi/eletters?lookup = by_date&days = 9999#74: August 15, 2003.

Bernstein PS, Gellermann W. (2003b): Author response: assessment of the Raman method of measuring human macular pigment (II). *Invest Ophthalmol Vis Sci* [serial online] Available at http://www.iovs.org/cgi/eletters?lookup = by_date&days = 9999#94: December 30, 2003.

Bernstein PS, Zhao DY, Sharifzadeh M, Ermakov IV, Gellermann W. (2004): Resonance Raman measurement of macular carotenoids in the living human eye. *Arch Biochem Biophys* 430: 163–169.

Bhosale P, Zhao DY, Bernstein PS. (2007): HPLC measurement of ocular carotenoid levels in human donor eyes in the lutein supplementation era. *Invest Ophthalmol Vis Sci* 48: 543–549.

Bohm F, Edge R, Truscott G. (2012): Interactions of dietary carotenoids with activated (singlet) oxygen and free radicals: potential effects for human health. *Mol Nutr Food Res* 56: 205–216.

Bone RA, Landrum JT, Cains A. (1992): Optical density spectra of the macular pigment *in vivo* and *in vitro*. *Vision Res* 32: 105–110.

Bone RA, Landrum JT, Cao Y, Howard AN, Alvarez-Calderon F. (2007): Macular pigment response to a supplement containing meso-zeaxanthin, lutein and zeaxanthin. *Nutr Metab (Lond)* 4: 12.

Bone RA, Landrum JT, Fernandez L, Tarsis SL. (1988): Analysis of the macular pigment by HPLC: retinal distribution and age study. *Invest Ophthalmol Vis Sci* 29: 843–849.

Bone RA, Landrum JT, Friedes LM, Gomez CM, Kilburn MD, Menendez E, Vidal I, Wang W. (1997): Distribution of lutein and zeaxanthin stereoisomers in the human retina. *Exp Eye Res* 64: 211–218.

Bone RA, Landrum JT, Guerra LH, Ruiz CA. (2003): Lutein and zeaxanthin dietary supplements raise macular pigment density and serum concentrations of these carotenoids in humans. *J Nutr* 133: 992–998.

Bone RA, Landrum JT, Hime GW, Cains A, Zamor J. (1993): Stereochemistry of the human macular carotenoids. *Invest Ophthalmol Vis Sci* 34: 2033–2040.

Bone RA, Landrum JT, Mayne ST, Gomez CM, Tibor SE, Twaroska EE. (2001): Macular pigment in donor eyes with and without AMD: a case-control study. *Invest Ophthalmol Vis Sci* 42: 235–240.

Borel P, de Edelenyi FS, Vincent-Baudry S, Malezet-Desmoulin C, Margotat A, Lyan B, Gorrand JM, Meunier N, Drouault-Holowacz S, Bieuvelet S. (2011): Genetic variants in BCMO1 and CD36 are associated with plasma lutein concentrations and macular pigment optical density in humans. *Ann Med* 43: 47–59.

Broekmans WMR, Berendschot TTJM, Klöpping WA, de Vries AJ, Goldbohm RA, Tijssen CC, Karplus M, van Poppel G. (2002): Macular pigment density in relation to serum and adipose tissue concentrations of lutein and serum concentrations of zeaxanthin. *Am J Clin Nutr* 76: 595–603.

Brown MJ, Ferruzzi MG, Nguyen ML, Cooper DA, Eldridge AL, Schwartz SJ, White WS. (2004): Carotenoid bioavailability is higher from salads ingested with full-fat than with fat-reduced salad dressings as measured with electrochemical detection. *Am J Clin Nutr* 80: 396–403.

Burke JD, Curran-Celentano J, Wenzel AJ. (2005): Diet and serum carotenoid concentrations affect macular pigment optical density in adults 45 years and older. *J Nutr* 135: 1208–1214.

Carpentier S, Knaus M, Suh M. (2009): Associations between lutein, zeaxanthin, and age-related macular degeneration: an overview. *Crit Rev Food Sci Nutr* 49: 313–326.

Chakravarthy U, Augood C, Bentham GC, de Jong PTVM, Rahu M, Seland J, Soubrane G, et al. (2007): Cigarette smoking and age-related macular degeneration in the EUREYE study. *Ophthalmology* 114: 1157–1163.

Chamberlain M, Baird P, Dirani M, Guymer R. (2006): Unraveling a complex genetic disease: age-related macular degeneration. *Surv Ophthalmol* 51: 576–586.

Chen SF, Chang Y, Wu JC. (2001): The spatial distribution of macular pigment in humans. *Curr Eye Res* 23: 422–434.

Cho E, Seddon JM, Rosner B, Willett WC, Hankinson SE. (2004): Prospective study of intake of fruits, vegetables, vitamins, and carotenoids and risk of age-related maculopathy. *Arch Ophthalmol* 122: 883–892.

Chong EW, Wong TY, Kreis AJ, Simpson JA, Guymer RH. (2007): Dietary antioxidants and primary prevention of age related macular degeneration: systematic review and meta-analysis. *BMJ* 335: 755.

Chung HY, Rasmussen HM, Johnson EJ. (2004): Lutein bioavailability is higher from lutein-enriched eggs than from supplements and spinach in men. *J Nutr* 134: 1887–1893.

Ciulla TA, Curran-Celentano J, Cooper DA, Hammond BR, Danis RP, Pratt LM, Riccardi KA, Filloon TG. (2001): Macular pigment optical density in a midwestern sample. *Ophthalmology* 108: 730–737.

Ciulla TA, Hammond JR. (2004): Macular pigment density and aging, assessed in the normal elderly and those with cataracts and age-related macular degeneration. *Am J Ophthalmol* 138: 582–587.

Clemons TE, Klein R, Seddon JM, Ferris FL, The Age-Related Eye Disease Study Research Group. (2005): Risk factors for the incidence of advanced age-related macular degeneration in the age-related eye disease study (AREDS): AREDS report no. 19. *Ophthalmology* 112: 533–539.

Connolly EE, Beatty S, Loughman J, Howard AN, Louw MS, Nolan JM. (2011): Supplementation with all three macular carotenoids: response, stability, and safety. *Invest Ophthalmol Vis Sci* 52: 9207–9217.

Curran-Celentano J, Hammond BR, Ciulla TA, Cooper DA, Pratt LM, Danis RB. (2001): Relation between dietary intake, serum concentrations, and retinal concentrations of lutein and zeaxanthin in adults in a midwest population. *Am J Clin Nutr* 74: 796–802.

Dallongeville J, Marecaux N, Fruchart JC, Amouyel P. (1998): Cigarette smoking is associated with unhealthy patterns of nutrient intake: a meta-analysis. *J Nutr* 128: 1450–1457.

Davies NP, Morland AB. (2002): Color matching in diabetes: optical density of the crystalline lens and macular pigments. *Invest Ophthalmol Vis Sci* 43: 281–289.

Davies NP, Morland AB. (2004): Macular pigments: their characteristics and putative role. *Prog Retin Eye Res* 23: 533–559.

Delcourt C, Carriere I, Delage M, Barberger-Gateau P, Schalch W, The Pola Study Group. (2006): Plasma lutein and zeaxanthin and other carotenoids as modifiable risk factors for age-related maculopathy and cataract: the POLA Study. *Invest Ophthalmol Vis Sci* 47: 2329–2335.

Delori FC, Goger DG, Hammond BR, Snodderly DM, Burns SA. (2001): Macular pigment density measured by autofluorescence spectrometry: comparison with reflectometry and heterochromatic flicker photometry. *J Opt Soc Am A* 18: 1212–1230.

DeMarco P, Pokorny J, Smith VC. (1992): Full-spectrum cone sensitivity functions for X-chromosome-linked anomalous trichromats. *J Opt Soc Am A* 9: 1465–1476.

Despriet DD, Klaver CCW, Witteman JC, Bergen AAB, Kardys I, de Maat MP, Boekhoorn SS, et al. (2006): Complement factor H polymorphism, complement activators, and risk of age-related macular degeneration. *JAMA* 296: 301–309.

Dietzel M, Zeimer M, Heimes B, Claes B, Pauleikhoff D, Hense HW. (2011): Determinants of macular pigment optical density and its relation to age-related maculopathy: results from the Muenster Aging and Retina Study (MARS). *Invest Ophthalmol Vis Sci* 52: 3452–3457.

Evans JR. (2001): Risk factors for age-related macular degeneration. *Prog Retin Eye Res* 20: 227–253.

Evans JR, Fletcher AE, Wormald RP. (2005): 28,000 cases of age related macular degeneration causing visual loss in people aged 75 years and above in the United Kingdom may be attributable to smoking. *Br J Ophthalmol* 89: 550–553.

Fagerness JA, Maller JB, Neale BM, Reynolds RC, Daly MJ, Seddon JM. (2009): Variation near complement factor I is associated with risk of advanced AMD. *Eur J Hum Genet* 17: 100–104.

Fletcher AE, Bentham GC, Agnew M, Young IS, Augood C, Chakravarthy U, de Jong PTVM, et al. (2008): Sunlight exposure, antioxidants, and age-related macular degeneration. *Arch Ophthalmol* 126: 1396–1403.

Gale CR, Hall NF, Phillips DI, Martyn CN. (2003): Lutein and zeaxanthin status and risk of age-related macular degeneration. *Invest Ophthalmol Vis Sci* 44: 2461–2465.

Gellermann W, Bernstein PS. (2004): Noninvasive detection of macular pigments in the human eye. *J Biomed Opt* 9: 75–78.

Gellermann W, Ermakov IV, Ermakova MR, McClane RW, Zhao DY, Bernstein PS. (2002): In vivo resonant Raman measurement of macular carotenoid pigments in the young and the aging human retina. *J Opt Soc Am A-Opt Image Sci Vis* 19: 1172–1186.

Granado F, Olmedilla B, Herrero C, Perez-Sacristan B, Blanco I, Blazquez S. (2006): Bioavailability of carotenoids and tocopherols from broccoli: *in vivo* and *in vitro* assessment. *Exp Biol Med* 231: 1733–1738.

Greene C, Waters D, Clark R, Contois J, Fernandez M. (2006): Plasma LDL and HDL characteristics and carotenoid content are positively influenced by egg consumption in an elderly population. *Nutr Metab (Lond)* 3: 6.

Hammond BR, Caruso-Avery M. (2000): Macular pigment optical density in a southwestern sample. *Invest Ophthalmol Vis Sci* 41: 1492–1497.

Hammond BR, Ciulla TA, Snodderly DM. (2002): Macular pigment density is reduced in obese subjects. *Invest Ophthalmol Vis Sci* 43: 47–50.

Hammond BR, Curran-Celentano J, Judd S, Fuld K, Krinsky NI, Wooten BR, Snodderly DM. (1996a): Sex differences in macular pigment optical density: relation to plasma carotenoid concentrations and dietary patterns. *Vision Res* 36: 2001–2012.

Hammond BR, Johnson EJ, Russell RM, Krinsky NI, Yeum KJ, Edwards RB, Snodderly DM. (1997): Dietary modification of human macular pigment density. *Invest Ophthalmol Vis Sci* 38: 1795–1801.

Hammond BR, Wooten BR, Snodderly DM. (1996b): Cigarette smoking and retinal carotenoids: implications for age-related macular degeneration. *Vision Res* 36: 3003–3009.

Handelman GJ, Dratz EA, Reay CC, van Kuijk JG. (1988): Carotenoids in the human macula and whole retina. *Invest Ophthalmol Vis Sci* 29: 850–855.

Hecker LA, Edwards AO. (2010): Genetic control of complement activation in humans and age related macular degeneration. *Adv Exp Med Biol* 703: 49–62.

Hogg RE, Ong EL, Chamberlain M, Dirani M, Baird PN, Guymer RH, Fitzke F. (2012): Heritability of the spatial distribution and peak density of macular pigment: a classical twin study. *Eye* 26: 1217–1225.

Hollyfield JG. (2010): Age-related macular degeneration: the molecular link between oxidative damage, tissue-specific inflammation and outer retinal disease: the proctor lecture. *Invest Ophthalmol Vis Sci* 51: 1276–1281.

Hyman L, Neborsky R. (2002): Risk factors for age-related macular degeneration: an update. *Curr Opin Ophthalmol* 13: 171–175.

Iannaccone A, Mura M, Gallaher KT, Johnson EJ, Todd WA, Kenyon E, Harris TL, et al. (2007): Macular pigment optical density in the elderly: findings in a large biracial midsouth population sample. *Invest Ophthalmol Vis Sci* 48: 1458–1465.

Izumi-Nagai K, Nagai N, Ohgami K, Satofuka S, Ozawa Y, Tsubota K, Umezawa K, Ohno S, Oike Y, Ishida S. (2007): Macular pigment lutein is antiinflammatory in preventing choroidal neovascularization. *Arterioscler Thromb Vasc Biol* 27: 2555–2562.

Jahn C, Brinkmann C, Mössner A, Wüstemeyer H, Schnurrbusch U, Wolf S. (2006): Seasonal fluctuations and influence of nutrition on macular pigment density. *Ophthalmologe* 103: 136–140.

Jahn C, Wüstemeyer H, Brinkmann C, Trautmann S, Moessner A, Wolf S. (2005): Macular pigment density in age-related maculopathy. *Graefes Arch Clin Exp Ophthalmol* 243: 222–227.

Johnson EJ, Chung HY, Caldarella SM, Snodderly DM. (2008): The influence of supplemental lutein and docosahexaenoic acid on serum, lipoproteins, and macular pigmentation. *Am J Clin Nutr* 87: 1521–1529.

Johnson EJ, Hammond BR, Yeum KJ, Qin J, Wang XD, Castaneda C, Snodderly DM, Russell RM. (2000): Relation among serum and tissue concentrations of lutein and zeaxanthin and macular pigment density. *Am J Clin Nutr* 71: 1555–1562.

Johnson EJ, Neuringer M, Russell RM, Schalch W, Snodderly DM. (2005): Nutritional manipulation of primate retinas, III: effects of lutein or zeaxanthin supplementation on adipose tissue and retina of xanthophyll-free monkeys. *Invest Ophthalmol Vis Sci* 46: 692–702.

Khachik F, Bernstein PS, Garland DL. (1997): Identification of lutein and zeaxanthin oxidation products in human and monkey retinas. *Invest Ophthalmol Vis Sci* 38: 1802–1811.

Khandhadia S, Cipriani V, Yates JRW, Lotery AJ. (2012): Age-related macular degeneration and the complement system. *Immunobiology* 217: 127–146.

Kijlstra A, Tian Y, Kelly ER, Berendschot TTJM. (2012): Lutein: more than just a filter for blue light. *Prog Retin Eye Res* 31: 303–315.

Kilbride PE, Alexander KR, Fishman M, Fishman GA. (1989): Human macular pigment assessed by imaging fundus reflectometry. *Vision Res* 29: 663–674.

Kim SR, Nakanishi K, Itagaki Y, Sparrow JR. (2006): Photooxidation of A2-PE, a photoreceptor outer segment fluorophore, and protection by lutein and zeaxanthin. *Exp Eye Res* 82: 828–839.

Kirby ML, Harrison M, Beatty S, Greene I, Nolan JM. (2009): Changes in macular pigment optical density and serum concentrations of lutein and zeaxanthin in response to weight loss. *Invest Ophthalmol Vis Sci* 50: 1709.

Klaver CCW, Assink JJ, Vingerling JR, Hofman A, de Jong PTVM. (1997): Smoking is also associated with age-related macular degeneration in persons aged 85 years and older: The Rotterdam Study. *Arch Ophthalmol* 115: 945.

Klein R, Cruickshanks KJ, Nash SD, Krantz EM, Nieto FJ, Huang GH, Pankow JS, Klein BEK. (2010): The prevalence of age-related macular degeneration and associated risk factors. *Arch Ophthalmol* 128: 750–758.

Klein R, Peto T, Bird A, Vannewkirk MR. (2004): The epidemiology of age-related macular degeneration. *Am J Ophthalmol* 137: 486–495.

Koh HH, Murray IJ, Nolan D, Carden D, Feather J, Beatty S. (2004): Plasma and macular responses to lutein supplement in subjects with and without age-related maculopathy: a pilot study. *Exp Eye Res* 79: 21–27.

Köpcke W, Schalch W, LUXEA-Study Group. (2005): Changes in macular pigment optical density following repeated dosing with lutein, zeaxanthin, or their combination in healthy volunteers—results of the LUXEA study. *Invest Ophthalmol Vis Sci* 46: 1768.

Landrum JT, Bone RA. (2004): Dietary lutein & zeaxanthin: reducing the risk for macular degeneration. *Agro Food Ind Hi Tech* 15: 22–25.

Landrum JT, Bone RA, Chen Y, Herrero C, Llerena CM, Twarowska E. (1999): Carotenoids in the human retina. *Pure Appl Chem* 71: 2237–2244.

Landrum JT, Bone RA, Joa H, Kilburn MD, Moore LL, Sprague KE. (1997a): A one year study of the macular pigment: the effect of 140 days of a lutein supplement. *Exp Eye Res* 65: 57–62.

Landrum JT, Bone RA, Kilburn MD. (1997b): The macular pigment: a possible role in protection from age-related macular degeneration. *Adv Pharmacol* 38: 537–556.

Lien EL, Hammond BR. (2011): Nutritional influences on visual development and function. *Prog Retin Eye Res* 30: 188–203.

Liew SHM, Gilbert CE, Spector TD, Mellerio J, Marshall J, van Kuijk FJGM, Beatty S, Fitzke F, Hammond CJ. (2005): Heritability of macular pigment: a twin study. *Invest Ophthalmol Vis Sci* 46: 4430–4436.

Loane E, Beatty S, Nolan JM. (2009): The relationship between lutein, zeaxanthin, serum lipoproteins and macular pigment optical density. *Invest Ophthalmol Vis Sci* 50: 1710.

Loane E, Kelliher C, Beatty S, Nolan JM. (2008): The rationale and evidence base for a protective role of macular pigment in age-related maculopathy. *Br J Ophthalmol* 92: 1163–1168.

Loane E, Nolan JM, Beatty S. (2010): The respective relationships between lipoprotein profile, macular pigment optical density, and serum concentrations of lutein and zeaxanthin. *Invest Ophthalmol Vis Sci* 51: 5897–5905.

Maiani G, Caston MJ, Catasta G, Toti E, Cambrodon IG, Bysted A, Granado-Lorencio F, et al. (2008): Carotenoids: actual knowledge on food sources, intakes, stability and bioavailability and their protective role in humans. *Mol Nutr Food Res* 53: S194–S218.

Mares-Perlman JA. (1999): Too soon for lutein supplements. *Am J Clin Nutr* 70: 431–432.

Mares-Perlman JA, Millen AE, Ficek TL, Hankinson SE. (2002): The body of evidence to support a protective role for lutein and zeaxanthin in delaying chronic disease. Overview. *J Nutr* 132: 518S–524S.

Mellerio J, Ahmadi-Lari S, van Kuijk F, Pauleikhoff D, Bird A, Marshall J. (2002): A portable instrument for measuring macular pigment with central fixation. *Curr Eye Res* 25: 37–47.

Meyers KJ, Johnson EJ, Iyengar SK, Igo RP, Snodderly M, Klein ML, Bernstein PS, Millen AE, Hageman GS, Mares JA. (2012): Genetic determinants of macular pigment optical density in the carotenoids in age-related eye diseases study (CAREDS). *Invest Ophthalmol Vis Sci* 53: 1323.

Miyazaki M, Nakamura H, Kubo M, Kiyohara Y, Oshima Y, Ishibashi T, Nose Y. (2003): Risk factors for age related maculopathy in a Japanese population: the Hisayama study. *Br J Ophthalmol* 87: 469–472.

Moeini HA, Masoudpour H, Ghanbari H. (2005): A study of the relation between body mass index and the incidence of age related macular degeneration. *Br J Ophthalmol* 89: 964–966.

Mozaffarieh M, Sacu S, Wedrich A. (2003): The role of the carotenoids, lutein and zeaxanthin, in protecting against age-related macular degeneration: a review based on controversial evidence. *Nutr J* 2: 20.

Neelam K, O'Gorman N, Nolan J, O'Donovan O, Wong HB, Eong KGA, Beatty S. (2005): Measurement of macular pigment: Raman spectroscopy versus heterochromatic flicker photometry. *Invest Ophthalmol Vis Sci* 46: 1023–1032.

Neuringer M, Sandstrom MM, Johnson EJ, Snodderly DM. (2004): Nutritional manipulation of primate retinas, I: effects of lutein or zeaxanthin supplements on serum and macular pigment in xanthophyll-free rhesus monkeys. *Invest Ophthalmol Vis Sci* 45: 3234–3243.

Nolan J, O'Donovan O, Kavanagh H, Stack J, Harrison M, Muldoon A, Mellerio J, Beatty S. (2004): Macular pigment and percentage of body fat. *Invest Ophthalmol Vis Sci* 45: 3940–3950.

Nolan JM, O'Reilly P, Loughman J, Loane E, Connolly EE, Stack J, Beatty S. (2009): Macular pigment levels increase following blue-light filtering intraocular lens implantation. *Invest Ophthalmol Vis Sci* 50: 1721.

Nolan JM, Stack J, Mellerio J, Godhinio M, O'Donovan O, Neelam K, Beatty S. (2006): Monthly consistency of macular pigment optical density and serum concentrations of lutein and zeaxanthin. *Curr Eye Res* 31: 199–213.

Nolan JM, Stack J, O'Connell E, Beatty S. (2007a): The relationships between macular pigment optical density and its constituent carotenoids in diet and serum. *Invest Ophthalmol Vis Sci* 48: 571–582.

Nolan JM, Stack J, O'Donovan O, Loane E, Beatty S. (2007b): Risk factors for age-related maculopathy are associated with a relative lack of macular pigment. *Exp Eye Res* 84: 61–74.

Nolan JM, Stringham JM, Beatty S, Snodderly DM. (2008): Spatial profile of macular pigment and its relationship to foveal architecture. *Invest Ophthalmol Vis Sci* 49: 2134–2142.

Obana A, Tanito M, Gohto Y, Gellermann W, Okazaki S, Ohira A. (2011): Macular pigment changes in pseudophakic eyes quantified with resonance Raman spectroscopy. *Ophthalmology* 118: 1852–1858.

O'Connell ED, Nolan JM, Stack J, Greenberg D, Kyle J, Maddock L, Beatty S. (2008): Diet and risk factors for age-related maculopathy. *Am J Clin Nutr* 87: 712–722.

Raman R, Biswas S, Gupta A, Kulothungan V, Sharma T. (2012): Association of macular pigment optical density with risk factors for wet age-related macular degeneration in the Indian population. *Eye* 26: 950–957.

Raman R, Rajan R, Biswas S, Vaitheeswaran K, Sharma T. (2011): Macular pigment optical density in a South Indian population. *Invest Ophthalmol Vis Sci* 52: 7910–7916.

Reynolds R, Rosner B, Seddon JM. (2010): Serum lipid biomarkers and hepatic lipase gene associations with age-related macular degeneration. *Ophthalmol* 117: 1989–1995.

Richer S, Devenport J, Lang JC. (2007): LAST II: differential temporal responses of macular pigment optical density in patients with atrophic age-related macular degeneration to dietary supplementation with xanthophylls. *Optometry* 78: 213–219.

Richer S, Stiles W, Statkute L, Pulido J, Franskowski J, Rudy D, Pei K, Tsipusrky M, Nyland J. (2004): Double-masked, placebo-controlled, randomized trial of lutein and antioxidant supplementation in the intervention of atrophic age-related macular degeneration: the Veterans LAST study (Lutein Antioxidant Supplementation Trial). *Optometry* 75: 216–230.

Roodenburg AJ, Leenen R, van het Hof KH, Weststrate JA, Tijburg LB. (2000): Amount of fat in the diet affects bioavailability of lutein esters but not of alpha-carotene, beta-carotene, and vitamin E in humans. *Am J Clin Nutr* 71: 1187–1193.

Rougier MB, Delyfer MN, Korobelnik JF. (2008): [Measuring macular pigment *in vivo*]. *J Fr Ophtalmol* 31: 445–453.

Rougier MB, Delyfer MN, Korobelnik JF. (2011): Macular pigment distribution in Stargardt macular disease. *J Fr Ophtalmol* 34: 287–293.

Sasamoto Y, Gomi F, Sawa M, Tsujikawa M, Nishida K. (2011): Effect of 1-year lutein supplementation on macular pigment optical density and visual function. *Graefes Arch Clin Exp Ophthalmol* 249: 1847–1854.

Schaefer EJ, Foster DM, Zech LA, Lindgren FT, Brewer HB, Jr., Levy RI. (1983): The effects of estrogen administration on plasma lipoprotein metabolism in premenopausal females. *J Clin Endocrinol Metab* 57: 262–267.

Schalch W, Cohn W, Barker FM, Kopcke W, Mellerio J, Bird AC, Robson AG, Fitzke FF, van Kuijk FJ. (2007): Xanthophyll accumulation in the human retina during supplementation with lutein or zeaxanthin—the LUXEA (LUtein Xanthophyll Eye Accumulation) study. *Arch Biochem Biophys* 458: 128–135.

Seddon JM, Ajani UA, Sperduto RD, Hiller R, Blair N, Burton TC, Farber MD, Gragoudas ES, Haller J, Miller DT. (1994): Dietary carotenoids, vitamins A, C, and E, and advanced age-related macular degeneration. Eye disease case-control study group. *JAMA* 272: 1413–1420.

Seddon JM, George S, Rosner B. (2006): Cigarette smoking, fish consumption, omega-3 fatty acid intake, and associations with age-related macular degeneration: the US Twin Study of Age-Related Macular Degeneration. *Arch Ophthalmol* 124: 995–1001.

Seddon JM, Reynolds R, Yu Y, Daly MJ, Rosner B. (2011): Risk models for progression to advanced age-related macular degeneration using demographic, environmental, genetic, and ocular factors. *Ophthalmology* 118: 2203–2211.

Sharpe LT, Stockman A, Knau H, Jägle H. (1998): Macular pigment densities derived from central and peripheral spectral sensitivity differences. *Vision Res* 38: 3233–3239.

Snellen EL, Verbeek AL, van den Hoogen GW, Cruysberg JR, Hoyng CB. (2002): Neovascular age-related macular degeneration and its relationship to antioxidant intake. *Acta Ophthalmol Scand* 80: 368–371.

Snodderly DM, Chung HC, Caldarella SM, Johnson EJ. (2005): The influence of supplemental lutein and docosahexaenoic acid on their serum levels and on macular pigment. *Invest Ophthalmol Vis Sci* 46: 1766.

Sommerburg O, Keunen JEE, Bird AC, van Kuijk FJGM. (1998): Fruits and vegetables that are sources for lutein and zeaxanthin: the macular pigment in human eyes. *Br J Ophthalmol* 82: 907–910.

Sparrow JR, Boulton M. (2005): RPE lipofuscin and its role in retinal pathobiology. *Exp Eye Res* 80: 595–606.

Sparrow JR, Nakanishi K, Parish CA. (2000): The lipofuscin fluorophore A2E mediates blue light-induced damage to retinal pigmented epithelial cells. *Invest Ophthalmol Vis Sci* 41: 1981–1989.

Stahl W, Sies H. (2012): Photoprotection by dietary carotenoids: concept, mechanisms, evidence and future development. *Mol Nutr Food Res* 56: 287–295.

Tang CY, Yip HS, Poon MY, Yau WL, Yap MK. (2004): Macular pigment optical density in young Chinese adults. *Ophthalmic Physiol Opt* 24: 586–593.

Tanito M, Obana A, Gohto Y, Okazaki S, Gellermann W, Ohira A. (2012): Macular pigment density changes in Japanese individuals supplemented with lutein or zeaxanthin: quantification via resonance Raman spectrophotometry and autofluorescence imaging. *Jpn J Ophthalmol* 56: 488–496.

Tanito M, Obana A, Okazaki S, Ohira A, Gellermann W. (2009): Change of macular pigment density quantified with resonance Raman spectrophotometry and autofluorescence imaging in normal subjects supplemented with oral lutein or zeaxanthin. *Invest Ophthalmol Vis Sci* 50: 1716.

Thomson LR, Toyoda Y, Delori FC, Garnett KM, Wong ZY, Nichols CR, Cheng KM, Craft NE, Kathleen DC. (2002): Long term dietary supplementation with zeaxanthin reduces photoreceptor death in light-damaged Japanese quail. *Exp Eye Res* 75: 529–542.

Thurnham DI. (2007): Macular zeaxanthins and lutein—a review of dietary sources and bioavailability and some relationships with macular pigment optical density and age-related macular disease. *Nutr Res Rev* 20: 163–179.

Tomany SC, Wang JJ, van Leeuwen R, Klein R, Mitchell P, Vingerling JR, Klein BEK, Smith W, de Jong PTVM. (2004): Risk factors for incident age-related macular degeneration: pooled findings from 3 continents. *Ophthalmology* 111: 1280–1287.

Trieschmann M, Beatty S, Nolan JM, Hense HW, Heimes B, Austermann U, Fobker M, Pauleikhoff D. (2007): Changes in macular pigment optical density and serum concentrations of its constituent carotenoids following supplemental lutein and zeaxanthin: the LUNA study. *Exp Eye Res* 84: 718–728.

Trieschmann M, Heimes B, Hense HW, Pauleikhoff D. (2006): Macular pigment optical density measurement in autofluorescence imaging: comparison of one- and two-wavelength methods. *Graefes Arch Clin Exp Ophthalmol* 244: 1565–1574.

Tsika C, Tsilimbaris MK, Makridaki M, Kontadakis G, Plainis S, Moschandreas J. (2011): Assessment of macular pigment optical density (MPOD) in patients with unilateral wet age-related macular degeneration (AMD). *Acta Ophthalmol* 89: e573–e578.

Tsika CI, Kontadakis G, Makridaki M, Plainis S, Moschandreas J, Pallikaris IG, Tsilimbaris MK. (2009): Assessment of macular pigment optical density in patients with unilateral wet AMD. *Invest Ophthalmol Vis Sci* 50: 1720.

van der Veen RLP, Berendschot TTJM, Hendrikse F, Carden D, Makridaki M, Murray IJ. (2009): A new desktop instrument for measuring macular pigment optical density based on a novel technique for setting flicker thresholds. *Ophthalmic Physiol Opt* 29: 127–137.

van Leeuwen R, Klaver CCW, Vingerling JR, Hofman A, de Jong PTVM. (2003): Epidemiology of age-related maculopathy: a review. *Eur J Epidemiol* 18: 845–854.

Vos JJ. (1972): Literature review of human macular absorption in the visible and its consequence for the cone receptor primaries. Report Institute for Perception TNO 17, Soesterberg, the Netherlands.

Wang JJ, Foran S, Smith W, Mitchell P. (2003): Risk of age-related macular degeneration in eyes with macular drusen or hyperpigmentation: the Blue Mountains Eye Study cohort. *Arch Ophthalmol* 121: 658–663.

Waters D, Clark RM, Greene CM, Contois JH, Fernandez ML. (2007): Change in plasma lutein after egg consumption is positively associated with plasma cholesterol and lipoprotein size but negatively correlated with body size in postmenopausal women. *J Nutr* 137: 959–963.

Wenzel AJ, Fuld K, Stringham JM. (2003): Light exposure and macular pigment optical density. *Invest Ophthalmol Vis Sci* 44: 306–309.

Wenzel AJ, Gerweck C, Barbato D, Nicolosi RJ, Handelman GJ, Curran-Celentano J. (2006): A 12-wk egg intervention increases serum zeaxanthin and macular pigment optical density in women. *J Nutr* 136: 2568–2573.

Wenzel AJ, Sheehan JP, Burke JD, Lefsrud MG, Curran-Celentano J. (2007): Dietary intake and serum concentrations of lutein and zeaxanthin, but not macular pigment optical density, are related in spouses. *Nutr Res* 27: 462–469.

Werner JS, Donnelly SK, Kliegl R. (1987): Aging and human macular pigment density. Appended with translations from the work of Max Schultze and Ewald Hering. *Vision Res* 27: 257–268.

Winkler BS, Boulton ME, Gottsch JD, Sternberg P. (1999): Oxidative damage and age-related macular degeneration. *Mol Vis* 5: 32.

Wooten BR, Hammond BR. (2003a): Assessment of the Raman method of measuring human macular pigment. *Invest Ophthalmol Vis Sci* [serial online] Available at http://www.iovs.org/cgi/eletters?lookup = by_date&days = 9999#73: August 15, 2003.

Wooten BR, Hammond BR. (2003b): Assessment of the Raman method of measuring human macular pigment (II). *Invest Ophthalmol Vis Sci* [serial online] Available at http://www.iovs.org/cgi/eletters?lookup = by_date&days = 9999#92: December 30, 2003.

Wüstemeyer H, Mößner A, Jahn C, Wolf S. (2003): Macular pigment density in healthy subjects quantified with a modified confocal scanning laser ophthalmoscope. *Graefes Arch Clin Exp Ophthalmol* 241: 647–651.

Yu J, Johnson EJ, Shang F, Lim A, Zhou H, Cui L, Xu J, et al. (2012): Measurement of macular pigment optical density in a healthy Chinese population sample. *Invest Ophthalmol Vis Sci* 53: 2106–2111.

Zagers NPA. (2004). Foveal reflection analyzer. On the spectral and directional reflectance of the retina. PhD thesis, defended at Utrecht University, the Netherlands.

Zagers NPA, van Norren D. (2003): Photoreceptors act as spectrally flat reflectors. *Invest Ophthalmol Vis Sci* 44: 2873.

Zeimer M, Hense HW, Heimes B, Austermann U, Fobker M, Pauleikhoff D. (2009): [The macular pigment: short- and intermediate-term changes of macular pigment optical density following supplementation with lutein and zeaxanthin and co-antioxidants. The LUNA Study]. *Ophthalmologe* 106: 29–36.

Zerbib J, Seddon JM, Richard F, Reynolds R, Leveziel N, Benlian P, Borel P, et al. (2009): rs5888 variant of SCARB1 gene is a possible susceptibility factor for age-related macular degeneration. *PLoS One* 4: e7341.

Zhang X, Hargitai J, Tammur J, Hutchinson A, Allikmets R, Chang S, Gouras P. (2002): Macular pigment and visual acuity in Stargardt macular dystrophy. *Graefes Arch Clin Exp Ophthalmol* 240: 802–809.

Zheng W, Zhang Z, Jiang K, Zhu J, He G, Ke B. (2012): Macular pigment optical density and its relationship with refractive status and foveal thickness in Chinese school-aged children. *Curr Eye Res* 38: 168–173, January 2013.

# 3 Epidemiology and Aetiopathogenesis of Age-Related Macular Degeneration

*Sobha Sivaprasad and Phil Hykin*

## CONTENTS

| | | |
|---|---|---|
| 3.1 | Introduction | 41 |
| 3.2 | Prevalence of AMD | 42 |
| 3.3 | Prevalence of Early AMD | 42 |
| 3.4 | Prevalence of Late AMD | 43 |
| 3.5 | Incidence of AMD | 44 |
| 3.6 | Risk Factors for AMD | 45 |
| 3.7 | Nonmodifiable Risk Factors | 45 |
| 3.8 | Modifiable Risk Factors | 48 |
| 3.9 | Risk Models | 49 |
| 3.10 | Pathogenesis of AMD | 49 |
| | 3.10.1 Genetics | 49 |
| | 3.10.2 Subclinical Inflammation | 50 |
| | 3.10.3 Oxidative Stress | 50 |
| | 3.10.4 Hydrodynamic Changes | 50 |
| | 3.10.5 Hemodynamic Changes | 51 |
| | 3.10.6 Angiogenesis | 51 |
| References | | 52 |

## 3.1 INTRODUCTION

Age-related macular degeneration (AMD) is a progressive and degenerative disease of the retina predominantly affecting the macula, a specialized part of the retina responsible for central and detailed vision (Hirsch et al. 1989). The disease presented in various phenotypes and several classifications have been developed to describe the severity grades of this condition (e.g., Bird et al. 1995; Ferris et al. 2005; Klein et al. 1991). In this publication, we define AMD as either "early AMD" or "late AMD." Early AMD, which does not usually affect vision, is defined as atrophic and/ or hypertrophic changes of the retinal pigment epithelium (RPE) as well as drusen formation beneath the RPE. Late AMD is subdivided into either geographic atrophy (GA) (dry) or neovascular AMD (wet) (Bird et al. 1995), and it is the leading cause of irreversible loss of central vision loss among the elderly in the western world (Congdon et al. 2004).

GA causes degeneration and thinning of the RPE, weakening its ability to nourish, and remove waste products from, the neural retina.

Neovascular AMD is characterized by the growth of abnormal blood vessels from the choroid, which penetrate Bruch's membrane and sometimes the RPE (Bhutto et al. 2006). If left untreated, the leakage results in subretinal and/or retinal scarring, and associated photoreceptor damage with consequential, irreversible loss of central vision (Wang et al. 2008).

Neovascular AMD accounts for up to 90% of blindness caused by AMD and the onset and progression of visual loss is more rapid compared to GA.

## 3.2 PREVALENCE OF AMD

There are many population-based studies that have reported on the prevalence of AMD (see Table 3.1). Only studies that used fundus imaging and international classification systems are included in this table. However, there are several differences between these studies that include variations in diagnostic procedures, definitions of AMD, and age of study population.

## 3.3 PREVALENCE OF EARLY AMD

It has been difficult to compare studies with respect to their findings on the prevalence of early AMD due to differences in how these studies have defined early AMD. For example, in the meta-analyses of studies conducted in the United States by Friedman et al. (2004), early AMD was defined as the presence of large drusen (>125 µm).

In those aged 40–49 years, the prevalence was 1.5% whereas in those aged 80 years or older the prevalence was 25%. In the European Eye (EUREYE) Study, the prevalence of large drusen in people aged 65 years or over across seven European countries ranged from 13.1% to 18.4% with similar increased prevalence in people aged 80 years and over (Augood et al. 2006). A meta-analyses of all the Asian studies show that age-specific pooled prevalence estimates of early AMD is slightly lower than those of European ancestry 6.8 (95% confidence interval [CI] 4.6–8.9) versus 8.8 (95% CI 3.8–13.8) (Kawasaki et al. 2010).

In the European studies, the prevalence of late AMD ranges from 1.65% to 3.5%. A meta-analysis of age-specific prevalence of late AMD by Friedman et al. (2004) showed no significant differences between population groups of European descent around the world. Similarly, in the EUREYE Study that included seven European countries, no significant differences were found among participating countries (Augood et al. 2006). A recent meta-analyses of age and gender variations of late AMD estimated prevalence of late AMD as 1.4% at 70 years of age, rising to 5.6% at age 80 and 20% at age 90 (Rudnicka et al. 2012).

## TABLE 3.1
### Prevalence of Late AMD Based on Population-Based Studies

| Study | Year Study Conducted | Number of Subjects | Mean Age (Years) | Prevalence of Late AMD (%) |
|---|---|---|---|---|
| **United States** | | | | |
| Baltimore Eye Survey (Friedman et al. 1999) | 1985–1988 | 5308 | 64.5 | 1.23 |
| Beaver Dam Study (Klein et al. 1992) | 1988–1990 | 4771 | 60.5 | 1.64 |
| Salisbury Eye Evaluation Project (Bressler et al. 2008) | 1993–1995 | 2387 | 73.3 | 1.75 |
| Atherosclerosis Risk in Communities Studies (Klein et al. 1999) | 1993–1995 | 11532 | 60.5 | 0.2 |
| Cardiovascular Health Study (Klein et al. 2003) | 1997–1998 | 2361 | 60 | 1.3 |
| Multiethnic Study of Atherosclerosis (Klein et al. 2006) | 2000–2002 | 6176 | 64.5 | 0.6 |
| **Europe** | | | | |
| Rotterdam Eye Study (Vingerling et al. 1995) | 1990–1993 | 6718 | 68.9 | 1.65 |
| POLA Study (Delcourt et al. 1998) | 1995–1997 | 2196 | 70.4 | 1.9 |
| Reykjavik Study (Jonasson et al. 2003) | 1996 | 1045 | 63.9 | 3.5 |
| Speedwell Eye Study (Ngai et al. 2011) | 1997 | 934 men | 71 | 0.5 |
| Thessaloniki Eye Study (Topouzis et al. 2006) | 2000 | 2554 | 72 | 2.5 |
| Oslo Macular Study (Bjornsson et al. 2006) | 2002 | 459 | 66.3 | 2.8 |
| EUREYE (Augood et al. 2006) | 2001–2002 | 4753 | 74.5 | 3.3 |
| **Australia** | | | | |
| Blue Mountain Eye Study (Mitchell et al. 1995) | 1992–1994 | 3654 | 66.4 | 2.06 |
| Visual Impairment Study (Van Newkirk, 2000) | 1992–1996 | 4744 | 60.2 | 0.68 |

## 3.4 PREVALENCE OF LATE AMD

The prevalence of late AMD in other ethnic groups appears to be much lower than in the white population. The estimated prevalence of late AMD in African American populations, derived from multiethnic population-based studies conducted in the United States, ranges from 0.04% to 0.3%, suggesting that the prevalence of late AMD is at least twofold lower in African Americans than in the white population (Bressler et al. 2008; Friedman et al. 1999; Klein et al. 1999; Klein, Marino et al. 2003; Klein et al. 2006). There are no studies from Africa. Nevertheless, the Barbados Eye Study (Schachat et al. 1995) estimated the prevalence of late AMD in people of African origin living in Barbados to be 0.56%, similar to those observed in African Americans.

Hispanics also have a lower prevalence of late AMD. Four epidemiological studies have estimated the prevalence of late AMD in this group (Cruickshanks et al. 1997; Klein et al. 2006; Munoz et al. 2005; Varma et al. 2004). The prevalence ranged between 0.09 and 0.5% again suggesting a lower risk of late AMD, similar to the African Americans.

However, the prevalence of late AMD in Asian populations is similar to those of European descent. In Japan, the prevalence is estimated at 0.8% (Kawasaki et al. 2008; Oshima et al. 2001). When considering the Chinese population, the Multiethnic Study of Atherosclerosis in United States (Klein et al. 2006) revealed a prevalence of 1% in the Chinese cohort, similar to the Shihpay study in Taiwan (Chen et al. 2008) and the Handan Eye Study (Yang et al. 2011) that showed a prevalence of 1.2%. In contrast, the Beijing Eye Study from mainland China (Li et al. 2008) estimated the prevalence of late AMD to be 0.2%. The reasons for these differences remain unclear. Studies on other Asian ethnic groups such as the Malays in the Malay Eye Study from Singapore (Kawasaki et al. 2008) also indicate similar prevalence of late AMD when compared to the Caucasians (0.7%). The Thailand National Survey of Visual Impairment in 2006–2007 observed a prevalence of 0.3% (Jenchitr et al. 2011). Five epidemiological studies from the Indian subcontinent also show prevalence of late AMD to range between 0.6% and 1.9% (Jonas et al. 2012; Krishnaiah et al. 2005; Krishnan et al. 2010; Nangia et al. 2011; Nirmalan et al. 2004).

Other ethnic-specific studies include the AMD prevalence study in Oklahoma Indians (Butt et al. 2011) and the Inuits in Greenland (Anderson et al. 2008). In Oklahoma Indians, the prevalence of late AMD was 0.8%. However, the prevalence of late AMD in Inuits is highest in the world, with a prevalence of 9.1% consisting mainly of neovascular AMD.

## 3.5 INCIDENCE OF AMD

Incidence is the number of new cases identified during a specified period. So the incidence of early AMD was defined by the presence of either soft indistinct drusen, or the presence of RPE depigmentation, or increased retinal pigment together with any type of drusen at follow-up when none of these lesions was present at baseline in the Beaver Dam Eye Study. The incidence of late AMD was defined by the appearance of either exudative macular degeneration or pure GA at follow-up when neither lesion was present at baseline. A few population-based studies have reported the incidence of AMD. The Beaver Dam Eye Study estimated the 15-year cumulative incidence of early and late AMD to be 14.3% and 3.1%, respectively. The incidence was significantly higher in subjects aged 75 years or over compared to those aged between 65 and 74 (Klein et al. 2007). Similar rates were observed in the Blue Mountains Eye Study (Wang et al. 2007). In Europe, the Reykjavik Study (Jonasson et al. 2005), Pathologies Oculaires Liées à (POLA) Study (Delcourt et al. 2005), and the Rotterdam Study (van Leeuwen et al. 2000) also showed similar incidence of early AMD with an exponential increase in late AMD with age. A 95% credible interval (CrI) represents the range of values within which the true prevalence is expected to lie with 95% probability. In a recent study in the United Kingdom, the overall number of incident cases in those aged over 50 years for late AMD, GA and neovascular AMD in women was 4.1 (95% CrI 2.4 to

6.8), 2.4 (95% CrI 1.5 to 3.9), and 2.3 (95% CrI 1.4 to 4.0) per 1000 per year; in men it was 2.6 (95% CrI 1.5 to 4.4), 1.7 (95% CrI 1.0 to 2.8) and 1.4 (95% CrI 0.8 to 2.4), respectively (Owen et al. 2012).

## 3.6 RISK FACTORS FOR AMD

The risk factors for AMD are now well known. Established risk factors for AMD include age, cigarette smoking, family history of AMD, and disease in the fellow eye. Putative risk factors for AMD include nutrition, light exposure, sex, ethnicity, obesity, and cardiovascular disease. In this chapter, we will discuss these risk factors with respect to those nonmodifiable and modifiable, and focus on those risk factors that the literature suggests are most important.

## 3.7 NONMODIFIABLE RISK FACTORS

Increasing age is the most important risk factor for AMD as confirmed in all the epidemiological studies (Tomany et al. 2004).

In the pooled analysis of the major population-based studies, the overall prevalence of AMD was similar in white men and women except for the age group older than 80 years (16.4% in women compared with 11.9% in men). However, women are more at risk of neovascular AMD than men (Rudnika et al. 2012). It is possible that the higher rates observed for women at older ages may reflect survival bias. On the contrary, in the Asian population, higher prevalence for early and late AMD were observed in men compared to women, and it is largely attributed to the substantially higher proportion of male smokers (Kawasaki et al. 2010).

Another very important risk factor for AMD is genetics. Recent progress in AMD genetics has established alleles as well as haplotypes on chromosome 1 in Complement Factor H (CFH) and on chromosome 10 in Age-Related Maculopathy Susceptibility 2 (ARMS2, formerly LOC387715/HtrA Serine Peptidase 1 [HTRA1]), as having large influences on the risk for all AMD subtypes in populations of various ethnicities (Edwards et al. 2005; Hageman et al. 2005; Haines et al. 2005).

CFH is a key regulator of the alternate complement pathway as well as innate immunity. CFH is localized to 1q32. Several independent investigators have shown the single nucleotide polymorphism (SNP) rs1061170 in CFH, encoding a tyrosine to histidine change (Y402H), to be associated with increased risk of both early and late stages of AMD (both neovascular and GA) (Edwards et al. 2005; Hageman et al. 2005; Haines et al. 2005). This variant confers a twofold higher risk of late AMD per copy in individuals of European descent and is not modified by smoking (Sofat et al. 2012). However, there is a lack of association of this gene in non-Europeans suggesting that this SNP may be more a biomarker than a cause of the disease. Combination of rare alleles of other adjacent genes, Complement Component 2 (C2) and Complement Factor B (CFB) on 6p21.3 has a protective effect against AMD, especially in those who are homozygous for the Y402H variant of CFH. Another important component of the complement pathway, Complement Component 3 (C3), is also associated with a higher risk of late AMD (Yates et al. 2007). The most significantly associated SNP, rs2230199, is nonsynonymous and results in an amino acid change from arginine to glycine. The association of CFH and its related pathway genes with AMD susceptibility is a major breakthrough in understanding the pathogenesis of this disease. However, despite the large influence of these genes on AMD risk, the combination of these genes alone is insufficient to correctly predict the development and progression of this disease, thereby suggesting that environmental and lifestyle factors have an important role to play in risk of AMD (DeAngelis et al. 2008).

Genes that reside on the long arm of chromosome 10 (10q26) have been the most strongly associated with neovascular AMD risk and it contains the genes PLEKHA1, ARMS2 (formerly LOC387715), and HTRA1. PLEKHA1 (Pleckstrin homology domain-containing family A member 1) is believed to participate in phospholipid binding and more indirectly in the immune response. The ARMS2 gene is located between PLEKHA1 and HTRA1 in the 10q26 locus. ARMS2 is expressed in the retina and may be located in the mitochondria. Within ARMS2, rs10490924, encoding the A69S change, has been associated with certain AMD phenotypes including early onset of disease and larger choroidal neovascularization lesions. Unlike CFH genotypes, this variation has been consistently associated with AMD risk across various ethnicities. HTRA1, the third gene in the 10q26, is highly conserved among species and has several variants that have frequently been found to be associated with AMD, the most consistent one being the SNP rs11200638. Given that these three genes within the 10q26 are in high linkage disequilibrium, it is difficult to distinguish between the individual contributions of each of these genes to overall AMD susceptibility. In 2005, CFH and ARMS2 were identified as the major risk genes for AMD, which together have been estimated to account for over 50% of risk for developing this disease. (Edwards et al. 2005; Gotoh et al. 2004; Hageman et al. 2005; Haines et al. 2005; Francis et al. 2008; Jakobsdottir et al. 2005; Rivera et al. 2005; Seddon et al. 2007).

Other genes that have been associated with AMD are summarized in Table 3.2.

### TABLE 3.2
### Genes Associated with Age-Related Macular Degeneration

| Chromosome | Authors | Loci | Mechanism of Action |
|---|---|---|---|
| 1q | Allikmets et al. 1997, 2000; Guymer et al. 2001 | ABCA4 | Lipid transport |
| | Abecasis et al. 2004; Hayashi et al. 2004; Iyengar et al. 2004 | HMCN1 | Extracellular matrix remodeling |
| | Zhang et al. 2008 | F13B | |
| | Edwards et al. 2005; Haines et al. 2005; Hageman et al. 2005 | CFH | Inflammation/Immune system |
| | Abarrategui-Garrido et al. 2009; Abrera-Abeleda et al. 2006; Hageman et al. 2006; Hughes et al. 2006; Moore et al. 2010 | CHR1-5 | Inflammation/Immune system |
| 3p | Tuo et al. 2004 | CX3CR1 | Inflammation/Immune system |
| 4q | Fagerness et al. 2009 | CFI | Inflammation/Immune system |
| | Allikmets et al. 2009; Edwards et al. 2008, 2009; Cho et al. 2009; Yang et al. 2008; Lewin 2009 | TLR3 | Inflammation/Immune system |

## TABLE 3.2 (*Continued*)
## Genes Associated with Age-Related Macular Degeneration

| Chromosome | Authors | Loci | Mechanism of Action |
|---|---|---|---|
| 6p | Gold et al. 2006; Francis et al. 2009 | CFB, C2 | Inflammation/Immune system |
|  | Haines et al. 2006; Churchill et al. 2006; Fang et al. 2009 | VEGF A | Angiogenesis |
| 6q | Ayyagari et al. 2001; Conley et al. 2005 | ELOVL4 | Lipid metabolism |
|  | Kimura et al. 2000; Esfandiary et al. 2005; Gotoh et al. 2008; Kondo et al. 2009 | SOD2 | Angiogenesis, oxidative stress |
| 7q | Baird, Chu et al. 2004; Ikeda et al. 2001 | PON1 | Inflammation/Immune system |
| 9q | Boekholdt et al. 2003; Zareparsi et al. 2005 | TLR4 | Inflammation/Immune system/ Lipid metabolism |
| 9p | Haines et al. 2006 | VLDLR | Lipid metabolism |
| 10q | Tuo et al. 2006 | ERCC6 | DNA repair |
|  | Jakobsdottir et al. 2005, 2008; Conley et al. 2006 | PLEKHA1 | Inflammation/Immune system/ Extracellular matrix |
|  | DeAngelis et al. 2008; Farwick et al. 2009; Francis et al. 2009; Fritsche et al. 2008 | ARMS2 | Inflammation/Immune system/ Extracellular matrix |
|  | Andreoli et al. 2009; Riveria et al. 2005; Brantley et al. 2007; Jiang et al. 2009; Kanda et al. 2007; Wong et al. 2008. | LOC387715 | Inflammation/Immune system/ Extracellular matrix |
| 12p | Haines et al. 2006 | LPR | Lipid metabolism |
| 14q | Lotery et al. 2006; Stone et al. 2004 | FBLN5 | Extracellular matrix remodeling |
| 15q | Lau et al. 2008; Zhu et al. 2006 | RORA | Inflammation/Immune system/ Lipid metabolism |
| 17q | Hamdi et al. 2002 | ACE | Blood pressure |
| 19q | Baird et al. 2004; Klaver et al. 1998; Losonczy et al. 2011; Souied et al. 1998; Pang et al. 2000 | APOE | Lipid metabolism |
| 19p | Francis et al. 2009; Maller et al. 2007; Park et al. 2009; Pei et al. 2009; Yates et al. 2007 | C3 | Inflammation/Immune system |
| 20q | Fiotti et al. 2005 | MMP-9 | Extracellular matrix remodeling |
| 22q | De La Paz et al. 1997; Felbor et al. 1997 | TIMP3 | Extracellular matrix remodeling |
|  | Hosaka et al. 1998 | SYN3 | Trafficking |

## 3.8 MODIFIABLE RISK FACTORS

Smoking is the most important modifiable risk factor for AMD (Thornton et al. 2005). The risk of late AMD is increased threefold in current smokers compared to nonsmokers irrespective of ethnic group. Moreover, a dose-response relationship is also observed (Khan et al. 2006; Seddon et al. 1996). Passive smoking is also associated with increased risk for AMD (Khan et al. 2006). Past smokers who have last smoked 20 years previously carry similar risks to nonsmokers. The risk of smoking is higher in those with the LOC387715 gene (Francis et al. 2007; Schmidt et al. 2006). The exact mechanisms by which smoking increases the risk of AMD is unclear but oxidative stress, inflammation, compromised choroidal circulation, decreased macular pigment, and decreased bioavailability of carotenoids may all contribute to this susceptibility.

Several other risk factors have been implicated in the pathogenesis of AMD. However, the results are inconsistent. Obese individuals are found to be at twice the risk of developing AMD compared to those with normal body mass index (Johnson 2005). Eating habits have been explored and high intake of animal fats is related to increased risk of AMD while intake of fish products and long chain omega-3 fatty acids are believed to be protective (Chong et al. 2008; Seddon et al. 2006a; Chong et al. 2007). Low serum antioxidant levels, for example vitamins C and E, may also negatively influence AMD (Delcourt et al. 1999; Fletcher et al. 2008). The putative protective role of the macular carotenoids lutein, zeaxanthin and *meso*-zeaxanthin for AMD continue to be explored in epidemiological studies (further discussion in Chapters 4 and 7) (Delcourt 2010; Ma et al. 2012; Sabour-Pickett et al. 2012). Increases in inflammatory markers in blood such as elevated levels of white blood cells, fibrinogen, oxidized low-density lipoproteins, cholesterol, and C-reactive protein are all associated with increased risk of AMD (Klein et al. 2003, 2009; Seddon et al. 2004; van Leeuwen et al. 2003). Sunlight and blue-light exposure have been studied in a number of reports and do not seem to be a major contributory factor in the progression of AMD, except in subjects with low antioxidant levels, suggesting insufficient defence mechanisms against the reactive oxygen species (ROS) (van Leeuwen et al. 2005; Cruickshanks et al. 2001; Taylor 1990; Fletcher et al. 2008; Delcourt 2001; West 1998).

Conflicting reports have been published on the association between AMD and cardiovascular factors including systemic hypertension (Klein et al. 2007; Smith et al. 1998; Vine et al. 2005). One hypothesis is that the vascular changes in the choriocapillaris may impair the supply of nutrients and decrease the outflow of waste products. However, systemic vascular determinants may not reflect the vascular status of the eye.

A possible protective role of dark-colored irides has not been confirmed (Mitchell et al. 1998; Topuzis 2009; West et al. 1989). Hypermetropia is consistently associated with increased risk of AMD irrespective of ethnic groups (Jonas et al. 2012; Lavanya et al. 2010). There is a probable relationship between cataract and AMD, as they both result, in part, from the aging process, and the contribution of oxidative stress (Klein 2005). Several investigators have reported on the link between cataract surgery and the progression of AMD with inconsistent results (Chew et al. 2009; Dong et al. 2009). Late AMD in one eye increases the risk of development of late AMD in the contralateral eye by 10% per year (Age-Related Eye Disease Study Research Group, 2001; Kassoff et al. 2001)

## 3.9 RISK MODELS

It is clear that there are many different risk factors for AMD. However, it is important that we discuss and understand the role of these risk factors for AMD not only as individual risk factors but when combined with others. For example, Seddon et al. (2011) evaluated risk models for progression to late AMD using demographic, environmental, genetic, and ocular factors. The model included rates of progression of AMD up to a 12-year period, AMD status at baseline, macular drusen size in both eyes at baseline, six genetic variants, and demographic and environmental factors. The presence of drusen and increasing drusen size in one or both eyes were strong risk factors for progression to late AMD. Individuals with late AMD in one eye at baseline had a seven- to ten-fold greater hazard of progression in the fellow eye compared with those without late AMD at baseline, even after controlling for genetic and other factors. Among individuals with the same baseline drusen phenotype and AMD status, the addition of demographic, lifestyle, and genetic factors was able to differentiate those who were at low, medium, and high risk of progression. Several other models are also being explored (Hageman et al. 2011; Klein et al. 2011; Yu et al. 2012). These models will potentially allow the use of personalized medicine to predict progression to late AMD. Also, such prediction models are now being used by eyecare professionals, whereby they input all AMD risk factors for a given individual, which provide a customized and personalized risk prediction for that individual, along with information on how that individual can reduce his/her risk of developing AMD in later life (www.sightrisk.com).

## 3.10 PATHOGENESIS OF AMD

The pathogenesis of AMD is not well understood. Many possible causative factors have been implicated including genetics, cumulative oxidative stress, ischaemia, and chronic inflammation (Ambati et al. 2003).

### 3.10.1 Genetics

As mentioned earlier, genetics play a significant role in the pathogenesis of the disease. The Y402H variant of CFH gene contributes to almost half of all cases of AMD independent of AMD classification and is not modified by smoking. Several other variants of both susceptibility and protective genes are shown in Table 3.2. However, causal links to the disease phenotype have not yet been established. The lack of association of these genes with AMD in the Japanese population indicates that these genotypes may indeed be a marker of the phenotype rather than a cause of the disease (Gotoh et al. 2006; Mori et al. 2010). Future larger population-based studies in Asia may provide better insight into the role of these variants. The late onset of this genetic disease also points toward a gene–environmental interaction consequent to cumulative exposure to stressors (Robman et al. 2010; Seddon et al. 2006, 2010).

### 3.10.2 SUBCLINICAL INFLAMMATION

Chronic inflammation at the level of RPE and Bruch's membrane plays a central role in AMD (Anderson et al. 2002; Hageman et al. 2001). There is sufficient evidence from the genetics of the disease that the complement system is involved in AMD (Anderson et al. 2010). Histopathological and proteome analyses have shown the presence of multiple inflammatory components in drusen including components of the complement pathway (Crabb et al. 2002). Hageman et al. (2001) proposed that drusen may be debris from degenerated RPE cells that form a nidus for an inflammatory cascade mediated partly through the complement system. McGeer and McGeer (2004) described this inflammation-induced tissue damage and progressive dysfunction as autotoxicity. As discussed earlier, RPE dysfunction may be the primary initiating factor for inflammation (Donoso et al. 2006). Further immunological evidence in the pathobiology of early AMD arises from observations of specific autoimmunity. The sera of AMD patients contain a high level of antiretinal autoantibodies (Morohoshi et al. 2012). However, the exact role of autoimmunity needs to be clarified. There is also sufficient evidence to suggest that choroidal neovascularization (CNV) development is mediated through inflammatory processes. Interestingly, some infections may also increase the risk for AMD. Chlamydia pneumoniae (Shen et al. 2009) and cytomegalovirus IgG antibody titres (Miller et al. 2004) have been reported to have an association with neovascular AMD lending weight to the postulation of an infectious process that may initiate the disease or its progression. However, the reports are conflicting (Khandhadia et al. 2012).

### 3.10.3 OXIDATIVE STRESS

A very well-researched environmental interaction is oxidative stress (Beatty et al. 2000). The retina is particularly susceptible to oxidative damage because of the high oxygen demand, exposure to visible light, and high proportion of polyunsaturated fatty acids in the photoreceptor outer segments. ROS are released mainly as by-products of oxygen metabolism. In addition, photochemical retinal injury and oxidative damage of the outer segments and the presence of lipofuscin also contribute to oxidative stress. Macular pigment is believed to limit retinal oxidative damage by absorbing incoming short wavelength (blue) light and/or quenching ROS. Several studies have convincingly shown that high plasma levels of lutein, zeaxanthin, and *meso*-zeaxanthin and the corresponding increase in macular pigment confer reduced risk of AMD (Delcourt 2010; Ma et al. 2012; Sabour-Pickett et al. 2012). However, we await large-scale studies such as the Age-Related Eye Disease Study (AREDS) II, Central Retinal Enrichment Supplementation Trials AMD, and other studies on the role of the macular carotenoids (lutein, zeaxanthin, and *meso*-zeaxanthin) that will provide a better understanding on the effect of nutritional antioxidant supplements on the onset and natural course of early AMD.

### 3.10.4 HYDRODYNAMIC CHANGES

Bruch's membrane is sandwiched between the choriocapillaris and the RPE, and is crucial for the transport of nutrients and waste products. Age-related changes of the Bruch's membrane, such as the progressive increase in lipid content, presence of advanced glycation end products, increase in heparan sulfate, decreased solubility of collagen (Karwatowski 1995), and the remodeling of the extracellular matrix, results in an increase in the diffusional path length and consequent exponential decline in the hydraulic conductivity especially at the macula (Curcio et al. 2011; Hewitt et al. 1989; Hussain et al. 2001; Pauleikhoff et al. 1990). The metabolically active RPE continuously phagocytose the outer segments of the photoreceptors throughout life. With increasing age, a decrease in lysosomal enzymes may account for incomplete phagolysosomal degradation and result in abnormal deposits in Bruch's membrane, further impeding the transport across this pentalaminar matrix (Cingle et al. 1996; Haimovici et al. 2002; Steinmetz et al. 1993).

### 3.10.5 Hemodynamic Changes

Several investigators have reported a decrease in choroidal blood flow in patients with AMD. A prolonged choroidal filling phase and the angiographic appearance of choroidal perfusion are observed in some eyes suggesting pathology in the choriocapillaris (Pauleikhoff et al. 1999). This is further substantiated by a decrease in the volume of choriocapillaris in AMD (Ramrattan et al. 1994). Prolonged choroidal filling also correlates with diffuse thickening of Bruch's membrane, elevated scotopic thresholds, and slow dark adaptation (Chen, JC 1992). A hemodynamic model proposed by Friedman (1997) attributes these findings to increased resistance of the choroidal circulation due to decreased compliance of the sclera and choroidal vessels.

### 3.10.6 Angiogenesis

CNV is the hallmark of neovascular AMD. Although the main driver for angiogenesis is hypoxia, other elements such as inflammation and oxidative stress also contribute significantly. Histopathological reports of surgically excised CNV of patients with AMD indicate the presence of both vascular elements as well as nonvascular components such as inflammatory cells, fibroblasts, and local retinal tissues such as RPE cells and glial cells (Kvanta et al. 1996). Although the vascular endothelial growth factor (VEGF) (Ferrara 2004) is the key inciting stimulus for the development of CNV, inflammatory cells such as macrophages are also crucial in determining the size and severity of the CNV. An imbalance of the angiogenic and antiangiogenic agents toward angiogenesis occurs in the initial stage of CNV development. For example, upregulation of angiopoietin-2, platelet-derived growth factor, basic fibroblast growth factor (FGF-2), transforming growth factor-beta (TGF-β), and hepatocyte growth factor and matrix metalloproteinases (MMP-2 and MMP-9) all amplify the angiogenic processes in CNV development and progression. Over time, the balance tilts to the antiangiogenic phase with an increase in fibrosis and involution of the CNV. The most important players at this stage are TGF-β and tissue inhibitors of MMPs (TIMP-3).

As VEGF is the key stimulus of CNV, the current treatments of neovascular AMD are focused on the frequent intravitreal injections of anti-VEGF agents such as ranibizumab, bevacizumab, and aflibercept (Rosenfeld et al. 2006). However, the burden of therapy and the collateral effects of atrophy and fibrosis point toward the need for combination therapies targeting different mechanisms.

In conclusion, significant advances in the understanding of the etiology and pathogenesis of AMD have led to the development of anti-VEGF agents for neovascular AMD. Several therapeutic interventions targeting oxidative stress, formation of drusen, accumulation of lipofuscin, local inflammation, and reactive gliosis are in the developmental phase for GA. Clearly, more studies are warranted for better understanding of the disease and identification of risk models that could be used to prevent the development and progression of this potentially blinding condition.

## REFERENCES

Abarrategui-Garrido C, Martínez-Barricarte R, López-Trascasa M, deCórdoba SR, Sánchez-Corral P. Characterization of complement factor H-related (CFHR) proteins in plasma reveals novel genetic variations of CFHR1 associated with atypical hemolytic uremic syndrome. *Blood*. 2009;114(19):4261–4271.

Abecasis GR, Yashar BM, Zhao Y, et al. Age-related macular degeneration: A high-resolution genome scan for susceptibility loci in a population enriched for late-stage disease. *Am. J. Hum. Genet*. 2004;74(3):482–494.

Abrera-Abeleda MA, Nishimura C, Smith JLH, et al. Variations in the complement regulatory genes factor H (CFH) and factor H related 5 (CFHR5) are associated with membranoproliferative glomerulonephritis type II (dense deposit disease). *J. Med. Genet*. 2006;43(7):582–589.

Age-Related Eye Disease Study Research Group. A randomized, placebo-controlled, clinical trial of high-dose supplementation with vitamins C and E, beta carotene, and zinc for age-related macular degeneration and vision loss: AREDS report no. 8. *Arch. Ophthalmol*. 2001;119(10):1417–1436.

Allikmets R, Bergen AA, Dean M, et al. Geographic atrophy in age-related macular degeneration and TLR3. *N. Engl. J. Med*. 2009;360(21):2252–2254; author reply 2255–2256.

Allikmets R, Shroyer NF, Singh N, et al. Mutation of the Stargardt disease gene (ABCR) in age-related macular degeneration. *Science* 1997;277(5333):1805–1807.

Ambati J, Ambati BK, Yoo SH, Ianchulev S, Adamis AP. Age-related macular degeneration: Etiology, pathogenesis, and therapeutic strategies. *Surv. Ophthalmol*. 2003;48:257–293.

Anderson DH, Mullins RF, Hageman GS, Johnson LV. A role for local inflammation in the formation of drusen in the aging eye. *Am. J. Ophthalmol*. 2002;134(3):411–431.

Anderson DH, Radeke MJ, Gallo NB, et al. The pivotal role of the complement system in aging and age-related macular degeneration: Hypothesis re-visited. *Prog. Retin. Eye Res*. 2010;29(2):95–112.

Andersen MV, Rosenberg T, la Cour M, Kiilgaard JF, Prause JU, Alsbirk PH, Borch-Johnsen K, Peto T, Carstensen B, Bird AC. Prevalence of age-related maculopathy and age-related macular degeneration among the inuit in Greenland. The Greenland Inuit Eye Study. *Ophthalmology*. 2008 Apr;115(4):700–707.

Andreoli MT, Morrison MA, Kim BJ, et al. Comprehensive analysis of complement factor H and LOC387715/ARMS2/HTRA1 variants with respect to phenotype in advanced age-related macular degeneration. *Am. J. Ophthalmol*. 2009;148(6):869–874.

Augood CA, Vingerling JR, de Jong PT, et al. Prevalence of age-related maculopathy in older Europeans: the European Eye Study (EUREYE). *Arch. Ophthalmol*. 2006 Apr;124(4):529–535.

Ayyagari R, Zhang K, Hutchinson A, et al. Evaluation of the ELOVL4 gene in patients with age-related macular degeneration. *Ophthalmic Genet*. 2001;22(4):233–239.

Baird PN, Chu D, Guida E, Vu HTV, Guymer R. Association of the M55L and Q192R paraoxonase gene polymorphisms with age-related macular degeneration. *Am. J. Ophthalmol.* 2004;138(4):665–666.

Baird PN, Guida E, Chu DT, Vu HTV, Guymer RH. The epsilon2 and epsilon4 alleles of the apolipoprotein gene are associated with age-related macular degeneration. *Invest. Ophthalmol. Vis. Sci.* 2004;45(5):1311–1315.

Beatty S, Koh H, Phil M, Henson D, Boulton M. The role of oxidative stress in the pathogenesis of age-related macular degeneration. *Surv. Ophthalmol.* 2000 Sep–Oct;45(2):115–134.

Bird AC, Bressler NM, Bressler SB, et al. An international classification and grading system for age-related maculopathy and age-related macular degeneration. The International ARM Epidemiological Study Group. *Surv. Ophthalmol.* 1995 Mar–Apr;39(5):367–374.

Bhutto I, Lutty G. Understanding age-related macular degeneration (AMD): Relationships between the photoreceptor/retinal pigment epithelium/Bruch's membrane/choriocapillaris complex. *Mol Aspects Med.* 2012 Aug;33(4):295–317.

Björnsson OM, Syrdalen P, Bird AC, Peto T, Kinge B. The prevalence of age-related maculopathy (ARM) in an urban Norwegian population: The Oslo Macular study. *Acta. Ophthalmol. Scand.* 2006 Oct;84(5):636–641.

Boekholdt SM, Agema WRP, Peters RJG, et al. Variants of toll-like receptor 4 modify the efficacy of statin therapy and the risk of cardiovascular events. *Circulation.* 2003;107(19):2416–2421.

Brantley MA, Fang AM, King JM, et al. Association of complement factor H and LOC387715 genotypes with response of exudative age-related macular degeneration to intravitreal bevacizumab. *Ophthalmol.* 2007;114(12):2168–2173.

Bressler SB, Muñoz B, Solomon SD, West SK, Salisbury Eye Evaluation (SEE) Study Team. Racial differences in the prevalence of age-related macular degeneration: the Salisbury Eye Evaluation (SEE) Project. *Arch. Ophthalmol.* 2008;126(2):241–245.

Butt AL, Lee ET, Klein R, et al. Prevalence and risks factors of age-related macular degeneration in Oklahoma Indians: the Vision Keepers Study. *Ophthalmol.* 2011 Jul;118(7):1380–1385.

Chen JC, Fitzke FW, Pauleikhoff D, Bird AC. Functional loss in age-related Bruch's membrane change with choroidal perfusion defect. *Invest. Ophthalmol. Vis. Sci.* 1992;33:334–340.

Chen SJ, Cheng CY, Peng KL, et al. Prevalence and associated risk factors of age-related macular degeneration in an elderly Chinese population in Taiwan: The Shihpai Eye Study. *Invest. Ophthalmol. Vis. Sci.* 2008 Jul;49(7):3126–3133.

Chew EY, Sperduto RD, Milton RC, et al. Risk of advanced age-related macular degeneration after cataract surgery in the Age-Related Eye Disease Study: AREDS report 25. *Ophthalmol.* 2009 Feb;116(2):297–303.

Cho Y, Wang JJ, Chew EY, et al. Toll-like receptor polymorphisms and age-related macular degeneration: Replication in three case-control samples. *Invest. Ophthalmol. Vis. Sci.* 2009;50(12):5614–5618.

Chong EW, Wong TY, Kreis AJ, Simpson JA, Guymer RH. Dietary antioxidants and primary prevention of age related macular degeneration: Systematic review and meta-analysis. *BMJ.* 2007 Oct 13;335(7623):755.

Chong EW, Kreis AJ, Wong TY, Simpson JA, Guymer RH. Dietary omega-3 fatty acid and fish intake in the primary prevention of age-related macular degeneration: A systematic review and meta-analysis. *Arch Ophthalmol.* 2008 Jun;126(6):826–33.

Churchill AJ, Carter JG, Lovell HC, et al. VEGF polymorphisms are associated with neovascular age-related macular degeneration. *Hum. Mol. Genet.* 2006;15(19):2955–2961.

Cingle KA, Kalski RS, Bruner WE, O'Brien CM, Erhard P, Wyszynski RE. Age-related changes of glycosidases in human retinal pigment epithelium. *Curr. Eye. Res.* 1996;15:433–438.

Congdon N, O'Colmain B, Klaver CC, et al. Causes and prevalence of visual impairment among adults in the United States. *Arch. Ophthalmol.* 2004;122(4):477–485.

Conley YP, Jakobsdottir J, Mah T, et al. CFH, ELOVL4, PLEKHA1 and LOC387715 genes and susceptibility to age-related maculopathy: AREDS and CHS cohorts and meta-analyses. *Hum. Mol. Genet.* 2006;15(21):3206–3218.

Conley YP, Thalamuthu A, Jakobsdottir J, et al. Candidate gene analysis suggests a role for fatty acid biosynthesis and regulation of the complement system in the etiology of age-related maculopathy. *Hum. Mol. Genet.* 2005;14(14):1991–2002.

Crabb JW, Miyagi M, Gu X, et al. Drusen proteome analysis: An approach to the etiology of age-related macular degeneration. *Proc. Natl. Acad. Sci. U.S.A.* 2002;99(23):14682–14687.

Cruickshanks KJ, Hamman RF, Klein R, Nondahl DM, Shetterly SM, Colorado-Wisconsin Study of Age-Related Maculopathy. The prevalence of age-related maculopathy by geographic region and ethnicity. *Arch. Ophthalmol.* 1997;115(2):242–250.

Cruickshanks KJ, Klein R, Klein BE, Nondahl DM. Sunlight and the 5-year incidence of early age-related maculopathy: The Beaver Dam Eye Study. *Arch. Ophthalmol.* 2001;119(2):246–250.

Curcio CA, Johnson M, Rudolf M, Huang JD. The oil spill in ageing Bruch membrane. *Br. J. Ophthalmol.* 2011 Dec;95(12):1638–1645.

DeAngelis M, Ji F. Genetics of age-related macular degeneration. In: *Principles & Practice of Ophthalmology*, 3rd ed., DM Albert, FA Jakobiec editors. Philadelphia, PA: Saunders, 2008, 1881–1900.

Deangelis MM, Ji F, Adams S, et al. Alleles in the HtrA serine peptidase 1 gene alter the risk of neovascular age-related macular degeneration. *Ophthalmol.* 2008;115(7):1209–1215.e7.

De La Paz MA, Pericak-Vance MA, Lennon F, Haines JL, Seddon JM. Exclusion of TIMP3 as a candidate locus in age-related macular degeneration. *Invest. Ophthalmol. Vis. Sci.* 1997;38(6):1060–1065.

Delcourt C, Carrière I, Ponton-Sanchez A, et al. Light exposure and the risk of age-related macular degeneration: The Pathologies Oculaires Liées à l'Age (POLA) study. *Arch. Ophthalmol.* 2001;119(10):1463–1468.

Delcourt C, Cristol JP, Léger CL, Descomps B, Papoz L. Associations of antioxidant enzymes with cataract and age-related macular degeneration. The POLA Study. Pathologies Oculaires Liées à l'Age. *Ophthalmol.* 1999 Feb;106(2):215–222.

Delcourt C, Lacroux A, Carrière I; POLA Study Group. The three-year incidence of age-related macular degeneration: The "Pathologies Oculaires Liées à l'Age" (POLA) prospective study. *Am J Ophthalmol.* 2005 Nov;140(5):924–6.

Delcourt C. Epidemiology of AMD in age related macular degeneration. In: *Age Related Macular Degneration*, 1st ed., R Silva, F Bandello editors. PA: GER group, Thea Portugal, Portugal, 2010, 11–20.

Dong LM, Stark WJ, Jefferys JL, et al. Progression of age-related macular degeneration after cataract surgery. *Arch. Ophthalmol.* 2009 Nov;127(11):1412–1419.

Donoso LA, Kim D, Frost A, Callahan A, Hageman G. The role of inflammation in the pathogenesis of age-related macular degeneration. *Surv. Ophthalmol.* 2006;51(2):137–152.

Edwards AO, Chen D, Fridley BL, et al. Toll-like receptor polymorphisms and age-related macular degeneration. *Invest. Ophthalmol. Vis. Sci.* 2008;49(4):1652–1659.

Edwards AO, Ritter R, Abel KJ, Manning A, Panhuysen C, Farrer LA. Complement factor H polymorphism and age-related macular degeneration. *Science.* 2005;308(5720):421–424.

Edwards AO, Swaroop A, Seddon JM. Geographic atrophy in age-related macular degeneration and TLR3. *N. Engl. J. Med.* 2009;360(21):2254–2255; author reply 2255–2256.

Esfandiary H, Chakravarthy U, Patterson C, Young I, Hughes AE. Association study of detoxification genes in age related macular degeneration. *Br. J. Ophthalmol.* 2005;89(4):470–474.

Fagerness JA, Maller JB, Neale BM, Reynolds RC, Daly MJ, Seddon JM. Variation near complement factor I is associated with risk of advanced AMD. *Eur. J. Hum. Genet.* 2009;17(1):100–104.

Fang AM, Lee AY, Kulkarni M, Osborn MP, Brantley MA. Polymorphisms in the VEGFA and VEGFR-2 genes and neovascular age-related macular degeneration. *Mol. Vis.* 2009;15:2710–2719.

Farwick A, Dasch B, Weber BHF, Pauleikhoff D, Stoll M, Hense HW. Variations in five genes and the severity of age-related macular degeneration: Results from the Muenster aging and retina study. *Eye (Lond).* 2009;23(12):2238–2244.

Felbor U, Doepner D, Schneider U, Zrenner E, Weber BH. Evaluation of the gene encoding the tissue inhibitor of metalloproteinases-3 in various maculopathies. *Invest. Ophthalmol. Vis. Sci.* 1997;38(6):1054–1059.

Ferrara N. Vascular endothelial growth factor: basic science and clinical progress. *Endocr. Rev.* 2004;25(4):581–611.

Ferris FL, Davis MD, Clemons TE, et al. A simplified severity scale for age-related macular degeneration: AREDS Report No. 18. *Arch. Ophthalmol.* 2005;123(11):1570–1574.

Fiotti N, Pedio M, Battaglia Parodi M, et al. MMP-9 microsatellite polymorphism and susceptibility to exudative form of age-related macular degeneration. *Genet. Med.* 2005 Apr;7(4):272–277.

Fletcher AE, Bentham GC, Agnew M, et al. Sunlight exposure, antioxidants, and age-related macular degeneration. *Arch. Ophthalmol.* 2008 Oct;126(10):1396–1403.

Francis PJ, George S, Schultz DW, et al. The LOC387715 gene, smoking, body mass index, environmental associations with advanced age-related macular degeneration. *Hum. Hered.* 2007;63(3–4):212–218.

Francis PJ, Hamon SC, Ott J, Weleber RG, Klein ML. Polymorphisms in C2, CFB and C3 are associated with progression to advanced age related macular degeneration associated with visual loss. *J. Med. Genet.* 2009;46(5):300–307.

Francis PJ, Zhang H, Dewan A, Hoh J, Klein ML. Joint effects of polymorphisms in the HTRA1, LOC387715/ARMS2, and CFH genes on AMD in a Caucasian population. *Mol. Vis.* 2008;14:1395–1400.

Friedman DS, Katz J, Bressler NM, Rahmani B, Tielsch JM. Racial differences in the prevalence of age-related macular degeneration: The Baltimore Eye Survey. *Ophthalmol.* 1999;106(6):1049–1055.

Friedman DS, O'Colmain BJ, Muñoz B, et al. Prevalence of age-related macular degeneration in the United States. *Arch. Ophthalmol.* 2004;122(4):564–572.

Friedman E. A hemodynamic model of the pathogenesis of age-related macular degeneration. *Am. J. Ophthalmol.* 1997;124:677–682.

Fritsche LG, Loenhardt T, Janssen A, et al. Age-related macular degeneration is associated with an unstable ARMS2 (LOC387715) mRNA. *Nat. Genet.* 2008;40(7):892–896.

Gold B, Merriam JE, Zernant J, et al. Variation in factor B (BF) and complement component 2 (C2) genes is associated with age-related macular degeneration. *Nat. Genet.* 2006;38(4):458–462.

Gotoh N, Kuroiwa S, Kikuchi T, et al. Apolipoprotein E polymorphisms in Japanese patients with polypoidal choroidal vasculopathy and exudative age-related macular degeneration. *Am. J. Ophthalmol.* 2004;138:567–573.

Gotoh N, Yamada R, Hiratani H, et al. No association between complement factor H gene polymorphism and exudative age-related macular degeneration in Japanese. *Hum. Genet.* 2006;120(1):139–143.

Gotoh N, Yamada R, Matsuda F, Yoshimura N, Iida T. Manganese superoxide dismutase gene (SOD2) polymorphism and exudative age-related macular degeneration in the Japanese population. *Am. J. Ophthalmol.* 2008;146(1):146; author reply 146–147.

Guymer RH, Héon E, Lotery AJ, et al. Variation of codons 1961 and 2177 of the Stargardt disease gene is not associated with age-related macular degeneration. *Arch. Ophthalmol.* 2001;119(5):745–751.

Hageman GS, Anderson DH, Johnson LV, et al. A common haplotype in the complement regulatory gene factor H (HF1/CFH) predisposes individuals to age-related macular degeneration. *Proc. Natl. Acad. Sci. U.S.A.* 2005;102(20):7227–7232.

Hageman GS, Gehrs K, Lejnine S, et al. Clinical validation of a genetic model to estimate the risk of developing choroidal neovascular age-related macular degeneration. *Hum. Genomics.* 2011 Jul;5(5):420–440.

Hageman GS, Hancox LS, Taiber AJ, et al. Extended haplotypes in the complement factor H (CFH) and CFH-related (CFHR) family of genes protect against age-related macular degeneration: Characterization, ethnic distribution and evolutionary implications. *Ann. Med.* 2006;38(8):592–604.

Hageman GS, Luthert PJ, VictorChong NH, Johnson LV, Anderson DH, Mullins RF. An integrated hypothesis that considers drusen as biomarkers of immune-mediated processes at the RPE-Bruch's membrane interface in aging and age-related macular degeneration. *Prog. Retin. Eye. Res.* 2001;20(6):705–732.

Haimovici R, Owens SL, Fitzke FW, Bird AC. Dark adaptation in age-related macular degeneration: Relationship to the fellow eye. *Graefes. Arch. Clin. Exp. Ophthalmol.* 2002; 240:90–95.

Haines JL, Hauser MA, Schmidt S, et al. Complement factor H variant increases the risk of age-related macular degeneration. *Science.* 2005;308(5720):419–421.

Haines JL, Schnetz-Boutaud N, Schmidt S, et al. Functional candidate genes in age-related macular degeneration: Significant association with VEGF, VLDLR, and LRP6. *Invest. Ophthalmol. Vis. Sci.* 2006;47(1):329–335.

Hamdi HK, Reznik J, Castellon R, et al. Alu DNA polymorphism in ACE gene is protective for age-related macular degeneration. *Biochem. Biophys. Res. Commun.* 2002; 295(3):668–672.

Hayashi M, Merriam JE, Klaver CCW, et al. Evaluation of the ARMD1 locus on 1q25–31 in patients with age-related maculopathy: Genetic variation in laminin genes and in exon 104 of HEMICENTIN-1. *Ophthalmic. Genet.* 2004;25(2):111–119.

Hewitt AT, Nakazawa K, Newsome DA. Analysis of newly synthesized Bruch's membrane proteoglycans. *Invest. Ophthalmol. Vis. Sci.* 1989;30:478–486.

Hirsch J, Curcio CA. The spatial resolution capacity of human foveal retina. *Vision Res.* 1989;29(9):1095–101.

Hosaka M, Südhof TC. Synapsin III, a novel synapsin with an unusual regulation by Ca2+. *J. Biol. Chem.* 1998;273(22):13371–13374.

Hughes AE, Orr N, Esfandiary H, Diaz-Torres M, Goodship T, Chakravarthy U. A common CFH haplotype, with deletion of CFHR1 and CFHR3, is associated with lower risk of age-related macular degeneration. *Nat. Genet.* 2006;38(10):1173–1177.

Hussain AA, Starita C, Marshall J. A new drug-based strategy for intervention in age-related macular degeneration (AMD). *Invest. Ophthalmol. Vis. Sci.* 2001;42:S223.

Ikeda T, Obayashi H, Hasegawa G, et al. Paraoxonase gene polymorphisms and plasma oxidized low-density lipoprotein level as possible risk factors for exudative age-related macular degeneration. *Am. J. Ophthalmol.* 2001;132(2):191–195.

Iyengar SK, Song D, Klein BEK, et al. Dissection of genomewide-scan data in extended families reveals a major locus and oligogenic susceptibility for age-related macular degeneration. *Am. J. Hum. Genet.* 2004;74(1):20–39.

Jakobsdottir J, Conley YP, Weeks DE, Ferrell RE, Gorin MB. C2 and CFB genes in age-related maculopathy and joint action with CFH and LOC387715 genes. *PLoS One.* 2008;3(5):e2199.

Jakobsdottir J, Conley YP, Weeks DE, Mah TS, Ferrell RE, Gorin MB. Susceptibility genes for age-related maculopathy on chromosome 10q26. *Am. J. Hum. Genet.* 2005;77:389–407.

Jenchitr W, Ruamviboonsuk P, Sanmee A, Pokawattana N. Prevalence of age-related macular degeneration in Thailand. *Ophthalmic. Epidemiol.* 2011 Feb;18(1):48–52.

Jiang H, Qu Y, Dang G, et al. Analyses of single nucleotide polymorphisms and haplotype linkage of LOC387715 and the HTRA1 gene in exudative age-related macular degeneration in a Chinese cohort. *Retina.* 2009;29(7):974–979.

Johnson EJ. Obesity, lutein metabolism, and age-related macular degeneration: A web of connections. *Nutr. Rev.* 2005;63(1):9–15.

Jonas JB, Nangia V, Kulkarni M, Gupta R, Khare A. Associations of early age-related macular degeneration with ocular and general parameters. The central India eyes and medical study. *Acta. Ophthalmol.* 2012 May;90(3):e185–91.

Jonasson F, Arnarsson A, Peto T, Sasaki H, Sasaki K, Bird AC. 5-year incidence of age-related maculopathy in the Reykjavik Eye Study. *Ophthalmology.* 2005 Jan;112(1):132–8.

Jonasson F, Arnarsson A, Sasaki H, Peto T, Sasaki K, Bird AC. The prevalence of age-related maculopathy in Iceland: Reykjavik Eye Study. *Arch. Ophthalmol.* 2003 Mar;121(3):379–85.

Kanda A, Chen W, Othman M, et al. A variant of mitochondrial protein LOC387715/ARMS2, not HTRA1, is strongly associated with age-related macular degeneration. *Proc. Natl. Acad. Sci. U.S.A.* 2007;104(41):16227–16232.

Karwatowski WS, Jeffries TE, Duance VC, Albon J, Bailey AJ, Easty DL. Preparation of Bruch's membrane and analysis of the age-related changes in the structural collagens. *Br. J. Ophthalmol.* 1995;79:944–952.

Kassoff A, Kassoff J, Buehler J, et al. A randomized, placebo-controlled, clinical trial of high-dose supplementation with vitamins C and E, beta carotene, and zinc for age-related macular degeneration and vision loss—AREDS Report No. 8. *Arch. Ophthalmol.* 2001;119:1417–1436.

Kawasaki R, Wang JJ, Aung T, et al. Prevalence of age-related macular degeneration in a Malay population: The Singapore Malay Eye Study. *Ophthalmol.* 2008 Oct;115(10):1735–1741.

Kawasaki R, Yasuda M, Song SJ, et al. The prevalence of age-related macular degeneration in Asians: a systematic review and meta-analysis. *Ophthalmol.* 2010 May;117(5):921–927.

Kemp CM, Jacobson SG, Cideciyan AV, Kimura AE, Sheffield VC, Stone EM. RDS gene mutations causing retinitis pigmentosa or macular degeneration lead to the same abnormality in photoreceptor function. *Invest. Ophthalmol. Vis. Sci.* 1994;35(8):3154–3162.

Khan JC, Thurlby DA, Shahid H, et al. Smoking and age related macular degeneration: The number of pack years of cigarette smoking is a major determinant of risk for both geographic atrophy and choroidal neovascularisation. *Br. J. Ophthalmol.* 2006 Jan;90(1):75–80.

Khandhadia S, Foster S, Cree A, et al. Chlamydia infection status, genotype, and age-related macular degeneration. *Mol. Vis.* 2012;18:29–37.

Kimura K, Isashiki Y, Sonoda S, Kakiuchi-Matsumoto T, Ohba N. Genetic association of manganese superoxide dismutase with exudative age-related macular degeneration. *Am. J. Ophthalmol.* 2000;130(6):769–773.

Klein R, Clegg L, Cooper LS, et al. Prevalence of age-related maculopathy in the Atherosclerosis Risk in Communities Study. *Arch. Ophthalmol.* 1999;117(9):1203–1210.

Klein R, Davis MD, Magli YL, Segal P, Klein BE, Hubbard L. The Wisconsin age-related maculopathy grading system. *Ophthalmol.* 1991;98(7):1128–1134.

Klein R, Klein BE, Knudtson MD, et al. Prevalence of age-related macular degeneration in 4 racial/ethnic groups in the multi-ethnic study of atherosclerosis. *Ophthalmol.* 2006;113(3):373–380.

Klein R, Klein BE, Knudtson MD, Meuer SM, Swift M, Gangnon RE. The fifteen-year cumulative incidence of age-related macular degeneration: The Beaver Dam Eye Study. *Ophthalmol.* 2007;114:253–262.

Klein R, Klein BE, Linton KL. Prevalence of age-related maculopathy. The Beaver Dam Eye Study. *Ophthalmol.* 1992;99(6):933–943.

Klein R, Klein BE, Marino EK, et al. Early age-related maculopathy in the cardiovascular health study. *Ophthalmol.* 2003;110(1):25–33.

Klein R, Klein BE, Myers CE. Risk assessment models for late age-related macular degeneration. *Arch. Ophthalmol.* 2011 Dec;129(12):1605–1606.

Klein R, Klein BEK, Tomany SC, Danforth LG, Cruickshanks KJ. Relation of statin use to the 5-year incidence and progression of age-related maculopathy. *Arch. Ophthalmol.* 2003;121(8):1151–1155.

Klein R, Knudtson MD, Lee KE, Klein BEK. Serum cystatin C level, kidney disease markers, and incidence of age-related macular degeneration: The Beaver Dam Eye Study. *Arch. Ophthalmol.* 2009;127(2):193–199.

Klein R. Epidemiology of age-related macular degeneration. In: *Macular Degeneration: Science and Medicine in Practice*, PL Penfold, JM Provis. New York, NY: Springer-Verlag, 2005, 79–121.

Klein R. Overview of progress in the epidemiology of age-related macular degeneration. *Ophthalmic. Epidemiol.* 2007;14(4):184–187.

Kondo N, Bessho H, Honda S, Negi A. SOD2 gene polymorphisms in neovascular age-related macular degeneration and polypoidal choroidal vasculopathy. *Mol. Vis.* 2009; 15:1819–1826.

Krishnaiah S, Das TP, Kovai V, Rao GN. Associated factors for age-related maculopathy in the adult population in southern India: The Andhra Pradesh Eye Disease Study. *Br. J. Ophthalmol.* 2009 Sep;93(9):1146–1150.

Krishnaiah S, Das T, Nirmalan PK, Nutheti R, Shamanna BR, Rao GN, Thomas R. Risk factors for age-related macular degeneration: Findings from the Andhra Pradesh eye disease study in South India. *Invest. Ophthalmol Vis Sci.* 2005 Dec;46(12):4442–9. PubMed PMID: 16303932.

Krishnan T, Ravindran RD, Murthy GV, et al. Prevalence of early and late age-related macular degeneration in India: The INDEYE Study. *Invest. Ophthalmol. Vis. Sci.* 2010 Feb; 51(2):701–707.

Kvanta A, Algvere PV, Berglin L, Seregard S. Subfoveal fibrovascular membranes in age-related macular degeneration express vascular endothelial growth factor. *Invest. Ophthalmol. Vis. Sci.* 1996;37(9):1929–1934.

Lau P, Fitzsimmons RL, Raichur S, Wang SC, Lechtken A, Muscat GE. The orphan nuclear receptor, RORalpha, regulates gene expression that controls lipid metabolism: Staggerer (SG/SG) mice are resistant to diet-induced obesity. *J. Biol. Chem.* 2008;283(26):18411–18421.

Lavanya R, Kawasaki R, Tay WT, et al. Hyperopic refractive error and shorter axial length are associated with age-related macular degeneration: The Singapore Malay Eye Study. *Invest. Ophthalmol. Vis. Sci.* 2010 Dec;51(12):6247–6252.

Lewin AS. Geographic atrophy in age-related macular degeneration and TLR3. *N. Engl. J. Med.* 2009;360(21):2251; author reply 2255–2256.

Li Y, Xu L, Jonas JB, Yang H, Ma Y, Li J. Prevalence of age-related maculopathy in the adult population in China: The Beijing Eye Study. *Am. J. Ophthalmol.* 2006 Nov;142(5):788–793.

Li Y, Xu L, Wang YX, You QS, Yang H, Jonas JB. Prevalence of age-related maculopathy in the adult population in China: The Beijing eye study. *Am J Ophthalmol.* 2008 Aug;146(2):329.

Losonczy G, Fekete Á, Vokó Z, Takács L, Káldi I, Ajzner É, Kasza M, Vajas A, Berta A, Balogh I. Analysis of complement factor H Y402H, LOC387715, HTRA1 polymorphisms and ApoE alleles with susceptibility to age-related macular degeneration in Hungarian patients. *Acta Ophthalmol.* 2011 May;89(3):255–62.

Lotery AJ, Baas D, Ridley C, et al. Reduced secretion of fibulin 5 in age-related macular degeneration and cutis laxa. *Hum. Mutat.* 2006;27(6):568–574.

Ma L, Dou HL, Wu YQ, et al. Lutein and zeaxanthin intake and the risk of age-related macular degeneration: A systematic review and meta-analysis. *Br. J. Nutr.* 2012 Feb; 107(3):350–359.

Maller JB, Fagerness JA, Reynolds RC, Neale BM, Daly MJ, Seddon JM. Variation in complement factor 3 is associated with risk of age-related macular degeneration. *Nat. Genet.* 2007;39(10):1200–1201.

McGeer PL, McGeer EG. Inflammation and the degenerative diseases of aging. *Ann. N Y Acad. Sci.* 2004 Dec;1035:104–116.

Miller DM, Espinosa-Heidmann DG, Legra J, et al. The association of prior cytomegalovirus infection with neovascular age-related macular degeneration. *Am. J. Ophthalmol.* 2004 Sep;138(3):323–328.

Mitchell P, Smith W, Attebo K, Wang JJ. Prevalence of age-related maculopathy in Australia. The Blue Mountains Eye Study. *Ophthalmol.* 1995;102(10):1450–1460.

Mitchell P, Smith W, Wang JJ. Iris color, skin sun sensitivity, and age-related maculopathy. The Blue Mountains Eye Study. *Ophthalmol.* 1998 Aug;105(8):1359–1363.

Moore I, Strain L, Pappworth I, et al. Association of factor H autoantibodies with deletions of CFHR1, CFHR3, CFHR4, and with mutations in CFH, CFI, CD46, and C3 in patients with atypical hemolytic uremic syndrome. *Blood.* 2010;115(2):379–387.

Mori K, Horie-Inoue K, Gehlbach PL, et al. Phenotype and genotype characteristics of age-related macular degeneration in a Japanese population. *Ophthalmol.* 2010; 117(5):928–938.

Morohoshi K, Patel N, Ohbayashi M, et al. Serum autoantibody biomarkers for age-related macular degeneration and possible regulators of neovascularization. *Exp. Mol. Pathol.* 2012 Feb;92(1):64–73.

Muñoz B, Klein R, Rodriguez J, Snyder R, West SK. Prevalence of age-related macular degeneration in a population-based sample of Hispanic people in Arizona: Proyecto VER. *Arch. Ophthalmol.* 2005;123(11):1575–1580.

Nangia V, Jonas JB, Kulkarni M, Matin A. Prevalence of age-related macular degeneration in rural central India: The Central India Eye and Medical Study. *Retina.* 2011 Jun;31(6):1179–1185.

Ngai LY, Stocks N, Sparrow JM, et al. The prevalence and analysis of risk factors for age-related macular degeneration: 18-year follow-up data from the Speedwell Eye Study, United Kingdom. *Eye (Lond).* 2011 Jun;25(6):784–793.

Nirmalan PK, Katz J, Robin AL, Tielsch JM, Namperumalsamy P, Kim R, Narendran V, Ramakrishnan R, Krishnadas R, Thulasiraj RD, Suan E. Prevalence of vitreoretinal disorders in a rural population of southern India: The Aravind Comprehensive Eye Study. *Arch Ophthalmol.* 2004 Apr;122(4):581–6. PubMed PMID: 15078677.

Oshima Y, Ishibashi T, Murata T, Tahara Y, Kiyohara Y, Kubota T. Prevalence of age related maculopathy in a representative Japanese population: The Hisayama study. *Br. J. Ophthalmol.* 2001 Oct;85(10):1153–1157.

Owen CG, Jarrar Z, Wormald R, Cook DG, Fletcher AE, Rudnicka AR. The estimated prevalence and incidence of late stage age related macular degeneration in the UK. *Br J Ophthalmol.* 2012 May;96(5):752–6.

Owen CG, Jarrar Z, Wormald R, Cook DG, Fletcher AE, Rudnicka AR. The estimated prevalence and incidence of late stage age related macular degeneration in the UK. *Br. J. Ophthalmol.* 2012 Feb 13.

Pang CP, Baum L, Chan WM, Lau TC, Poon PM, Lam DS. The apolipoprotein E epsilon4 allele is unlikely to be a major risk factor of age-related macular degeneration in Chinese. *Ophthalmologica.* 2000;214(4):289–291.

Park KH, Fridley BL, Ryu E, Tosakulwong N, Edwards AO. Complement component 3 (C3) haplotypes and risk of advanced age-related macular degeneration. *Invest. Ophthalmol. Vis. Sci.* 2009;50(7):3386–3393.

Pauleikhoff D, Harper CA, Marshall J, Bird AC. Aging changes in Bruch's membrane. A histochemical and morphologic study. *Ophthalmol.* 1990 Feb;97(2):171–178.

Pauleikhoff D, Spital G, Radermacher M, Brumm GA, Lommatzsch A, Bird AC. A fluorescein and indocyanine green angiographic study of choriocapillaris in age-related macular disease. *Arch. Ophthalmol.* 1999;117:1353–1358.

Pei X, Li X, Bao Y, et al. Association of c3 gene polymorphisms with neovascular age-related macular degeneration in a Chinese population. *Curr. Eye Res.* 2009;34(8):615–622.

Ramrattan RS, van der Schaft TL, Mooy CM, de Bruijn WC, Mulder PG, de Jong PT. Morphometric analysis of Bruch's membrane, the choriocapillaris, and the choroid in aging. *Invest. Ophthalmol. Vis. Sci.* 1994;35:2857–2864.

Rivera A, Fisher SA, Fritsche LG, et al. Hypothetical LOC387715 is a second major susceptibility gene for age-related macular degeneration, contributing independently of complement factor H to disease risk. *Hum. Mol. Genet.* 2005;14(21):3227–3236.

Robman L, Baird PN, Dimitrov PN, Richardson AJ, Guymer RH. C-reactive protein levels and complement factor H polymorphism interaction in age-related macular degeneration and its progression. *Ophthalmol.* 2010;117(10):1982–1988.

Rosenfeld PJ, Brown DM, Heier JS, et al. Ranibizumab for neovascular age-related macular degeneration. *N. Engl. J. Med.* 2006;355(14):1419–1431.

Rudnicka AR, Jarrar Z, Wormald R, Cook DG, Fletcher A, Owen CG. Age and gender variations in age-related macular degeneration prevalence in populations of European ancestry: A meta-analysis. *Ophthalmol.* 2012 Mar;119(3):571–580.

Sabour-Pickett S, Nolan JM, Loughman J, Beatty S. A review of the evidence germane to the putative protective role of the macular carotenoids for age-related macular degeneration. *Mol. Nutr. Food Res.* 2012 Feb;56(2):270–286.

Schachat AP, Hyman L, Leske MC, Connell AM, Wu SY. Features of age-related macular degeneration in a black population. The Barbados Eye Study Group. *Arch. Ophthalmol.* 1995;113(6):728–735.

Schmidt S, Hauser MA, Scott WK, et al. Cigarette smoking strongly modifies the association of LOC387715 and age-related macular degeneration. *Am. J. Hum. Genet.* 2006;78(5):852–864.

Seddon JM, Francis PJ, George S, Schultz DW, Rosner B, Klein ML. Association of CFH Y402H and LOC387715 A69S with progression of age-related macular degeneration. *JAMA.* 2007;297:1793–1800.

Seddon JM, Gensler G, Milton RC, Klein ML, Rifai N. Association between C-reactive protein and age-related macular degeneration. *JAMA.* 2004 Feb 11;291(6):704–10.

Seddon JM, Gensler G, Rosner B. C-reactive protein and CFH, ARMS2/HTRA1 gene variants are independently associated with risk of macular degeneration. *Ophthalmol.* 2010 Aug;117(8):1560–6. Epub 2010 Mar 26

Seddon JM, George S, Rosner B, Klein ML. CFH gene variant, Y402H, and smoking, body mass index, environmental associations with advanced age-related macular degeneration. *Hum. Hered.* 2006a;61(3):157–165.

Seddon JM, George S, Rosner B. Cigarette smoking, fish consumption, omega-3 fatty acid intake, and associations with age-related macular degeneration: The US Twin Study of Age-Related Macular Degeneration. *Arch. Ophthalmol.* 2006b Jul;124(7):995–1001.

Seddon JM, Reynolds R, Yu Y, Daly MJ, Rosner B. Risk models for progression to advanced age-related macular degeneration using demographic, environmental, genetic, and ocular factors. *Ophthalmol.* 2011 Nov;118(11):2203–2211.

Seddon JM, Willett WC, Speizer FE, Hankinson SE. A prospective study of cigarette smoking and age-related macular degeneration in women. *JAMA.* 1996 Oct 9;276(14):1141–6.

Shen D, Tuo J, Patel M, et al. Chlamydia pneumoniae infection, complement factor H variants and age-related macular degeneration. *Br. J. Ophthalmol.* 2009 Mar;93(3):405–408.

Smith W, Mitchell P, Leeder SR, Wang JJ. Plasma fibrinogen levels, other cardiovascular risk factors, and age-related maculopathy: The Blue Mountains Eye Study. *Arch. Ophthalmol.* 1998 May;116(5):583–587.

Sofat R, Casas JP, Webster AR, et al. Complement factor H genetic variant and age-related macular degeneration: Effect size, modifiers and relationship to disease subtype. *Int. J. Epidemiol.* 2012 Feb;41(1):250–262.

Souied EH, Benlian P, Amouyel P, et al. The epsilon4 allele of the apolipoprotein E gene as a potential protective factor for exudative age-related macular degeneration. *Am. J. Ophthalmol.* 1998;125(3):353–359.

Steinmetz RL, Haimovici R, Jubb C, Fitzke FW, Bird AC. Symptomatic abnormalities of dark adaptation in patients with age-related Bruch's membrane change. *Br. J. Ophthalmol.* 1993;77:549–554.

Stone EM, Braun TA, Russell SR, et al. Missense variations in the fibulin 5 gene and age-related macular degeneration. *N. Engl. J. Med.* 2004;351(4):346–353.

Taylor HR, Muñoz B, West S, Bressler NM, Bressler SB, Rosenthal FS. Visible light and risk of age-related macular degeneration. *Trans. Am. Ophthalmol. Soc.* 1990;88:163–173; discussion 173–178.

Thornton J, Edwards R, Mitchell P, Harrison RA, Buchan I, Kelly SP. Smoking and age-related macular degeneration: A review of association. *Eye (Lond).* 2005;19(9):935–944.

Tomany SC, Wang JJ, VanLeeuwen R, et al. Risk factors for incident age-related macular degeneration: Pooled findings from 3 continents. *Ophthalmol.* 2004;111(7):1280–1287.

Topouzis F, Anastasopoulos E, Augood C, et al. Association of diabetes with age-related macular degeneration in the EUREYE Study. *Br. J. Ophthalmol.* 2009 Aug;93(8):1037–1041.

Tuo J, Ning B, Bojanowski CM, et al. Synergic effect of polymorphisms in ERCC6 5' flanking region and complement factor H on age-related macular degeneration predisposition. *Proc. Natl. Acad. Sci. U.S.A.* 2006;103(24):9256–9261.

Tuo J, Smith BC, Bojanowski CM, et al. The involvement of sequence variation and expression of CX3CR1 in the pathogenesis of age-related macular degeneration. *FASEB J.* 2004;18(11):1297–1299.

van Leeuwen R, Boekhoorn S, Vingerling JR, et al. Dietary intake of antioxidants and risk of age-related macular degeneration. *JAMA.* 2005 Dec 28;294(24):3101–3107.

van Leeuwen R, Klaver CC, Vingerling JR, Hofman A, de Jong PT. Epidemiology of age-related maculopathy: a review. *Eur J Epidemiol.* 2003;18(9):845–54.

van Leeuwen R, Klaver CC, Vingerling JR, Hofman A, de Jong PT. The risk and natural course of age-related maculopathy: Follow-up at 6 1/2 years in the Rotterdam study. *Arch Ophthalmol.* 2003 Apr;121(4):519–26.

VanNewkirk MR, Nanjan MB, Wang JJ, Mitchell P, Taylor HR, McCarty CA. The prevalence of age-related maculopathy: The visual impairment project. *Ophthalmol.* 2000;107(8):1593–1600.

Varma R, Fraser-Bell S, Tan S, Klein R, Azen SP, Los Angeles Latino Eye Study Group. Prevalence of age-related macular degeneration in Latinos: The Los Angeles Latino Eye Study. *Ophthalmol.* 2004;111(7):1288–1297.

Vine AK, Stader J, Branham K, Musch DC, Swaroop A. Biomarkers of cardiovascular disease as risk factors for age-related macular degeneration. *Ophthalmol.* 2005; 112(12):2076–2080.

Vingerling JR, Dielemans I, Hofman A, et al. The prevalence of age-related maculopathy in the Rotterdam Study. *Ophthalmol.* 1995;102(2):205–210.

Wang G, Spencer KL, Court BL, et al. Localization of age-related macular degeneration-associated ARMS2 in cytosol, not mitochondria. *Invest. Ophthalmol. Vis. Sci.* 2009;50(7):3084–3090.

Wang JJ, Rochtchina E, Lee AJ, Chia EM, Smith W, Cumming RG, Mitchell P. Ten-year incidence and progression of age-related maculopathy: The blue Mountains Eye Study. *Ophthalmology*. 2007 Jan;114(1):92–8.

Wang JJ, Ross RJ, Tuo J, et al. The LOC387715 polymorphism, inflammatory markers, smoking, and age-related macular degeneration. A population-based case-control study. *Ophthalmol*. 2008;115(4):693–699.

West SK, Duncan DD, Muñoz B, et al. Sunlight exposure and risk of lens opacities in a population-based study: The Salisbury Eye Evaluation project. *JAMA*. 1998;280(8):714–718.

West SK, Rosenthal FS, Bressler NM, Bressler SB, Munoz B, Fine SL, Taylor HR. Exposure to sunlight and other risk factors for age-related macular degeneration. *Arch Ophthalmol*. 1989 Jun;107(6):875–9.

Yang K, Liang YB, Gao LQ, et al. Prevalence of age-related macular degeneration in a rural Chinese population: the Handan Eye Study. *Ophthalmol*. 2011 Jul;118(7):1395–1401.

Yang Z, Stratton C, Francis PJ, et al. Toll-like receptor 3 and geographic atrophy in age-related macular degeneration. *N. Engl. J. Med*. 2008;359(14):1456–1463.

Yates J, Sepp T, Matharu BK, et al. Complement C3 variant and the risk of age-related macular degeneration. *N. Engl. J. Med*. 2007;357(6):553–561.

Yu Y, Reynolds R, Rosner B, Daly MJ, Seddon JM. Prospective assessment of genetic effects on progression to different stages of age-related macular degeneration using multi-state Markov models. *Invest. Ophthalmol. Vis. Sci*. 2012 Mar 21;53(3):1548–56.

Zareparsi S, Buraczynska M, Branham KEH, et al. Toll-like receptor 4 variant D299G is associated with susceptibility to age-related macular degeneration. *Hum. Mol. Genet*. 2005;14(11):1449–1455.

Zhang H, Morrison MA, Dewan A, et al. The NEI/NCBI dbGAP database: Genotypes and haplotypes that may specifically predispose to risk of neovascular age-related macular degeneration. *BMC Med. Genet*. 2008;9:51.

Zhu Y, McAvoy S, Kuhn R, Smith DI. RORA, a large common fragile site gene, is involved in cellular stress response. *Oncogene*. 2006;25(20):2901–2908.

Zurdel J, Finckh U, Menzer G, Nitsch RM, Richard G. CST3 genotype associated with exudative age related macular degeneration. *Br. J. Ophthalmol*. 2002;86(2):214–219.

# 4 Relationships of Lutein and Zeaxanthin to Age-Related Macular Degeneration
## *Epidemiological Evidence*

*Julie A. Mares*

### CONTENTS

4.1 Introduction .................................................................................................. 63
4.2 Evidence for Relationships with Antioxidants .............................................. 64
4.3 Evidence for Relationships with Dietary L and Z ........................................ 65
4.4 Evidence for Relationships with Macular Pigment Optical Density ............. 68
4.5 Summary and Conclusions ........................................................................... 69
Acknowledgments .................................................................................................. 70
References .............................................................................................................. 70

### 4.1 INTRODUCTION

Epidemiological studies comprise one of several types of scientific approaches that can be used to understand whether the macular carotenoids, lutein and zeaxanthin/meso-zeaxanthin (L and Z), might protect against the development and worsening of age-related macular degeneration (AMD) in the general population. These are generally observations made in large samples of free-living populations, who often represent a variety of genetic and environmental attributes and may be followed over longer periods of time than are possible in clinical trials. Thus, results from epidemiological studies, together with results from other study types in animals, cells, and people, provide a more complete body of evidence for making sound public health or clinical recommendations than results from single types of studies.

To date, there is strong evidence from experiments with cells and in animals that L and Z protect against physiological processes that promote AMD. (This is further discussed in Chapters 1 through 3 and 6) Evidence from clinical trials over one to five years suggests that L and Z supplements, with or without other ingredients, may benefit vision in patients with AMD over the short term. (Richer et al. 2004; Piermarocchi et al. 2011; Richer et al. 2011; Mac et al. 2012; Ma et al. 2012).

Furthermore, trials which are randomized and placebo-controlled provide evidence that benefit is not explained by other differences in people assigned to different treatment groups. However, the results of individual small trials can be explained by chance or publication bias, although this likelihood decreases as the number of different trials resulting in benefit increases. Yet, results are limited to the specific supplement formulations tested. Because these trials are costly, they are generally conducted over short periods and in small samples of people who may not be fully representative of the general population of people at risk for AMD; typically, participants are healthier and already have some clinical signs of AMD. Moreover, trials are not sufficiently large to adequately assess long-term risks.

Epidemiological studies can provide early clues about benefit and fill in the gaps in evidence not covered by other study types. Adding the results of epidemiological studies to the larger body of evidence permits *broader inferences in the general population* about the likelihood that the intake of L and Z from foods and/or supplements might be extended to longer periods of time than can be tested in trials. Evidence from observational epidemiological studies can facilitate the evaluation of associations *over wider ranges of lutein and zeaxanthin intakes* than can be practically tested in trials. When fundus photographs are collected, epidemiological studies can uniquely evaluate relationships of long-term diet and supplement use to the onset of early stages of AMD. Large epidemiological studies can also inform researchers about the *possibility that benefits or risks differ in people who have different lifestyles* (e.g., smokers versus nonsmokers, or people with otherwise good versus poor diets) or genotypes. This chapter provides an overview of evidence from epidemiological studies of relationships between lutein and zeaxanthin in the diet, blood, and macular pigment and AMD. Gaps in existing knowledge and suggestions for future research directions are also discussed.

## 4.2 EVIDENCE FOR RELATIONSHIPS WITH ANTIOXIDANTS

L and Z are two of several dietary constituents (such as vitamins C and E and a variety of food phytochemicals), which function as antioxidants; which is to say, they protect against oxidative damage because of light and metabolic processes in which oxygen radical/nitrogen radicals and singlet oxygen are side products. A large body of evidence from observational studies suggests that intake of one or more antioxidant nutrients from foods or supplements is related to lower rates of AMD, but the specific type of antioxidant related to AMD varies across studies (Mares et al. 2012). In two large clinical trials (AREDS Study Group 2001 and 2013) and other small trials (Richer et al. 2004; Piermarocchio et al. 2011; Richer et al. 2011), antioxidant supplements of various types slowed the loss of vision in people who already had AMD. In one (the Age-Related Eye Disease Study [AREDS 2001]), there was also evidence that this treatment reduces the age-related oxidation of cysteine in the blood, which supports the possibility that the benefit is due to a reduction in oxidative stress (Moriarty-Craige 2005).

Despite this body of evidence to support the benefit of antioxidants against the progression of AMD, there is insufficient evidence that specific antioxidant supplements prevent or delay the onset of AMD (Evans and Henshaw 2009). Such evidence is difficult to gather in clinical trials which are usually of short duration. Observational studies of single antioxidant nutrients in foods and supplements in relation to prevention or delay of AMD are conflicting (Chiu and Taylor 2007; Chong et al. 2007; Evans and Henshaw 2009; Krishnadev et al. 2010; Mares et al. 2012). However, this may be, in part, because of an attempt to evaluate single antioxidants in isolation from others. Investigators in the Rotterdam Eye Study (van Leeuwen et al. 2005) studied relationships of diets rich in a combination of antioxidants to incident AMD. They observed a 35% lower risk of incident AMD among the 10% of participants who were consuming diets rich in several antioxidants (i.e., diets with above-the-median intake of all four antioxidants in AREDS supplements) compared with above-median intake for only one antioxidant.

## 4.3 EVIDENCE FOR RELATIONSHIPS WITH DIETARY L AND Z

In addition to being able to scavenge free radicals like other antioxidants, L and Z may protect by virtue of their ability to absorb blue light that reaches the retina (discussed in Chapters 1, 2, and 6) and reduce inflammation (Izumi-Nagai et al. 2007; Sazacki et al. 2009; Bian et al. 2012). Over the past 3 years, consistency in the relationships of macular carotenoids to lower rates of *advanced* stages of AMD has been observed (Seddon et al. 1994; Mares-Perlman et al. 2001; SanGiovanni et al. 2007; Cho et al. 2008; Tan et al. 2008). However, there is a remaining need to determine whether these associations reflect L and Z, per se, or other aspects of diet or lifestyle linked to high intake of these carotenoids. Most evidence in population studies reflects the intake of L and Z from foods because supplemental L and Z have not been commonly used until the last decade. Secondary analyses in the AREDS2 study support a benefit of these specific carotenoids in supplements in slowing progression to neovascular AMD, in persons with low intake of L and Z from foods and when replacing beta-carotene with L and Z. There is be a remaining need to determine whether dietary or supplemental L and Z slows the development of earlier stages and to identify subgroups of people who might benefit most. If the results indicate no significant benefit over 5 years, then epidemiological studies might evaluate whether free-living people who have taken L and Z supplements for longer periods have lower rates of development on progression of AMD.

Slowing the onset of AMD in its early stages, rather than in late stages, may offer even more promise in preventing AMD-related vision loss and the consequent costs of treatment than slowing progression once the condition is clinically detectable (Mares 2006). This is because the occurrence of early AMD increases the risk of developing advanced AMD over a decade by 10–20 times (Klein et al. 2002; Wang et al. 2007). If early AMD is delayed, then some people may not develop visually-limiting AMD in their lifetimes. However, evidence that L and Z intake (or blood levels) is related to lower incidence or prevalence of early AMD is not as consistent as the evidence relating these carotenoids to advanced AMD (VandenLangenberg et al. 1998; Mares-Perlman et al. 2001; van Leeuwen et al. 2005; Moeller et al. 2006; Cho et al. 2008; Tan et al. 2008; Ma et al. 2012). Yet, it is premature to conclude from this that L and Z do not protect against early stages. There are several reasons that a protective relationship of L and Z to early AMD might have been missed in some studies. Results from the carotenoids in Age-Related Eye Disease Study (CAREDS) indicate that a protective association between dietary lutein or serum lutein and AMD is observed only after excluding women who have made marked dietary changes and who are less than 75 years of age, in whom the association is less likely to be influenced by survivor bias (Moeller et al. 2006). Stronger protective associations between dietary L and Z and AMD have been observed in younger subgroups of other large samples, or in samples that include young subjects (Mares-Perlman et al. 2001; Mares 2003). However, in many study samples, the majority of cases of early AMD occur in people who are older than 75 years. Over this age, the possibility of survivor bias and of recent dietary changes is greatest. This may have contributed to the inconsistency across study samples.

Another contributor to the inconsistency in associations of L and Z intake to early AMD may be misclassification of AMD in early stages. Results of studies which use self-reported AMD diagnoses to assess AMD (Cho et al. 2008) may fail to detect many people who have AMD in early stages. Younger portions of study samples (less than 65 years of age) may be less likely than those in the oldest age strata to routinely visit an ophthalmologist; so many cases are likely to be missed. In CAREDS between 2001 and 2004, 77% of women, aged between 55 and 84 with early AMD as determined from fundus photographs were unaware of its presence (Moeller et al. 2006). Missing cases increases the possibility of low were unaware of its presence statistical power to observe a protective association, if one exists. More importantly, health consciousness and health care seeking behavior are often linked, so that people who are more likely to visit an eye doctor are also more apt to have healthier diets (which would be higher in lutein). This could create a bias that would mask a protective relationship if one existed.

Compounding these potential biases is the possibility that protective associations with AMD may be difficult to observe in some previous epidemiological studies because of inability to adequately account for some personal attributes which modify a protective benefit. Genetics may modify a protective effect. In the Rotterdam Eye Study, high dietary intake of lutein alone was not associated with lower risk of incident AMD (van Leeuwen et al. 2005). However, lutein intake modified the risk associated with *CFH* Y402H genotype (Ho et al. 2011). Being in the high tertile for lutein intake was only associated with significantly reduced risk for incidence of AMD (most of whom had early AMD) in persons who were homozygous for the high-risk allele. In a separate study (Seddon et al. 2010), subjects with the TT genotype of a specific variant (rs10468017) in the hepatic lipase (*LIPC*) gene and lutein intake above the median had lower risk for advanced AMD (OR = 0.2 [0.1–0.5]) compared with those who had both high-risk alleles and lutein intake below the median. Whether this is a chance finding (the $p$-value for interaction was not significant [$p = 0.11$]) or would apply to earlier stages of AMD is unknown. Interactions between genotype and AMD have not been well studied. More research is needed.

Exposure to sunlight and vitamin D may also influence the protective benefit of L and Z, but have also not been well studied. Sunlight exposure presents both a risk, because of the oxidative stress and potential for photic damage that result from high exposures, and a benefit, providing the opportunity to make vitamin D in the skin, as a result of ultraviolet-B exposure. L and Z may minimize risk associated with high sunlight exposure. In the EurEye Study, sunlight exposure was only related to neovascular AMD in people who also had low dietary intake of four antioxidants, including zeaxanthin (Fletcher et al. 2008).

To observe a protective benefit of lutein in some studies among persons with high sunlight exposure, it may be necessary to account for vitamin D exposure. Increased vitamin D intake has been associated with lower risk for AMD in three previous studies (Parekh et al. 2007; Millen et al. 2011; Seddon et al. 2011). In CAREDS, sunlight exposure history was not initially related to early or late AMD (Mares, unpublished data). However, in women less than 75 years of age, after adjusting for vitamin D which was associated with a twofold lower risk for AMD (Millen et al. 2011), there was a trend toward an adverse association of sunlight exposure 20 years before AMD assessment in women in the lowest quartile for macular pigment optical density (multivariate adjusted OR = 2.8, 95%; CI 0.9–9.0; Mares, unpublished data). In addition, there was a significant interaction between status for vitamin D and intake of L and Z, with risk being lowest among persons with better status for both (Millen et al. 2011). Taken together, joint and interactive influences of sunlight, vitamin D and macular pigment are suggested by these few studies but have not been well explored. Such interactions, if real, may have reduced an ability to observe a protective benefit of L and Z on AMD in other studies, particularly in the early stages.

Another explanation for the inconsistency in epidemiological data could be that the ability to accumulate macular pigments may also depend on a person's ability to accumulate carotenoids from the diet or supplements. The reported macular response to supplementation with L and Z varies widely (from 50 to 95%) across studies. (Hammond et al. 1997; Aleman et al. 2001; Bone et al. 2003 and 2007; Johnson et al. 2000; Berendschott et al. 2000; Kvansakul et al 2006; Treischmann et al. 2007; Schalch et al. 2007; Nolan et al. 2011). In general, blood responses to oral carotenoids vary between individuals as well (Bowen et al. 1993; Bowen et al. 2002). The influences on the ability to take up carotenoids by the intestinal tract and the eye are largely unknown. Having high levels of obesity (Hammond et al. 2002; Burke et al. 2005; Nolan et al. 2007), diabetes (Mares et al. 2006; Moeller et al. 2009), or certain genetic variants (Loane et al. 2010; Borel et al. 2011; Loane et al. 2011) is related to lower macular pigment density. These may reflect influences on the ability to absorb and accumulate L and Z within the eye or conditions which increase the turnover of carotenoids in the eye due to inflammation and oxidative stress. They may also be markers for genetic or epigenetic patterns that promote the uptake of L and Z by the eye or other tissues, such as adipose.

Variants in genes, which could influence macular pigment optical density levels and modify relationships between lutein intake and AMD risk, might include those which are involved in reverse cholesterol transport, which appear to influence the ability to absorb xanthophylls into the gut or retina. An example is SR-B1, a membrane receptor for high-density lipoproteins, which mediates cholesterol efflux and carotenoid uptake in the intestine and retina (Reboul et al. 2005; During et al. 2008). Heterozygosity for a specific variant within *SRB1* was related to AMD in one sample of people without high-risk variants in the well-known AMD susceptibility genes *CFH* and *ARMS2* (Zerbib et al. 2009). Another example is variation in the gene which encodes for the adenosine binding cassette transporter protein 1 (the *ABCA1* gene) which plays a role in cholesterol metabolism, especially high-density lipoprotein (HDL-cholesterol). Homozygous or heterozygous carriers of the dysfunctional allele variants of *ABCA1* lead to low levels and unstable HDL particles in humans with Tangiers disease (Schaefer et al. 2010) as well as deficiencies in tissue carotenoids in chickens (Attie et al. 2002; Connor et al. 2007). Variants within cholesterol transport genes have been linked to the presence of AMD in some past studies (Zareparsi et al. 2005; Tikellis et al. 2007; SanGiovanni and Mehta 2009) and in a recent genome-wide association study of a very large number of advanced AMD cases (Chen et al. 2010).

Previous investigators have also found relationships between macular pigment optical density and variants in *BCM01*, a gene for the enzyme which cleaves carotenoids to make vitamin A (Borel et al. 2011). Although L and Z are not known substrates for *BCM01*, variants in this gene are related to low levels of L and Z in the blood (Ferrucci et al. 2009); this has been hypothesized to result from competition of provitamin A carotenoids with xanthophylls for absorption and transport (Lietz et al. 2012).

Common variants in *CD36* have also been related to macular pigment optical density (Borel et al. 2011). This gene encodes a transmembrane protein which is involved in lutein uptake by adipocytes (Moussa et al. 2011). However, expression of *CD36* is also influenced by diet (Llorente-Cortes et al. 2010) which might further modify the association between lutein status and AMD.

Genetic variants in specific lutein- or zeaxanthin-binding proteins in the retina might also influence macular pigment density or the concentration of these carotenoids elsewhere in the retina. These include *GSTP1* that binds zeaxanthin in the human macula (Bhosale et al. 2004) and *StARD3* which binds lutein in the human macula (Li et al. 2011). To date, variants in these genes have not yet been reported to be related to AMD risk, nor well studied.

Recent results in the CAREDS suggest that many common variants of the genes discussed above are related to the optical density of macular pigment in women. (Meyers 2013) Confirmation in independent study samples are needed.

## 4.4 EVIDENCE FOR RELATIONSHIPS WITH MACULAR PIGMENT OPTICAL DENSITY

Direct study of relationships of macular carotenoids to the occurrence of AMD circumvents the influence of unknown influences on uptake into the gut and eye, and may offer more direct evidence for a protective benefit of L and Z. The strong biological plausibility that macular pigment protects against degeneration is supported by the observation of lower levels of L and Z in autopsy specimens of donor eyes with AMD, compared with donor eyes in people without AMD (Boneet et al. 2001).

Clinical and epidemiological studies which have direct measures of macular pigment optical density are, to date, limited. Relationships of AMD to macular pigment optical density, measured noninvasively in living persons, have not been detected in cross-sectional studies (Berendschot et al. 2002; Bernstein et al. 2002; LaRowe et al. 2008; Obana et al. 2008; Dietzel et al. 2011). This may be due, in part, to bias as a result of recent lutein supplement use in people who have been diagnosed with or who have a family history for AMD, or due to survivor bias common in epidemiological studies of older people (discussed earlier). Significant (Bernstein et al. 2002) or marginally significant (LaRowe et al. 2008) trends for a protective association have been observed when such people are excluded from analyses. A trend in the direction of protection of macular pigment on AMD progression was observed in the only longitudinal study (Kanis et al. 2007) to date, but it was not statistically significant, nor well powered (only 27 people developed AMD over 10 years.) For this reason, larger longitudinal studies are needed to evaluate the magnitude of protection that higher macular pigment optical density may have on lowering risk for AMD.

Even if macular pigment is found to be related to lower risk for developing or worsening AMD in future studies, it will remain to be determined whether this is due to dietary intake of L and Z, or the many other components in fruits and vegetables or healthy lifestyles in people with lutein-rich diets that may slow the development of AMD. Women who had a combination of healthy lifestyles (fruit- and vegetable-rich diet, not smoking and high levels of physical activity) had higher macular pigment optical density than women who did not, despite having only slightly greater intake of L and Z (Mares et al. 2011). This suggests that a combination of lutein- and antioxidant-rich diets, physical activity, and not smoking, which could all contribute to lowering oxidative stress, might lead to enriched macular pigment. However, this has not been tested.

## 4.5 SUMMARY AND CONCLUSIONS

Relationships of dietary L and Z to advanced AMD have become consistent in recent longitudinal epidemiological studies and are suggested by results of several clinical trials. These lines of evidence add to the strong evidence for plausibility that these carotenoids protect the macula from oxidative stress, inflammation, and sunlight. The body of evidence to support a protective benefit of L and Z to advanced AMD has become moderately strong.

Whether these macular carotenoids protect against earlier stages of AMD is less clear from available epidemiological evidence. Results to date are conflicting, but likely to be biased. Understanding these associations may be enhanced by studying very large study samples, as is possible when several study samples are pooled at the individual level. Comprehensive measurement of nutritional, lifestyle, and genetic risks for AMD, to detect high risk conferred by the joint presence of these factors, will also permit better detection of protective benefit of L and Z on risk for AMD in some people, if it exists. This will allow for early intervention and the development of strategies to motivate people to take steps to lower the risk for developing this common and costly condition in their lifetimes.

## ACKNOWLEDGMENTS

Our work is supported by the Research to Prevent Blindness and, in the Carotenoids in CAREDS, by taxpayers, through National Institute of Health Grants: National Eye Institute Grants EY013018 and EY016886, and by the National Heart Lung and Blood Institute, which supports the Women's Health Initiative to which CAREDS is an ancillary study.

## REFERENCES

Aleman, T. S., J. L. Duncan, et al. (2001). "Macular pigment and lutein supplementation in retinitis pigmentosa and Usher syndrome." *Invest Ophthalmol Vis Sci* **42**(8): 1873–1881.

AREDS Group. (2001). A randomized, placebo-controlled, clinical trial of high-dose supplementation with vitamins C and E, beta carotene, and zinc for age-related macular degeneration and vision loss: AREDS report no. 8. *Arch Ophthalmol.* **119**(10): 1417–1436.

AREDS Group. (2013). Lutein + zeaxanthin and omega-3 fatty acids for age-related macular degeneration: the Age-Related Eye Disease Study 2 (AREDS2) randomized clinical trial. *JAMA.* **309**(19): 2005–2015.

Attie, A. D., Y. Hamon, et al. (2002). "Identification and functional analysis of a naturally occurring E89K mutation in the ABCA1 gene of the WHAM chicken." *J Lipid Res* **43**(10): 1610–1617.

Berendschot, T. T., R. A. Goldbohm, W. A. Klopping, J. van de Kraats, J. van Norel, D. van Norren. (2000). Influence of lutein supplementation on macular pigment, assessed with two objective techniques. *Investigative Ophthalmology & Visual Science.* **41**(11): 3322–3326.

Berendschot, T. T., J. J. Willemse-Assink, et al. (2002). "Macular pigment and melanin in age-related maculopathy in a general population." *Invest Ophthalmol Vis Sci* **43**(6): 1928–1932.

Bernstein, P. S., D. Y. Zhao, et al. (2002). "Resonance Raman measurement of macular carotenoids in normal subjects and in age-related macular degeneration patients." *Ophthalmology* **109**(10): 1780–1787.

Bhosale, P., A. J. Larson, et al. (2004). "Identification and characterization of a Pi isoform of glutathione S-transferase (GSTP1) as a zeaxanthin-binding protein in the macula of the human eye." *J Biol Chem* **279**(47): 49447–49454.

Bian, Q., S. Gao, J. Zhou, et al. (2012). Lutein and zeaxanthin supplementation reduces photo-oxidative damage and modulates the expression of inflammation-related genes in retinal pigment epithelial cells. *Free Radic Biol Med.* **53**(6): 1298–1307.

Bone, R. A., J. T. Landrum, et al. (2001). "Macular pigment in donor eyes with and without AMD: a case-control study." *Invest Ophthalmol Vis Sci* **42**(1): 235–240.

Bone, R. A., J. T. Landrum, Y. Cao, A. N. Howard, F. Alvarez-Calderon. (2007). Macular pigment response to a supplement containing meso-zeaxanthin, lutein and zeaxanthin. *Nutr Metab (Lond).* **4**: 12.

Bone, R. A., J. T. Landrum, et al. (2003). "Lutein and zeaxanthin dietary supplements raise macular pigment density and serum concentrations of these carotenoids in humans." *J Nutr* **133**(4): 992–998.

Borel, P., F. S. de Edelenyi, et al. (2011). "Genetic variants in BCMO1 and CD36 are associated with plasma lutein concentrations and macular pigment optical density in humans." *Ann Med* **43**(1): 47–59.

Bowen, P. E., V. Garg, et al. (1993). "Variability of serum carotenoids in response to controlled diets containing six servings of fruits and vegetables per day." *Ann NY Acad Sci* **691**: 241–243.

Bowen, P. E., S. M. Herbst-Espinosa, et al. (2002). "Esterification does not impair lutein bioavailability in humans." *J Nutr* **132**(12): 3668–3673.
Burke, J. D., J. Curran-Celentano, et al. (2005). "Diet and serum carotenoid concentrations affect macular pigment optical density in adults 45 years and older." *J Nutr* **135**(5): 1208–1214.
Chen, W., D. Stambolian, et al. (2010). "Genetic variants near TIMP3 and HDL-associated loci influence susceptibility to age-related macular degeneration." *Proc Natl Acad Sci USA* **In Press.**
Chew, E. Y., T. Clemons, et al. (2012). "The Age-Related Eye Disease Study 2 (AREDS2): study design and baseline characteristics (AREDS2 report number 1)." *Ophthalmology* **119**(11): 2282–2289.
Chiu, C. J. and A. Taylor. (2007). "Nutritional antioxidants and age-related cataract and maculopathy." *Exp Eye Res* **84**(2): 229–245.
Cho, E., S. E. Hankinson, et al. (2008). "Prospective study of lutein/zeaxanthin intake and risk of age-related macular degeneration." *Am J Clin Nutr* **87**(6): 1837–1843.
Chong, E. W., T. Y. Wong, et al. (2007). "Dietary antioxidants and primary prevention of age related macular degeneration: systematic review and meta-analysis." *BMJ* **335**(7623): 755.
Connor, W. E., P. B. Duell, et al. (2007). "The prime role of HDL to transport lutein into the retina: evidence from HDL-deficient WHAM chicks having a mutant ABCA1 transporter." *Invest Ophthalmol Vis Sci* **48**(9): 4226–4231.
Dietzel, M., M. Zeimer, et al. (2011). "Determinants of macular pigment optical density and its relation to age-related maculopathy: results from the Muenster Aging and Retina Study (MARS)." *Invest Ophthalmol Vis Sci* **52**(6): 3452–3457.
During, A., S. Doraiswamy, et al. (2008). "Xanthophylls are preferentially taken up compared with beta-carotene by retinal cells via a SRBI-dependent mechanism." *J Lipid Res* **49**(8): 1715–1724.
Evans, J. R. and K. Henshaw. (2008). "Antioxidant vitamin and mineral supplements for preventing age-related macular degeneration." *Cochrane Database Syst Rev* (1): 1–33, CD000253.
Ferrucci, L., J. R. Perry, et al. (2009). "Common variation in the beta-carotene 15,15'-monooxygenase 1 gene affects circulating levels of carotenoids: a genome-wide association study." *Am J Hum Genet* **84**(2): 123–133.
Fletcher, A. E., G. C. Bentham, et al. (2008). "Sunlight exposure, antioxidants, and age-related macular degeneration." *Arch Ophthalmol* **126**(10): 1396–1403.
Group, T. E. D. C. C. S. (1993). "Antioxidant status and neovascular age-related macular degeneration." *Arch Ophthalmol* **111**: 104–109.
Hammond, B. R., Jr., T. A. Ciulla, et al. (2002). "Macular pigment density is reduced in obese subjects." *Invest Ophthalmol Vis Sci* **43**(1): 47–50.
Hammond, B. R., Jr., E. J. Johnson, et al. (1997). "Dietary modification of human macular pigment density." *Invest Ophthalmol Vis Sci* **38**(9): 1795–1801.
Ho, L., R. van Leeuwen, et al. (2011). "Reducing the genetic risk of age-related macular degeneration with dietary antioxidants, zinc, and omega-3 fatty acids: the Rotterdam study." *Arch Ophthalmol* **129**(6): 758–766.
Izumi-Nagai, K., N. Nagai, K. Ohgami, et al. (2007). Macular pigment lutein is antiinflammatory in preventing choroidal neovascularization. *Arterioscler Thromb Vasc Biol.* **27**(12): 2555–2562.
Johnson, E. J., B. R. Hammond, K. J. Yeum, et al. (2000). Relation among serum and tissue concentrations of lutein and zeaxanthin and macular pigment density. *Am J Clin Nutr.* **71**(6): 1555–1562.
Kanis, M. J., T. T. Berendschot, et al. (2007). "Influence of macular pigment and melanin on incident early AMD in a white population." *Graefes Arch Clin Exp Ophthalmol* **245**(6): 767–773.

Klein, R., B. E. Klein, et al. (2002). "Ten-year incidence and progression of age-related maculopathy: The Beaver Dam Eye study." *Ophthalmology* **109**(10): 1767–1779.

Krishnadev, N., A. D. Meleth, et al. (2010). "Nutritional supplements for age-related macular degeneration." *Curr Opin Ophthalmol* **21**(3): 184–189.

Kvansakul, J., M. Rodriguez-Carmona, D. Edgar, et al. (2006). Supplementation with the carotenoids lutein or zeaxanthin improves human visual performance. *Ophthalmic & Physiological Optics.* **26**(4): 362–371.

LaRowe, T. L., J. A. Mares, et al. (2008). "Macular pigment density and age-related maculopathy in the Carotenoids in Age-Related Eye Disease Study. An ancillary study of the women's health initiative." *Ophthalmology* **115**(5): 876–883 e871.

Li, B., P. Vachali, et al. (2011). "Identification of StARD3 as a lutein-binding protein in the macula of the primate retina." *Biochemistry* **50**(13): 2541–2549.

Lietz, G., A. Oxley, et al. (2012). "Single nucleotide polymorphisms upstream from the beta-carotene 15,15′-monoxygenase gene influence provitamin A conversion efficiency in female volunteers." *J Nutr* **142**(1): 161S–165S.

Llorente-Cortes, V., R. Estruch, et al. (2010). "Effect of Mediterranean diet on the expression of pro-atherogenic genes in a population at high cardiovascular risk." *Atherosclerosis* **208**(2): 442–450.

Loane, E., G. J. McKay, et al. (2010). "Apolipoprotein E genotype is associated with macular pigment optical density." *Invest Ophthalmol Vis Sci* **51**(5): 2636–2643.

Loane, E., J. M. Nolan, et al. (2011). "The association between macular pigment optical density and CFH, ARMS2, C2/BF, and C3 genotype." *Exp Eye Res* **93**(5): 592–598.

Ma, L., H. L. Dou, et al. (2012a). "Improvement of retinal function in early age-related macular degeneration after lutein and zeaxanthin supplementation: a randomized, double-masked, placebo-controlled trial." *Am J Ophthalmol* **154**(4): 625–634 e621.

Ma, L., H. L. Dou, et al. (2012b). "Lutein and zeaxanthin intake and the risk of age-related macular degeneration: a systematic review and meta-analysis." *Br J Nutr* **107**(3): 350–359.

Ma, L., S. F. Yan, et al. (2012c). "Effect of lutein and zeaxanthin on macular pigment and visual function in patients with early age-related macular degeneration." *Ophthalmology* **119**(11): 2290–2297.

Mares, J. (2003). "Carotenoids and eye disease: epidemiologic evidence." In: *Carotenoids in Health & Disease*, N. I. Krinsky, Mayne S (eds.), Chapter 19, Marcel Dekker, New York.

Mares, J. A. (2006). "Potential value of antioxidant-rich foods in slowing age-related macular degeneration." *Arch Ophthalmol* **124**(9): 1339–1340.

Mares, J. A., T. L. LaRowe, et al. (2006). "Predictors of optical density of lutein and zeaxanthin in retinas of older women in the Carotenoids in Age-Related Eye Disease Study, an ancillary study of the Women's Health Initiative." *Am J Clin Nutr* **84**(5): 1107–1122.

Mares, J. A., A. E. Millen, et al. (2012). "Diet and supplements and the prevention and treatment of eye diseases." In: *Nutrition in the Prevention and Treatment of Disease*, A. M. Coulston, Boushey C L, Ferruzzi M (eds.), Elsevier, San Diego, CA.

Mares, J. A., R. P. Voland, et al. (2011). "Healthy lifestyles related to subsequent prevalence of age-related macular degeneration." *Arch Ophthalmol* **129**(4): 470–480.

Mares-Perlman, J. A., A. I. Fisher, et al. (2001). "Lutein and zeaxanthin in the diet and serum and their relation to age-related maculopathy in the third national health and nutrition examination survey." *Am J Epidemiol* **153**(5): 424–432.

Meyers, K. J., E. J. Johnson, P. S. Bernstein, et al. (2013). Genetic determinants of macular pigments in women of the Carotenoids in Age-Related Eye Disease Study. *Invest Ophthalmol Vis Sci.* **54**(3): 2333–2345.

Millen, A. E., R. Voland, et al. (2011). "Vitamin D status and early age-related macular degeneration in postmenopausal women." *Arch Ophthalmol* **129**: 481–489. DOI: 129/4/481 [pii] 10.1001/archophthalmol.2011.48.

Moeller, S. M., N. R. Mehta, et al. (2006). "Associations between intermediate age-related macular degeneration and lutein and zeaxanthin in the Carotenoids in Age-Related Eye Disease Study (CAREDS), an ancillary study of the Women's Health Initiative." *Arch Ophthalmol* **124**: 1–24.

Moeller, S. M., N. Parekh, et al. (2006). "Associations between intermediate age-related macular degeneration and lutein and zeaxanthin in the Carotenoids in Age-Related Eye Disease Study (CAREDS): ancillary study of the Women's Health Initiative." *Arch Ophthalmol* **124**(8): 1151–1162.

Moeller, S. M., R. Voland, et al. (2009). "Women's health initiative diet intervention did not increase macular pigment optical density in an ancillary study of a subsample of the women's health initiative." *J Nutr* **139**(9): 1692–1699.

Moriarty-Craige, S. E., J. Adkison, et al. (2005). "Antioxidant supplements prevent oxidation of cysteine/cystine redox in patients with age-related macular degeneration." *Am J Ophthalmol* **140**(6): 1020–1026.

Moussa, M., E. Gouranton, et al. (2011). "CD36 is involved in lycopene and lutein uptake by adipocytes and adipose tissue cultures." *Mol Nutr Food Res* **55**(4): 578–584.

Nolan, J. M., J. Loughman, M. C. Akkali, et al. (2011). The impact of macular pigment augmentation on visual performance in normal subjects: COMPASS. *Vision Res.* **51**(5): 459–469.

Nolan, J. M., J. Stack, et al. (2007). "The relationships between macular pigment optical density and its constituent carotenoids in diet and serum." *Invest Ophthalmol Vis Sci* **48**(2): 571–582.

Obana, A., T. Hiramitsu, et al. (2008). "Macular carotenoid levels of normal subjects and age-related maculopathy patients in a Japanese population." *Ophthalmology* **115**(1): 147–157.

Parekh, N., R. Chappell, et al. (2007). "Association between Vitamin D and age-related macular degeneration in the Third National Health and Nutrition Examination Survey 1988–94." *Arch Ophthalmol* **125**: 661–669.

Piermarocchi, S., S. Saviano, et al. (2011). "Carotenoids in Age-Related Maculopathy Italian Study (CARMIS): two-year results of a randomized study." *Eur J Ophthalmol* **22**(2):216–225.

Reboul, E., L. Abou, et al. (2005). "Lutein transport by Caco-2 TC-7 cells occurs partly by a facilitated process involving the scavenger receptor class B type I (SR-BI)." *Biochem J* **387**(Pt 2): 455–461.

Richer, S., W. Stiles, et al. (2004). "Double-masked, placebo-controlled, randomized trial of lutein and antioxidant supplementation in the intervention of atrophic age-related macular degeneration: the Veterans LAST study (Lutein Antioxidant Supplementation Trial)." *Optometry* **75**(4): 216–230.

Richer, S. P., W. Stiles, et al. (2011). "Randomized, double-blind, placebo-controlled study of zeaxanthin and visual function in patients with atrophic age-related macular degeneration: the Zeaxanthin and Visual Function Study (ZVF) FDA IND #78, 973." *Optometry* **82**(11): 667–680 e666.

SanGiovanni, J. P., E. Y. Chew, et al. (2007). "The relationship of dietary carotenoid and vitamin A, E, and C intake with age-related macular degeneration in a case-control study: AREDS Report No. 22." *Arch Ophthalmol* **125**(9): 1225–1232.

SanGiovanni, J. P. and S. Mehta. (2009). "Variation in lipid-associated genes as they relate to risk of advanced age-related macular degeneration." *World Rev Nutr Diet* **99**: 105–158.

Schaefer, E. J., R. D. Santos, et al. (2010). "Marked HDL deficiency and premature coronary heart disease." *Curr Opin Lipidol* **21**(4): 289–297.

Schalch, W., W. Cohn, F. M. Barker, et al. (2007). Xanthophyll accumulation in the human retina during supplementation with lutein or zeaxanthin—the LUXEA (LUtein Xanthophyll Eye Accumulation) study. *Arch Biochem Biophys.* **458**(2): 128–135.

Seddon, J., U. A. Ajani, et al. (1994). "Dietary carotenoids, vitamins A, C, and E, and advanced age related macular degeneration." *JAMA* **272**: 1413–1420.

Seddon, J. M., R. Reynolds, et al. (2010). "Associations of smoking, body mass index, dietary lutein, and the LIPC gene variant rs10468017 with advanced age-related macular degeneration." *Mol Vis* **16**: 2412–2424.

Seddon, J. M., R. Reynolds, et al. (2011). "Smoking, dietary betaine, methionine, and vitamin D in monozygotic twins with discordant macular degeneration: epigenetic implications." *Ophthalmology* **118**(7): 1386–1394.

Tan, J. S. L., J. J. Wang, et al. (2008). "Dietary antioxidants and the long-term incidence of age-related macular degeneration: The Blue Mountains Eye Study." *Ophthalmology* **115**(2): 334–341.

Tikellis, G., C. Sun, et al. (2007). "Apolipoprotein E gene and age-related maculopathy in older individuals: the cardiovascular health study." *Arch Ophthalmol* **125**(1): 68–73.

Trieschmann, M., S. Beatty, J. M. Nolan, et al. (2007). Changes in macular pigment optical density and serum concentrations of its constituent carotenoids following supplemental lutein and zeaxanthin: the LUNA study. *Exp Eye Res.* **84**(4): 718–728.

VandenLangenberg, G. M., J. A. Mares-Perlman, et al. (1998). "Associations between antioxidant and zinc intake and the 5-year incidence of early age-related maculopathy in the Beaver Dam Eye Study." *Am J Epidemiol* **148**(2): 204–214.

van Leeuwen, R., S. Boekhoorn, et al. (2005). "Dietary intake of antioxidants and risk of age-related macular degeneration." *JAMA* **294**(24): 3101–3107.

Wang, J. J., E. Rochtchina, et al. (2007). "Ten-year incidence and progression of age-related maculopathy: The Blue Mountains Eye Study." *Ophthalmology* **114**(1): 92–98.

Zareparsi, S., K. E. Branham, et al. (2005). "Strong association of the Y402H variant in complement factor H at 1q32 with susceptibility to age-related macular degeneration." *Am J Hum Genet* **77**(1): 149–153.

Zerbib, J., J. M. Seddon, et al. (2009). "rs5888 variant of SCARB1 gene is a possible susceptibility factor for age-related macular degeneration." *PLoS One* **4**(10): e7341.

# 5 Clinical Trials Investigating the Macular Carotenoids

*Sarah Sabour-Pickett, John M. Nolan, and Stephen Beatty*

## CONTENTS

- 5.1 Introduction ........................................................................................................ 75
- 5.2 Clinical Trial and Its Contribution to Evidence ............................................... 76
- 5.3 Clinical Trials Investigating Macular Carotenoids in Subjects with AMD ... 77
    - 5.3.1 Proof of Principle ................................................................................... 77
    - 5.3.2 Interventional Studies ............................................................................ 78
    - 5.3.3 Trials Awaiting Completion .................................................................. 81
- 5.4 Clinical Trials Investigating Macular Carotenoids in Normal Subjects ........ 81
    - 5.4.1 Compass .................................................................................................. 83
    - 5.4.2 MOST Vision .......................................................................................... 83
    - 5.4.3 Supplementation with the Macular Carotenoids in Subjects with an Atypical MPOD Spatial Profile ....................................................... 84
- 5.5 Observational Studies ........................................................................................ 84
    - 5.5.1 Serum and Retinal Response to Supplementation with the Macular Carotenoids ............................................................................. 87
    - 5.5.2 Moving Forward .................................................................................... 87
- 5.6 Conclusion .......................................................................................................... 88
- References .................................................................................................................. 88

## 5.1 INTRODUCTION

The anatomical (central retinal), biochemical (antioxidant), and optical (short-wavelength filtering) properties of macular pigment (MP) have generated interest in the biologically plausible rationale that MP may not only confer protection against age-related macular degeneration (AMD) but also optimize vision. The testing of this hypothesis has been a subject of growing interest over the past decade. This chapter reviews the evidence from the clinical trials that have investigated the role of macular carotenoid supplementation on macular pigment optical density (MPOD), on visual performance, and on the risk of disease progression in subjects afflicted with AMD. In addition, the more recent studies investigating the impact of supplementation with the macular carotenoids on visual performance in normal subjects (without disease) will be discussed.

## 5.2 CLINICAL TRIAL AND ITS CONTRIBUTION TO EVIDENCE

Clinical trials vary in size and complexity of design. Large, controlled studies provide statistical power and the ability to control for many confounding factors. Smaller, simpler study designs, on the other hand, are typically less costly and are of shorter duration. In practice, it is often smaller (pilot) trials and observational studies that precede and inform researchers about the range, nature, and magnitude of the responses that might be anticipated in a larger interventional trial. An observational study is one in which conclusions are drawn by observation alone, examples of which may include case-control and cohort studies. The randomized controlled trial (RCT) is the accepted "gold standard" clinical trial, where participants are assigned at random, rather than by conscious decision of clinician or patient, to receive one or more clinical interventions, typically compared with a control or placebo. The reliability of the RCT is strengthened when it is blinded or masked by "procedures that prevent study participants, caregivers, or outcome assessors from knowing which intervention was received" (Wood et al. 2008). Generally, the greater the sample size, the more reduced the likelihood of bias.

The quality of the evidence produced by a trial is usually evaluated on the basis of study design. Systematic reviews, or meta-analyses, of RCTs are widely accepted as providing the best evidence on the effects of preventative, or interventional, therapies and treatments in medicine (Sackett 1994) (see Table 5.1). However, the results of meta-analyses can be influenced by issues such as publication selection bias, heterogeneity of the studies being assessed, limitations in information availability, and the method of data analysis, thus potentially influencing the interpretation of the results (Bailar 1997; Walker et al. 2008). Similarly, and although RCTs are regarded as the gold standard in clinical research, they do have certain limitations (Black 1996). Occasionally inappropriate outcome measures and/or biased sample recruitment can occur. Given that studies involving humans are laden with ethical issues and, in many cases, may not be feasible, practical or, indeed, appropriate (Black 1996; Stephenson and Imrie 1998), many important findings have necessarily been the result of observational studies. Although lacking the benefit

### Table 5.1
### Hierarchy of Study Types

1. Systematic reviews and meta-analyses of RCTs
2. RCTs
3. Nonrandomized intervention studies
4. Observational studies
5. Nonexperimental studies
6. Expert opinion

of a control group, large numbers of observational outcomes can provide highly reliable conclusions (Mann 2003). The preference accorded to RCTs can, in some instances, result in the exclusion of evidence arising from other and valid sources. Experience demonstrates that studies with alternative designs should be seen as complementary to RCTs.

In addition, the capacity and resources of competing stakeholders (e.g., pharmaceutical companies or academic institutions) to generate and disseminate evidence has a profound influence on the prestige and volume of available published literature on a given subject (Rychetnik et al. 2002).

AMD is a slow, complex disorder. To effectively investigate the protective role of carotenoids in AMD, including a possible role for supplementation in preventing or delaying the onset of the condition, an RCT of considerable length (decades) would be required. Securing funding for such a study would be challenging indeed. In addition, the carotenoids, particularly lutein (L) and zeaxanthin (Z), are commonly found in a typical diet and are readily available in supplement form on the open market, complicating baseline parameters of interest (e.g., MPOD, serum concentrations of L and Z). Such issues further complicate any investigation with respect to the protective role of the macular carotenoids in AMD. Accepting these limitations inherent in the design of an RCT, it is evident that more limited approaches, including observational studies, will play an important role in expanding our understanding of the significance of carotenoids in retarding the onset and/or progression of AMD.

In this chapter, and with full appreciation of these and other challenges related to the design and interpretation of trials investigating the role of macular carotenoid supplements in ameliorating the natural course of AMD and their ability to enhance visual performance, we summarize the currently available evidence germane to this growing area of interest.

## 5.3 CLINICAL TRIALS INVESTIGATING MACULAR CAROTENOIDS IN SUBJECTS WITH AMD

### 5.3.1 Proof of Principle

In 2001, the National Eye Institute published the Age-Related Eye Disease Study (AREDS). This was a 5-year, randomized, double-blind, placebo-controlled trial of 4757 subjects. In brief, AREDS showed that supplementation with vitamins C and E, β-carotene, and zinc, in combination, resulted in a 25% risk reduction of progression from intermediate to advanced AMD. Of note, the AREDS formulation did not contain any of the macular carotenoids, primarily because these compounds were not available in supplement form at the inception of that study. This landmark work did, however, provide Level 1 evidence that supplemental dietary antioxidants were of benefit to patients with AMD. Evaluation of dietary data from this study population also provided observational results that showed that higher dietary intake of L/Z was independently associated with decreased likelihood of having the later stages of the disease (neovascular AMD, geographic atrophy, and large or extensive intermediate drusen) (SanGiovanni et al. 2007).

### 5.3.2 INTERVENTIONAL STUDIES

The evidence with respect to the antioxidant role of MPs has been robust and growing since the landmark 1994 nutritional study by Seddon et al. (1994) This accumulating evidence stimulated investigations into the effect of supplementation with MP's constituent carotenoids, in particular, in the protection against AMD and in the enhancement of visual performance. Landrum et al. (1997) was the first to demonstrate that MPOD (as assessed by heterochromatic flicker phototometry) could be augmented through dietary supplementation with L. Since then, there have been a number of interventional studies reporting on supplementation with macular carotenoids and its impact on AMD and visual performance (Table 5.2), ranging from case series to RCTs (Bartlett and Eperjesi 2007; Beatty et al. 2013; Jentsch et al. 2011; Nolan et al. 2012b; Olmedilla et al. 2001; Piermarocchi et al. 2011; Richer 1999; Richer et al. 2004, 2011; Sasamoto et al. 2011; Weigert et al. 2011).

In 2004, the LAST study (Lutein Antioxidant Supplementation Trial) was carried out in an attempt to evaluate the effect of L, either alone or in combination with coantioxidants, vitamins, and minerals, on the progression of atrophic AMD (Richer et al. 2004). This study was a prospective, 12-month, randomized, double-masked, placebo-controlled trial, involving 90 subjects with atrophic AMD. The subjects were assigned to one of three groups: Group 1 received L (10 mg) only; Group 2 received a broad-spectrum supplementation formula containing L (10 mg) as well as coantioxidants, vitamins, and minerals; Group 3 received a placebo. Results showed that the subjects in Groups 1 and 2 demonstrated an increase in mean MPOD as well as an improvement in visual acuity, contrast sensitivity, glare recovery, and visual distortion. This study, therefore, demonstrated that visual function is improved in patients with atrophic AMD following supplementation with either L alone or L in combination with coantioxidants, vitamins, and minerals.

The Carotenoids in Age-Related Maculopathy (CARMA) study was a randomized, double-blind, placebo-controlled clinical trial of L (12 mg) and Z (0.6 mg) supplementation with coantioxidants versus placebo in patients with AMD (Neelam et al. 2008). This study included 433 subjects, who were recruited and randomly assigned to the treatment or the placebo arms of the study. The primary outcome measure, corrected distance visual acuity (CDVA) at 1 year, did not differ significantly between the placebo and the intervention arms of the study. It was noted that CDVA was significantly better in the intervention arm of the study at 36 months follow-up. In addition, an increase in serum L was associated with significantly improved CDVA and slowing of progression along the AMD severity scale (Beatty et al. 2013). These results were found despite a relatively small sample size at 36 months ($n = 41$, 20 in the intervention group and 21 in the placebo group).

The optical, anatomical, and antioxidant properties of MP have generated a consensus that MP plays an important role in vision. Indeed, a number of studies have already demonstrated the positive cross-sectional association between MPOD and measures of visual performance, including visual acuity, contrast sensitivity,

## TABLE 5.2
### Interventional Studies Investigating the Effect of Supplementation with the Macular Carotenoids in Subjects with AMD

| Principal Author(s) | Study | Year | n | Study Design | Age | Carotenoids (mg) | VP Measures Used | Finding |
|---|---|---|---|---|---|---|---|---|
| Richer et al. | — | 1999 | 14 | Case series | 61–79 | L (14) | CDVA, CS, GR, Ams. | Improved VP |
| Olmedilla et al. | — | 2001 | 5 | Case series | 69–75 | L (15) | CDVA, GD | Improved VP |
| Richer et al. | LAST | 2004 | 90 | RCT | 68–82 | L (10) | CDVA, CS, Ams. | Improved VP[a] |
| Bartlett et al. | — | 2007 | 25 | RCT | 55–82 | L (6) | CS | No benefit |
| Weigart et al. | LISA | 2011 | 126 | RCT | 50–90 | L (20/10)[b] | CDVA, MDLT | Improved VP[a] |
| Richer et al. | ZVF | 2011 | 60 | RCT | 75(±10)[c] | Z (8) and L (9) | CDVA, low contrast CDVA, FSD, KVF, GR, CSF | Improved VP[a] |
| Sasamoto et al. | — | 2011 | 33 | Case series | 65(±9)[c] | L (6) | CS, ROS | Improved VP[a] |
| Piermarocchi et al. | CARMIS | 2011 | 145 | PRS | — | L (10) and Z (1) | CDVA, CS | Improved VP |
| Jentsch et al. | Lutega | 2011 | 172 | RCT | 50+ | L (10/20) and Z (1/2) | CDVA | Improved VP[a] |
| Beatty et al. | CARMA | 2012 | 433 | RCT | 50+ | L (12) and Z (0.6) | CDVA, CS | Improved VP[a] |
| Nolan et al. | MOST AMD | 2012 | 52 | Randomized trial | 67(±8)[c] | L (20), Z (2); MZ (10), L (10), Z (2); MZ (17), L (3), Z (2) | CDVA, CS, GD | Improved VP[a] |

*Note:* $n$ = number of subjects participating in study; Carotenoids, macular carotenoids assessed in the study; VP, visual performance; —, data unavailable; CS, contrast sensitivity; GR, glare recovery; Ams., Amsler grid assessment; GD, glare disability; LISA, Lutein Intervention Study Austria; MDLT, mean differential light threshold; ZVF, Zeaxanthin Visual Function; FSD, foveal shape discrimination; KVF, kinetic visual fields; CSF, contrast sensitivity function; ROS, retinotopic ocular sensitivity; CARMIS, Carotenoids in Age-Related Maculopathy Italian Study; PRS, prospective randomized study; CARMA, Carotenoids in Age-Related Maculopathy.

[a] MP measurements were obtained in the study.
[b] Twenty milligrams taken for first 3 months and 10 mg taken for remaining 3 months.
[c] Mean(±SD).

glare disability, photostress recovery, critical flicker fusion frequency, and color vision (Engles et al. 2007; Hammond and Wooten 2005; Kvansakul et al. 2006; Loughman et al. 2010; Stringham and Hammond 2007; Wooten and Hammond 2002). One might hypothesize, therefore, that an increase in MPOD will be paralleled by an improvement in vision. Indeed, Weigert et al. (2011) have recently shown that an increase in MPOD correlated significantly with a decrease in mean differential light threshold (assessed by microperimetry), suggesting that augmentation of MPOD enhances retinotopic ocular sensitivity. Psychophysical function is adversely affected in AMD (Neelam et al. 2009), and this is confounded by age-related decline in many aspects of visual function in the absence of macular pathology (Arundale 1978; Gittings and Fozard 1986; Klein et al. 1991). Therefore, a demonstrable improvement, or even stabilization of visual function in response to supplemental macular carotenoids in an older population with a known degenerative disease should be deemed beneficial. In this context, it is interesting to note that 9 of the 10 studies (see Table 5.2) investigating changes in visual performance following supplementation with macular carotenoids in AMD subjects have demonstrated an improvement in visual function. In the remaining study, consisting of only 25 subjects supplemented with 6 mg L (alone), vision did not deteriorate (Bartlett and Eperjesi 2007).

Studies are now underway investigating the potential of *meso*-zeaxanthin (MZ), the third and currently least explored macular carotenoid, with respect to the development or progression of AMD (see later). Of note, MZ accounts for about one-third of total MP (Bone et al. 1993), and of the three macular carotenoids, it is the most powerful antioxidant (Bhosale and Bernstein 2005). The *Meso*-zeaxanthin Ocular Supplementation Trial (MOST) is ongoing and designed to investigate the effect of three macular carotenoid formulations on MPOD, visual performance, and retinal morphology, in subjects with early AMD. Preliminary data have demonstrated an improvement in visual performance following supplementation at 12 and 24 months, in terms of contrast sensitivity at a range of spatial frequencies. The greatest effect is seen in subjects supplementing with a formulation that contains all three macular carotenoids (L, Z, and MZ). The observation supports the conclusion that such a formulation may be uniquely efficacious in augmenting MPOD across its spatial profile in the same study (Table 5.3).

### TABLE 5.3
### Mean (±SD) MPOD at Baseline and 12 Months in MOST

| Eccentricity (°) | Group 1 (20 mg L, 2 mg Z) | | | Group 2 (10 mg L, 10 mg MZ, 2 mg Z) | | | Group 3 (17 mg MZ, 3 mg L, 2 mg Z) | | |
|---|---|---|---|---|---|---|---|---|---|
| | Baseline | 12 Months | p | Baseline | 12 Months | p | Baseline | 12 Months | p |
| 0.25 | 0.50 ± 0.25 | 0.59 ± 0.30 | .077 | 0.50 ± 0.25 | 0.60 ± 0.21 | .005 | 0.46 ± 0.21 | 0.57 ± 0.20 | .010 |
| 0.5 | 0.38 ± 0.27 | 0.47 ± 0.27 | .055 | 0.42 ± 0.22 | 0.50 ± 0.19 | .005 | 0.35 ± 0.18 | 0.45 ± 0.21 | .014 |
| 1 | 0.27 ± 0.18 | 0.34 ± 0.16 | .083 | 0.27 ± 0.13 | 0.34 ± 0.17 | .005 | 0.24 ± 0.16 | 0.32 ± 0.16 | .019 |
| 1.75 | 0.16 ± 0.11 | 0.21 ± 0.09 | .018 | 0.14 ± 0.11 | 0.22 ± 0.12 | .002 | 0.11 ± 0.12 | 0.19 ± 0.10 | .009 |

### 5.3.3 TRIALS AWAITING COMPLETION

There are a number of trials underway investigating the role of L and Z in individuals with AMD. The AREDS 2 is an ongoing multicenter RCT ($n = 4203$) evaluating the impact of supplemental L and Z (and/or omega-3) on the progression of intermediate to advanced AMD and the influence of these supplements on visual acuity (Chew et al. 2012). Additionally, it seeks to assess whether modified forms of the original AREDS supplement, with reduced zinc and no β-carotene, work as effectively as the original supplement in reducing the risk of progression to advanced AMD.

AREDS 2 is expected to be completed in December 2012. The results of AREDS 2 will provide valuable and timely data on the potential role of antioxidants, including L and Z, in delaying AMD progression and expects to inform current professional practice with respect to the role of dietary modification and/or supplementation in patients with AMD. A limitation of the trial, however, rests on the fact that the measurement of MPOD has not been incorporated into the study design (with the exception of a small cohort [$n = 55$] in one study location [Bernstein et al. 2012]). Unfortunately, therefore, a finding that supplemental L and Z in AREDS 2 are not beneficial cannot be generalized to mean that MPOD augmentation is not beneficial. Further, it is likely that a very high proportion of participants in the U.S.-based AREDS 2 will have been supplementing with dietary antioxidants for years prior to entering the study (Bernstein et al. 2012). This confounding baseline characteristic in the study population complicates the interpretation of findings for all study groups. The modest 3-month washout period of subjects who reported supplementing is arguably insufficient to normalize these individuals to "nonsupplementing" participants. This compromises the trial's capacity to demonstrate a beneficial effect of supplementation. In addition, should there be a positive outcome in this respect, it is unlikely to reflect that which might have been observed had no study participants been taking supplements.

## 5.4 CLINICAL TRIALS INVESTIGATING MACULAR CAROTENOIDS IN NORMAL SUBJECTS

The optical properties of the MP, in the context of these pigments' accumulation exclusively in the macula, prompted the hypothesis that MP is important for visual performance. Indeed, there is a general consensus that MPOD is important for visual performance in the central retina. Many studies have reported on the cross-sectional relationship between MP and a plethora of visual performance parameters, and a number of trials have investigated the impact of supplementation with the macular carotenoids on visual performance in subjects without disease (see Table 5.4). MP achieves its role in this respect through its blue light–filtering properties at a preceptoral level, thereby attenuating chromatic aberration and light scatter (which are the results of defocus and scatter of blue visible light, respectively), consequentially enhancing contrast sensitivity and reducing the effects of glare. The dichroic properties of MP may further contribute to glare reduction because of consequential and preferential absorption of plane polarized light.

## TABLE 5.4
### Interventional Studies Assessing the Effects of the Macular Carotenoids on Visual Performance in Normal Subjects

| Principal Author(s) | Year | n | Placebo-Control | Carotenoids | Visual Performance Tests | Study Duration (Months) | Observed Visual Benefit Following Supplementation |
|---|---|---|---|---|---|---|---|
| Monje[a] | 1948 | 14 | No | L dipalmitate | Dark adaptation and scotopic VA | 2–6 | Yes[b] |
| Wustenberg[a] | 1951 | 7 | No | L dipalmitate | Dark adaptation | — | No |
| Klaes and Riegel | 1951 | — | No | L dipalmitate | Dark adaptation | — | Yes |
| Andreani and Volpi[a] | 1956 | 10 | No | L dipalmitate | Dark adaptation | — | Yes |
| Mosci[a] | 1956 | — | No | L dipalmitate | Light sensitivity | — | Yes |
| Hayano[a] | 1959 | — | No | L dipalmitate | Dark adaptation | — | Yes[c] |
| Wenzel | 2006 | 10 | Yes | 30 mg L + 2.7 mg Z | Photophobia | 3 | Yes |
| Rodriguez-Carmona | 2006 | 24 | Yes[d] | 10 mg/20 mg of L/Z/L + Z | B/Y color discrimination | 12 | No |
| Kvansakul | 2006 | 34 | Yes | 10 mg L/10 mg Z/combination | Mesopic CS | 6 | Yes |
| Bartlett and Eperjesi | 2008 | 29 | Yes | 6 mg L | VA (distance and near), CS, photostress recovery | 18 | No |
| Stringham and Hammond | 2008 | 40 | No | 10 mg L + 2 mg Z | Photostress recovery and grating visibility | 6 | Yes; both |
| Nolan | 2010 | 121 | Yes | 12 mg L + 1 mg Z | VA, CS, GD, photostress recovery | 12 | Yes; CS, GD |
| Loughman | 2012 | 36 | Yes | 10 mg L + 2 mg Z + 10 mg MZ/20 mg L + 2 mg Z | VA, CS, GD, photostress recovery | 6 | Yes; VA, CS, GD |

*Note: n*, number of subjects; Carotenoids, macular carotenoids investigated; VA, visual acuity; —, data not available; B/Y, blue/yellow; CS, contrast sensitivity; GD, glare disability.

[a] Data obtained from Nussbaum et al. (1981).
[b] Described as having a "transient" benefit.
[c] Proportional to serum L.
[d] For second 6 months of the study.

### 5.4.1 COMPASS

The Collaborative Optical Macular Pigment ASsessment Study (COMPASS), an RCT, was designed to investigate the impact of supplementation with macular carotenoids versus placebo, on MPOD and visual performance. One hundred and twenty-one normal subjects were recruited (age range: 18–41 years) to COMPASS. The active group consumed 12 mg of L and 1 mg of Z (but no MZ) every day for 12 months ($n = 61$), and the remainder of the subjects were on placebo. A range of psychophysical tests were used to assess visual performance, including visual acuity, contrast sensitivity, glare disability, and photostress recovery. Subjective visual function was determined by questionnaire, and MPOD was measured using customized heterochromatic flicker photometry. The results of this study showed that central MPOD increased significantly in the active group, whereas no such augmentation was demonstrable in the placebo group. Of note, however, this augmentation in MPOD (using an L-based formulation) was only observed at (and not prior to) the 12-month study visit. Although the increase in MPOD did not correlate, generally, with an improvement in visual performance, statistically significant differences in mesopic contrast sensitivity (with and without glare) were observed between those who had high MPOD and those who had low MPOD at 12 months, whereas this was not the case at baseline (Nolan et al. 2011).

### 5.4.2 MOST VISION

Optimal blue-light filtration is achieved in the presence of all three macular carotenoids (L, Z, and MZ) (Landrum and Bone 2001). The *Meso*-zeaxanthin Ocular Supplementation Vision Trial (MOST Vision) investigated the effect of supplemental macular carotenoids, including a formulation containing MZ, on visual performance in normal subjects (Loughman et al. 2012). Thirty-six recruited subjects were assigned to one of three groups: the first was given a high dose (20 mg) of L and 2 mg Z (Group 1); the second group was given 10 mg L, 10 mg MZ, and 2 mg Z (Group 2); and the third group was given placebo (Group 3), every day for six months. A statistically significant rise in MP was observed (determined three months following commencement of supplementation) only among subjects supplemented with a formulation containing all three macular carotenoids, including MZ (Group 2). Statistically significant improvements in visual acuity were observed at six months, but only for subjects in Group 2. Contrast sensitivity (under mesopic and photopic conditions) and glare disability under mesopic conditions were assessed using the Functional Acuity Analyzer™ at the following spatial frequencies: 1.5, 3, 6, 12, and 18 cycles per degree (cpd). Statistically significant improvements in contrast sensitivity were noted across a range of spatial frequencies, under photopic (3, 12, and 18 cpd) and mesopic conditions (1.5, 3, 12, and 18 cpd), again only among subjects supplemented with MZ (with a single exception of improved contrast sensitivity at a single spatial frequency [6 cpd] in the high L group [Group 1]). There were no statistically significant improvements in mesopic glare disability between baseline and 6 months in Groups 1 and 3. However, there was a demonstrable improvement in mesopic glare disability for subjects in Group 2 for all spatial frequencies tested (with the exception of 18 cpd).

### 5.4.3 SUPPLEMENTATION WITH THE MACULAR CAROTENOIDS IN SUBJECTS WITH AN ATYPICAL MPOD SPATIAL PROFILE

A study reporting on data collected from 828 subjects between the ages of 18 and 55 (and who, therefore, did not suffer from AMD) demonstrated that there was a relative lack of MP in association with tobacco use and with a family history of AMD. This study also reported an age-related decline in MP, suggesting that the risk that such variables represent for AMD may be attributable, at least in part, to a parallel lack of MP. In other words, this study showed that, prior to disease onset, known risk factors for AMD are independently associated with a relative lack of MP (Nolan et al. 2007).

Of the 828 subjects, a proportion (12%) exhibited a "central dip" (i.e., they did not exhibit the typical central peak that declines in a monotonic fashion from the foveal center) in their MPOD spatial profile. Interestingly, this central dip was associated with tobacco use and increasing age (Kirby et al. 2010), indicating that such atypical spatial profiles of MP are also and independently representative of risk for AMD. Given that MZ is the dominant carotenoid in the foveal center, it has been hypothesized that the observed central dips in the MP spatial profile in about 12% of the population were attributable to a relative lack of this carotenoid. Since retinal MZ is formed from retinal L (but not retinal Z), the observed central dip in the MP spatial profile may be the result of an inability among these subjects to convert retinal L to MZ. Such subjects would require this carotenoid in supplemental form if they are to achieve a typical and desirable spatial profile characterized by a central peak with a monotonic decline from the foveal center.

A study was conducted to investigate the effect of supplementation on a group of subjects that exhibited a central dip in their MP spatial profile (Nolan et al. 2012a). Thirty-one subjects were assigned to one of three intervention groups, as follows: one given a 20 mg of L and 2 mg of Z (Group 1); the second group was given 10 mg L, 10 mg MZ, and 2 mg Z (Group 2); and the third group was supplemented 17 mg MZ and 3 mg L (Group 3). Subjects took one capsule a day for eight weeks. A significant increase in MPOD was not demonstrable among subjects supplemented with high doses of L (Group 1), at any eccentricity. Subjects supplemented with high doses of MZ (Group 3) exhibited significant increases in MPOD at the center of the MP spatial profile, but at no other eccentricity. Subjects in the combined carotenoid group (Group 2, containing L, Z, and MZ) exhibited a significant augmentation of MPOD at $0.25°$ and at $0.5°$ eccentricity, and a trend toward a rise in MP approaching statistical significance at all other eccentricities. The authors concluded that these atypical spatial profiles of MP, characterized by central dips, which are known to be associated with risk for AMD, can be normalized following supplementation with a formulation containing MZ, but not with a formulation that is lacking this carotenoid. It was concluded that augmentation across the spatial profile of MP required supplementation with a formulation containing all three macular carotenoids.

## 5.5 OBSERVATIONAL STUDIES

A large number of studies have investigated the relationship between dietary intake of the macular carotenoids and AMD (Table 5.5) (Flood et al. 2002; Mares-Perlman et al. 2001; Moeller et al. 2006; SanGiovanni et al. 2007; Seddon et al.

## TABLE 5.5
## Observational Studies Investigating the Relationship between the Macular Carotenoids and AMD

| Principal Author(s) | Study | Year | n | Study Design | Age | Carotenoids | Nutrient/AMD Relationship |
|---|---|---|---|---|---|---|---|
| | | | | Observational Dietary Studies | | | |
| Seddon et al. | EDCCS | 1994 | 356/520[a] | Case–Control | 55–80 | L&Z | Inverse |
| VandenLangenberg et al. | BDES | 1996 | 1968 | Cohort | 45–86 | L&Z | None |
| Mares-Perlman et al. | NHANES III | 2001 | 8222 | Cross-sectional | 40+ | L&Z | Inverse |
| Flood et al. | BMES | 2002 | 2335 | Cohort | 49+ | L&Z | None |
| Snellen et al. | — | 2002 | 72/66[a] | Case–Control | 60+ | L | Inverse |
| Moeller et al. | CAREDS | 2006 | 1787 | Cross-sectional | 50–79 | L&Z | None |
| San Giovanni et al. | AREDS | 2007 | 4519 | Case–Control | 60–80 | L&Z | Inverse |
| Tan et al. | BMES | 2008 | 2454 | Cohort | 49+ | L&Z | Inverse |
| Cho et al. | | 2008 | 66,993 | Cohort | 50+ | L&Z | None |
| Olea et al. | — | 2012 | 52 | Cross-sectional | Mean = 79 | L&Z | Inverse[b] |
| | | | | Observational Serum Studies | | | |
| — | EDCCS | 1993 | 421/615[a] | Case–Control | — | L&Z | Inverse |
| Mares-Perlman et al. | BDES | 1995 | 167/167[a] | Case–Control | 43–86 | L&Z | None |
| Mares-Perlman et al. | NHANES III | 2001 | 8222 | Cross-sectional | 40+ | L&Z | Inverse |
| Simonelli et al. | — | 2002 | 48/46[a] | Case–Control | Mean = 67 | L&Z | None |
| Gale et al. | — | 2003 | 380 | Cross-sectional | 66–75 | L&Z; L; Z | Inverse (Z only) |
| Cardinault et al. | — | 2005 | 34/21[a] | Case–Control | 72–74 | L; Z | None |

(*Continued*)

## TABLE 5.5 (Continued)
## Observational Studies Investigating the Relationship between the Macular Carotenoids and AMD

| Principal Author(s) | Study | Year | n | Study Design | Age | Carotenoids | Nutrient/AMD Relationship |
|---|---|---|---|---|---|---|---|
| Michikawa et al. | — | 2009 | 722 | Cross-sectional | 65+ | L&Z | Inverse[b] |
| Zhao et al. | — | 2011 | 263 | Cross-sectional | 50–88 | L&Z | Inverse[b] |

*Note:* n, number of subjects; Carotenoids, macular carotenoids assessed in the study; age, age range (years) of subjects in study; EDCCS, Eye Disease Case Control Study; BDES, Beaver Dam Eye Study; NHANES, National Health and Nutrition Examination Survey; —, data unavailable; BMES, Blue Mountains Eye Study; CAREDS, Carotenoids in Age-Related Eye Disease Study; POLA, Pathologies Oculaires Liées à l'Age; EES, European Eye Study.

[a] Cases/controls.
[b] For late stages of AMD.

1994; Snellen et al. 2002; Tan et al. 2008; VandenLangenberg et al. 1998). Of these 10 published observational studies, 6 reported that a high dietary intake of the carotenoids was associated with a reduced risk of AMD. The relationship between AMD and serum concentration of the macular carotenoids has also been investigated (Cardinault et al. 2005; Delcourt et al. 2006; Early Disease Case Control Study Group 1993; Fletcher et al. 2008; Gale et al. 2003; Mares-Perlman et al. 1995, 2001; Michikawa et al. 2009; Simonelli et al. 2002; Zhou et al. 2011), and of the 10 published studies in this respect, 7 have shown that low serum concentrations of the macular carotenoids are associated with increased risk of this condition.

### 5.5.1 SERUM AND RETINAL RESPONSE TO SUPPLEMENTATION WITH THE MACULAR CAROTENOIDS

There have been many published studies on serum and retinal response (i.e., MPOD) to supplementation with the macular carotenoids, in normal and in AMD subjects, and it is clear that serum carotenoid levels and MPOD rise in response to supplementation with L, Z, and MZ. However, it is important to point out that the magnitude of response is influenced by many factors, including the type of carotenoid used (i.e., L, Z, MZ, independently or in combination), the total amount of carotenoid present in the supplement (dose), the duration of supplementation (time), individual characteristics (e.g., adiposity), and baseline MP levels (Richer et al. 2007). It appears that a formulation containing all three macular carotenoids is the most efficacious in terms of achieving the highest combined concentration of the three carotenoids in the serum (Meagher et al. 2013).

The data suggest that supplementation with all three macular carotenoids may result in the greatest response in terms of MP augmentation and changes in its spatial profile (Loughman et al. 2012; Nolan et al. 2012a). *In vitro* analysis supports the conclusion that the antioxidant capacity of MP is maximized in the presence of all three macular carotenoids (Li et al. 2010). Since the objective of MP augmentation is to confer protection against (photo)-oxidative stress and AMD, supplementation with all three macular carotenoids appears to effectively have the greatest potential to (1) limit (photo)-oxidative retinal injury with a consequential reduction in risk of AMD development or progression and (2) maximally enhance visual performance.

### 5.5.2 MOVING FORWARD

There have been no published trials investigating the potential of the macular carotenoids to prevent the onset of AMD. This would require recruitment of subjects who are not afflicted with the condition and evaluating the incidence of AMD with respect to both dietary intake of the carotenoids and MPOD. The trial would need to be observational in design rather than interventional, as the study period would need to be no less than 15 years in duration following completion of recruitment. Of note, however, a study of such unique design is currently underway in Ireland. The Irish Longitudinal Study of Ageing (TILDA) (Whelan et al. 2013) is investigating health, lifestyles, and financial status of circa 8000 randomly selected people aged 50 years and older. A major component of this prospective cohort study is the investigation into the relationship between baseline MP levels and the prevalence and incidence of AMD (Nolan et al. 2010). MP measurements and retinal photographs are being obtained at three separate study visits (waves), commencing in year 1, year 4, and year 8, respectively. The first wave is complete and has generated some interesting MPOD data (Beatty et al. 2012; Nolan et al. 2010). This study will investigate, for the first time, whether baseline MP levels relate to the ultimate risk of developing AMD.

## 5.6 CONCLUSION

In summary, the properties of MP, namely its central retinal location, its prereceptoral filtration of damaging short-wavelength light, and its ability to quench free radicals, suggest that it plays a key role in the aetiopathogenesis of AMD and its progression. In addition, it appears to contribute to the optimization of visual performance (in subjects with and without disease). Level 1 evidence has demonstrated that supplemental dietary antioxidants reduce the risk of vision loss in AMD. Evidence of this quality for supplementation with the macular carotenoids is still lacking. The visual performance hypothesis of MP, on the other hand, is now generally accepted. A number of clinical trials have reported that macular carotenoid supplementation demonstrably enhances visual performance in subjects with and without disease. Clinical trials have repeatedly shown that dietary supplementation with the macular carotenoids (L, Z, and/or MZ) results in augmentation of MPOD. The best response in terms of augmentation, changes in spatial profile of the pigment, global fortification of the antioxidant defenses of the tissue to be protected and in terms of visual performance appears to be a supplement containing all three macular carotenoids.

## REFERENCES

Arundale, K. 1978. An investigation into the variation of human contrast sensitivity with age and ocular pathology. *Br. J. Ophthalmol.*, 62(4): 213–215.

Bailar, J.C., III. 1997. The promise and problems of meta-analysis. *N. Engl. J. Med.*, 337(8): 559–561.

Bartlett, H.E. and Eperjesi, F. 2007. Effect of lutein and antioxidant dietary supplementation on contrast sensitivity in age-related macular disease: A randomized controlled trial. *Eur. J. Clin. Nutr.*, 61(9): 1121–1127.

Beatty, S., Chakravarthy, U., Nolan, J.M., et al. 2013. Secondary outcomes in a clinical trial of carotenoids with coantioxidants versus placebo in early age-related macular degeneration. *Ophthalmology*, 120(3):600–606, http://www.ncbi.nlm.nih.gov/pubmed/23218821.

Beatty, S., Feeney, J., Kenny, R., et al. 2012. Education is positively associated with macular pigment: The Irish Longitudinal Study on Ageing (TILDA). *Invest. Ophthalmol. Vis. Sci.*, 53(12): 7855–7861.

Bernstein, P.S., Ahmed, F., Liu, A., et al. 2012. Macular pigment imaging in AREDS2 participants: An ancillary study of AREDS2 subjects enrolled at the Moran Eye Center. *Invest. Ophthalmol. Vis. Sci.*, 53(10): 6178–6186.

Bhosale, P. and Bernstein, P.S. 2005. Synergistic effects of zeaxanthin and its binding protein in the prevention of lipid membrane oxidation. *Biochim. Biophys. Acta*, 1740(2): 116–121.

Black, N. 1996. Why we need observational studies to evaluate the effectiveness of health care. *BMJ*, 312(7040): 1215–1218.

Bone, R.A., Landrum, J.T., Hime, G.W., et al. 1993. Stereochemistry of the human macular carotenoids. *Invest. Ophthalmol. Vis. Sci.*, 34(6): 2033–2040.

Cardinault, N., Abalain, J.H., Sairafi, B., et al. 2005. Lycopene but not lutein nor zeaxanthin decreases in serum and lipoproteins in age-related macular degeneration patients. *Clin. Chim. Acta*, 357(1): 34–42.

Chew, E.Y., Clemons, T., SanGiovanni, J.P., et al. 2012. The Age-Related Eye Disease Study 2 (AREDS2): Study design and baseline characteristics (AREDS2 report number 1). *Ophthalmology*, 119(11): 2282–2289.

Delcourt, C., Carriere, I., Delage, M., et al. 2006. Plasma lutein and zeaxanthin and other carotenoids as modifiable risk factors for age-related maculopathy and cataract: The POLA Study. *Invest. Ophthalmol. Vis. Sci.*, 47(6): 2329–2335.

Early Disease Case Control Study Group. 1993. Antioxidant status and neovascular age-related macular degeneration. Eye Disease Case-Control Study Group. *Arch. Ophthalmol.*, 111(1): 104–109.

Engles, M., Wooten, B., and Hammond, B. 2007. Macular pigment: A test of the acuity hypothesis. *Invest. Ophthalmol. Vis. Sci.*, 48(6): 2922–2931.

Fletcher, A.E., Bentham, G.C., Agnew, M., et al. 2008. Sunlight exposure, antioxidants, and age-related macular degeneration. *Arch. Ophthalmol.*, 126(10): 1396–1403.

Flood, V., Smith, W., Wang, J.J., et al. 2002. Dietary antioxidant intake and incidence of early age-related maculopathy: The Blue Mountains Eye Study. *Ophthalmology*, 109(12): 2272–2278.

Gale, C.R., Hall, N.F., Phillips, D.I., et al. 2003. Lutein and zeaxanthin status and risk of age-related macular degeneration. *Invest. Ophthalmol. Vis. Sci.*, 44(6): 2461–2465.

Gittings, N.S. and Fozard, J.L. 1986. Age related changes in visual acuity. *Exp. Gerontol.*, 21(4–5): 423–433.

Hammond, B.R., Jr., and Wooten, B.R. 2005. CFF thresholds: Relation to macular pigment optical density. *Ophthalmic. Physiol. Opt.*, 25(4): 315–319.

Jentsch, S., Schweitzer, D., Hammer, M., et al. 2011. The Lutega-Study: Lutein and omega-3-fatty acids and their relevance for macular pigment in patients with age-related macular degeneration (AMD). ARVO 2011, May 1–5, 2011, Fort Lauderdale, Florida.

Kirby, M.L., Beatty, S., Loane, E., et al. 2010. A central dip in the macular pigment spatial profile is associated with age and smoking. *Invest. Ophthalmol. Vis. Sci.*, 5: 6722–6728.

Klein, R., Klein, B.E., Linton, K.L., et al. 1991. The Beaver Dam Eye Study: Visual acuity. *Ophthalmology*, 98(8): 1310–1315.

Kvansakul, J., Rodriguez-Carmona, M., Edgar, D.F., et al. 2006. Supplementation with the carotenoids lutein or zeaxanthin improves human visual performance. *Ophthalmic. Physiol. Opt.*, 26(4): 362–371.

Landrum, J.T. and Bone, R.A. 2001. Lutein, zeaxanthin, and the macular pigment. *Arc. Biochem. Biophys.*, 385(1): 28–40.

Landrum, J.T., Bone, R.A., Joa, H.I.L.D., et al. 1997. A one year study of the macular pigment: The effect of 140 days of a lutein supplement. *Exp. Eye Res.*, 65(1): 57–62.

Li, B., Ahmed, F., and Bernstein, P.S. 2010. Studies on the singlet oxygen scavenging mechanism of human macular pigment. *Arch. Biochem. Biophys.*, 504(1): 56–60.

Loughman, J., Akkali, M.C., Beatty, S., et al. 2010. The relationship between macular pigment and visual performance. *Vision Res.*, 50(13): 1249–1256.

Loughman, J., Nolan, J.M., Howard, A.N., et al. 2012. The impact of macular pigment augmentation on visual performance using different carotenoid formulations. *Invest. Ophthalmol. Vis. Sci.*, 53(12): 7871–7880.

Mann, C.J. 2003. Observational research methods. Research design II: Cohort, cross sectional, and case-control studies. *Emerg. Med. J.*, 20(1): 54–60.

Mares-Perlman, J.A., Brady, W.E., Klein, R., et al. 1995. Serum antioxidants and age-related macular degeneration in a population-based case-control study. *Arch. Ophthalmol.*, 113(12): 1518–1523.

Mares-Perlman, J.A., Fisher, A.I., Klein, R., et al. 2001. Lutein and zeaxanthin in the diet and serum and their relation to age-related maculopathy in the third National Health and Nutrition Examination Survey. *Am. J. Epidemiol.*, 153(5): 424–432.

Meagher, K.A., Thurnham, D.I., Beatty, S., et al. 2013. Serum response to supplemental macular carotenoids in subjects with and without age-related macular degeneration. *Br. J. Nutr.*, 110(2): 289–300.

Michikawa, T., Ishida, S., Nishiwaki, Y., et al. 2009. Serum antioxidants and age-related macular degeneration among older Japanese. *Asia. Pac. J. Clin. Nutr.*, 18(1): 1–7.

Moeller, S.M., Parekh, N., Tinker, L., et al. 2006. Associations between intermediate age-related macular degeneration and lutein and zeaxanthin in the Carotenoids in Age-Related Eye Disease Study (CAREDS): Ancillary study of the Women's Health Initiative. *Arch. Ophthalmol.*, 124(8): 1151–1162.

Neelam, K., Hogg, R.E., Stevenson, M.R., et al. 2008. Carotenoids and co-antioxidants in age-related maculopathy: design and methods. *Ophthalmic. Epidemiol.*, 15(6): 389–401.

Neelam, K., Nolan, J., Chakravarthy, U., et al. 2009. Psychophysical function in age-related maculopathy. *Surv. Ophthalmol.*, 54(2): 167–210.

Nolan, J.M., Akkali, M.C., Loughman, J., et al. 2012a. Macular carotenoid supplementation in subjects with atypical spatial profiles of macular pigment. *Exp. Eye Res.*, 101: 9–15.

Nolan, J.M., Kenny, R., O'Regan, C., et al. 2010. Macular pigment optical density in an ageing Irish population: The Irish Longitudinal Study on Ageing. *Ophthalmic. Res.*, 44(2): 131–139.

Nolan, J.M., Loughman, J., Akkali, M.C., et al. 2011. The impact of macular pigment augmentation on visual performance in normal subjects: COMPASS. *Vision Res.*, 51(5): 459–469.

Nolan, J.M., Sabour-Pickett, S., Connolly, E., et al. 2012b. Macular pigment response to three different macular carotenoid interventions in patients with early age-related macular degeneration. *ARVO 2012*, Poster No. 3368/A513, May 8, 2012.

Nolan, J.M., Stack, J., O' Donovan O., et al. 2007. Risk factors for age-related maculopathy are associated with a relative lack of macular pigment. *Exp. Eye Res.*, 84(1): 61–74.

Nussbaum, J.J., Pruett, R.C., and Delori, F.C. 1981. Historic perspectives. Macular yellow pigment. The first 200 years. *Retina*, 1(4): 296–310.

Olmedilla, B., Granado, F., Blanco, I., et al. 2001. Lutein in patients with cataracts and age-related macular degeneration: A long-term supplementation study. *J. Sci. Food Agric.*, 81(9): 904–909.

Piermarocchi, S., Saviano, S., Parisi, V., et al. 2011. Carotenoids in Age-Related Maculopathy Italian Study (CARMIS): Two-year results of a randomized study. *Eur. J. Ophthalmol.*, 22(2): 216–225.

Richer, S. 1999. ARMD—Pilot (case series) environmental intervention data. *J. Am. Optom. Assoc.*, 70(1): 24–36.

Richer, S., Devenport, J., and Lang, J.C. 2007. LAST II: Differential temporal responses of macular pigment optical density in patients with atrophic age-related macular degeneration to dietary supplementation with xanthophylls. *Optometry*, 78(5): 213–219.

Richer, S.P., Stiles, W., Graham-Hoffman, K., et al. 2011. Randomized, double-blind, placebo-controlled study of zeaxanthin and visual function in patients with atrophic age-related macular degeneration: The Zeaxanthin and Visual Function Study (ZVF) FDA IND #78, 973. *Optometry*, 82(11): 667–680.

Richer, S., Stiles, W., Statkute, L., et al. 2004. Double-masked, placebo-controlled, randomized trial of lutein and antioxidant supplementation in the intervention of atrophic age-related macular degeneration: The Veterans LAST Study (Lutein Antioxidant Supplementation Trial). *Optometry*, 75(4): 216–230.

Rychetnik, L., Frommer, M., Hawe, P., et al. 2002. Criteria for evaluating evidence on public health interventions. *J. Epidemiol. Community Health*, 56(2): 119–127.

Sackett, D.L. 1994. The Cochrane Collaboration. *ACP J. Club*, 120(Suppl 3): A11.

SanGiovanni, J.P., Chew, E.Y., Clemons, T.E., et al. 2007. The relationship of dietary carotenoid and vitamin A, E, and C intake with age-related macular degeneration in a case-control study: AREDS Report No. 22. *Arch. Ophthalmol.*, 125(9): 1225–1232.

Sasamoto, Y., Gomi, F., Sawa, M., et al. 2011. Effect of 1-year lutein supplementation on macular pigment optical density and visual function. *Graefes Arch. Clin. Exp. Ophthalmol.*, 249(12): 1847–1854.

Seddon, J.M., Ajani, U.A., Sperduto, R.D., et al. 1994. Dietary carotenoids, vitamins A, C, and E, and advanced age-related macular degeneration. Eye Disease Case-Control Study Group. *JAMA*, 272(18): 1413–1420.

Simonelli, F., Zarrilli, F., Mazzeo, S., et al. 2002. Serum oxidative and antioxidant parameters in a group of Italian patients with age-related maculopathy. *Clin. Chim. Acta*, 320(1–2): 111–115.

Snellen, E.L., Verbeek, A.L., Van Den Hoogen, G.W., et al. 2002. Neovascular age-related macular degeneration and its relationship to antioxidant intake. *Acta. Ophthalmol. Scand.*, 80(4): 368–371.

Stephenson, J., and Imrie, J. 1998. Why do we need randomised controlled trials to assess behavioural interventions? *BMJ*, 316(7131): 611–613.

Stringham, J.M., and Hammond, B.R., Jr. 2007. The glare hypothesis of macular pigment function. *Optom. Vis. Sci.*, 84(9): 859–864.

Tan, J.S., Wang, J.J., Flood, V., et al. 2008. Dietary antioxidants and the long-term incidence of age-related macular degeneration: The Blue Mountains Eye Study. *Ophthalmology*, 115(2): 334–341.

VandenLangenberg, G.M., Mares-Perlman, J.A., Klein, R., et al. 1998. Associations between antioxidant and zinc intake and the 5-year incidence of early age-related maculopathy in the Beaver Dam Eye Study. *Am. J. Epidemiol.*, 148(2): 204–214.

Walker, E., Hernandez, A.V., and Kattan, M.W. 2008. Meta-analysis: Its strengths and limitations. *Cleve. Clin. J. Med.*, 75(6): 431–439.

Weigert, G., Kaya, S., Pemp, B., et al. 2011. Effects of lutein supplementation on macular pigment optical density and visual acuity in patients with age-related macular degeneration. *Invest. Ophthalmol. Vis. Sci.*, 52(11): 8174–8178.

Whelan, B.J., Savva, G.M. 2013. *The Design of the Irish Longitudinal Study on Ageing (TILDA)*. *J Am Geriatr Soc*, 61 Suppl 2: S265–268

Wood, L., Egger, M., Gluud, L.L., et al. 2008. Empirical evidence of bias in treatment effect estimates in controlled trials with different interventions and outcomes: Meta-epidemiological study. *BMJ*, 336(7644): 601–605.

Wooten, B.R., and Hammond, B.R. 2002. Macular pigment: Influences on visual acuity and visibility. *Prog. Retin. Eye Res.*, 21(2): 225–240.

Zhou, H., Zhao, X., Johnson, E.J., et al. 2011. Serum carotenoids and risk of age-related macular degeneration in a chinese population sample. *Invest. Ophthalmol. Vis. Sci.*, 52(7): 4338–4344.

# 6 The Promise of Molecular Genetics for Investigating the Influence of Macular Xanthophyllys on Advanced Age-Related Macular Degeneration

*John Paul SanGiovanni and Martha Neuringer*

## CONTENTS

| | | |
|---|---|---|
| 6.1 | Overview | 94 |
| 6.2 | Introduction | 95 |
| 6.3 | Promise of Molecular Genetics for Examining the Influence of MXs on AMD | 96 |
| | 6.3.1 Molecular Genetics of AMD and the Putative Influence MXs on AMD Pathogenesis | 96 |
| | 6.3.2 Relationships of AMD with DNA Sequence Variation | 98 |
| |     6.3.2.1 Group 1: Binding, Transport, and Accretion of Macular Xanthophylls | 99 |
| | 6.3.3 Group 2: Cleavage of Macular Xanthophylls | 110 |
| |     6.3.3.1 β-Carotene-15, 15′-Monooxygenase 1 | 111 |
| |     6.3.3.2 β-Carotene-9′, 10′-Oxygenase 2 | 112 |
| |     6.3.3.3 Lipoprotein Lipase | 112 |
| | 6.3.4 Group 3: Antioxidant Potential of MXs in DNA Preservation and Repair | 113 |
| | 6.3.5 Group 4: Influence on Adaptive Cellular Response | 113 |
| | 6.3.6 Group 5: Interference of Carotenoids with Growth Factors | 115 |
| | 6.3.7 Group 6: Diseases Associated with Low MX in the Retina | 115 |
| |     6.3.7.1 Sjögren–Larsson Syndrome (Aldehyde Dehydrogenase 3 Family Member A2) | 115 |

|  |  | 6.3.7.2 | Stargardt Disease (ATP-Binding Cassette Subfamily A Member 4) | 116 |
|---|---|---|---|---|
| 6.4 | Conclusion | | | 117 |
| 6.5 | Method | | | 117 |
| | 6.5.1 | Subjects and Study Design | | 118 |
| | | 6.5.1.1 | AREDS Cohort | 118 |
| | | 6.5.1.2 | NEI-AMD Cohorts | 118 |
| | 6.5.2 | Outcome Ascertainment | | 118 |
| | | 6.5.2.1 | AREDS Cohort | 118 |
| | | 6.5.2.2 | NEI-AMD Cohorts | 119 |
| | 6.5.3 | Genotyping | | 119 |
| | 6.5.4 | Statistical Analyses | | 120 |
| Acknowledgments | | | | 120 |
| References | | | | 120 |

## 6.1 OVERVIEW

Age-related macular degeneration (AMD) is the prime cause of vision loss in elderly people of Western European ancestry. Over 9 million U.S. residents are living with sight-threatening AMD. Biochemical analyses and *in vivo* imaging studies indicate that genetic, dietary, and environmental factors influence tissue concentrations of macular xanthophylls (MXs) within the retina. Photoreceptors, retinal cell types manifesting AMD pathology, have been shown to contain MXs. In this work, we comment on the putative role of the MXs (lutein, zeaxanthin, and *meso*-zeaxanthin) in AMD and report findings from our genome-wide analyses of AMD-associated genes that encode enzymes, transporters, ligands, and receptors influencing or influenced by MXs. Our cohorts contained a total of 1555 people with advanced AMD (AAMD) and 1170 of their peers who were both AMD-free and ≥65 years of age. Data from our 12-year prospective discovery sample (Age-Related Eye Disease Study [AREDS]) were used to investigate the age-, sex-, and smoking-adjusted relationships of AAMD (neovascular [NV] AMD and/or geographic atrophy [GA]) with allelic variation in 1060 single-nucleotide polymorphisms (SNPs) residing within 275 genes encoding products with the capacity to interact with MXs. We used age-, sex-, and smoking-adjusted meta-regression to analyze 2710 gene variants from these same genes within three university hospital-based cohorts (1177 people with AAMD and 1024 AMD-free controls) in our replication study. In both our discovery and replication cohorts, AMD-associated DNA sequence variants existed in genes encoding transporters and receptors reported to bind MXs in model systems (scavenger receptor class B type I [SCARB1], ATP-binding cassette transporter A1 [ABCA1], and tubulin gamma complex associated protein 3 [TUBGCP3]), nuclear hormone receptors that bind MXs (retinoic acid receptor gamma [RARG] and retinoic acid receptor beta [RARB]), growth factors influenced by carotenoids (insulin-like growth factor 1 receptor [IGF1R] and insulin-like growth factor binding protein 1 [IGFBP1]), and genes associated with both low levels of retinal MXs and inherited retinal degenerations (aldehyde dehydrogenase 3 family member A2 [*ALDH3A2*] and *ABCA4*). We have demonstrated that data from large-scale genotyping projects may be effectively applied to strengthen inferences and elucidate novel relationships on factors and processes influencing metabolism and metabolic fate of nutrients in the context of health and disease.

## 6.2 INTRODUCTION

The dietary carotenoids lutein ((3R,3′R,6′R)-β,ε-caroten-3,3′-diol) and zeaxanthin ((3R,3′R)-β,β-caroten-3,3′-diol) are primary constituents of macular pigment (Bone et al. 1985; Handelman et al. 1988) and have been examined for their influence on health and disease of the retina for over 200 years (Beatty et al. 1999; Nussbaum et al. 1981; Whitehead et al. 2006). These two nutrients and *meso*-zeaxanthin ((3R,3′S)-β, β-caroten-3,3′-diol), a metabolite of lutein (Johnson et al. 2005), are known collectively as MXs. There is biologic plausibility in the idea that lutein and zeaxanthin may influence nutrient–retinal disease relationships because these compounds (1) demonstrate intake-dependent and intake-modifiable accretion to the retina; (2) show preferential concentration and localization in retinal regions manifesting retinal pathology of AMD; and (3) participate in biophysical and biochemical events and processes that are implicated in regulation of pathogenesis and progression of retinal diseases. A number of large-scale human studies on AMD, a common and complex sight-threatening disease of the retina (Congdon et al. 2004; Friedman et al. 2004; Klein et al. 2011), have yielded evidence for associations of this disease with the intake of lutein and zeaxanthin and with MX status in serum and the retina (Cho et al. 2008; Delcourt et al. 2006; Gale et al. 2003; SanGiovanni et al. 2007; Seddon et al. 1994; Snellen et al. 2002; Tan et al. 2008; The Eye Disease Case-Control Study Group 1992).

In the human and monkey retina, high concentrations of MXs exist within areas manifesting high susceptibility to damage from environmental and metabolic stressors. The topographic distribution of macular pigment and concentration of MXs within certain retinal laminae may arguably explain how the human retina is capable of handling such stressors without appreciable loss of function, usually for at least six decades. Even in people with AMD, remarkable resilience of the MX-rich areas (especially the foveal center) is sometimes seen (Bone et al. 2003). Foveal sparing is observed in AMD (Sunness et al. 1999) and in a number of inherited macular diseases of high penetrance (reviewed in Aleman et al. 2007).

Primates cannot synthesize lutein and zeaxanthin de novo; we have adapted with a capacity for efficient MX uptake (Handelman et al. 1991; Neale et al. 2010), transport (Bhosale et al. 2004; Li et al. 2011), and retention (Beatty et al. 2000; Bone et al. 2003; Hammond et al. 1997a, 1997b; Zeimer et al. 2009) in the retina. Efficiency of these processes may be testimony to the physiologic significance of MXs in retinal health and disease. There is an unmet need to examine the means by which regulatory mechanisms and metabolic fate of MXs may influence MX–AMD associations. A number of unifying concepts, helpful in addressing this need, have emerged from research in the field over the past 2.5 decades and they are as follows:

1. MX concentration is increased 1000- to 10000-fold from the circulation to the retina (Bernstein et al. 2010; Krinsky et al. 2003) through active transport mechanisms involving specific binding proteins.
2. MXs are localized within retinal cells damaged in AMD (Rapp et al. 2000).

3. MXs demonstrate a capacity to act on processes implicated in AMD pathogenesis (Snodderly 1995).
4. Relationships of MX intake → MX status in the retina exist. MX status → retinal structure relationships have been observed in nonhuman primates (Leung et al. 2004, 2005) and rodents (Chucair et al. 2007).

We applied these four concepts to examine the putative role of MXs in AMD pathogenesis within the context of knowledge on genes encoding enzymes, transporters, and receptors that may influence or be influenced by MXs, their metabolites (e.g., apocarotenoids), and cofactors (e.g., nuclear receptors regulating transcription). Throughout the work, we discuss emerging research opportunities and barriers, knowledge gaps, and tools offering promise for meaningful investigation and inference in the field.

## 6.3 PROMISE OF MOLECULAR GENETICS FOR EXAMINING THE INFLUENCE OF MXS ON AMD

There is an unmet need to examine the molecular genetics of factors influencing regulatory mechanisms and metabolic fates of MXs as they may relate to processes implicated in AMD pathogenesis (SanGiovanni and Neuringer 2012). Findings may inform development of a "molecular phenotype" (a pattern of DNA variation) representing the capacity for an individual to produce transporters, receptors, enzymes, and hormones targeting or influenced by MXs, particularly to determine whether these genetic characteristics may reduce risk of AMD incidence or progression.

### 6.3.1 Molecular Genetics of AMD and the Putative Influence MXs on AMD Pathogenesis

AMD is a complex disease with a strong hereditary component (Swaroop et al. 2009). Aspects of retinal MX absorption, distribution, metabolism, and excretion are genetic or genetically linked as demonstrated by studies on twins (Klein et al. 1994; Seddon et al. 2005) and first-degree family members (Heiba et al. 1994; Klaver et al. 1998b; Seddon et al. 1997). Notably, estimates of retinal MX concentrations (macular pigment optical density [MPOD]) have shown stronger relationships among monozygotic than dizygotic twins (Liew et al. 2005). Findings from a large cohort of married people indicate that interspouse relationships exist for dietary intake and serum concentrations of MXs (Wenzel et al. 2007), but not for MPOD.

Most work on the molecular genetics of sight-threatening AMD has focused on DNA sequence variation in genes of the complement system (Chen et al. 2010) (a proteolytic pathway in plasma acting to mediate innate immunity through a series of proenzyme-to-protease conversions) and high-density lipoprotein (HDL) transport and metabolism (Chen et al. 2010; Neale et al. 2010). In the case of the HDL-related genes, the common inference is that altered cholesterol status or metabolism is a dominant pathogenic process in AMD. The epidemiologic evidence on this matter is equivocal, as dietary intake of cholesterol, saturated fat, and total fat has not yielded a clear risk profile in AMD. When considered with the studies on 3-hydroxy-3-methylglutaryl-CoA (HMG-CoA) reductase reductase inhibitors (statins) failing to consistently identify this drug class as a protective factor against progression to AAMD (Maguire et al. 2009; Peponis et al. 2010), it is clear that alternative or supplementary explanations should be considered. MXs are carried on HDL particles (Connor et al. 2007) that interact with numerous receptors and transporters involved in cholesterol metabolism. Epidemiologic studies are concordant in support of MX intake–NV AMD relationships (Cho et al. 2008; Delcourt et al. 2006; Gale et al. 2003; SanGiovanni et al. 2007; Seddon et al. 1994; Snellen et al. 2002; Tan et al. 2008; The Eye Disease Case-Control Study Group 1992). NV AMD, a form of the disease characterized by florid neovascularization of the choriocapillaris, is present in ~90% of people with AMD-related blinding events. Regarding the complement system, there are links between MX intake, complement genes, and AMD (Ho et al. 2011) suggesting interactions of diet with genetic AMD risk loci. Carriers of both AMD-associated complement factor H and age-related maculopathy susceptibility 2 (ARMS2/HTRA1) locus variants exhibit lower MPOD values than their peers (Loane et al. 2011). Our approach has allowed us to link the concepts of inflammation and HDL metabolism in AMD to genes encoding proteins associated with MX uptake, transport, accretion, and retention. It has also facilitated examination of putative influence of MX-associated genetic factors in AMD-associated processes of oxidative stress and cell–cell signaling.

We used original research reports and substantive reviews to identify molecules influencing or influenced by MXs, their metabolites (products of MX cleavage and biosynthesis), and cofactors (transcription factors that form complexes with MXs) to compile a compendium of 275 genes. These genes were classified into functional groups (described in Section 6.3.2) for multivariable association testing with AAMD. We used the *Retina Central* database at the University of Regensburg (http://www.retinacentral.org/), the *NEIBank* database at the National Institutes of Health (http://neibank.nei.nih.gov/index.shtml), and findings from immunolocalization studies in whole retinal sections to determine whether AMD-associated members of our gene set are expressed in the primate retina. Gene–disease analysis involved the examination of DNA sequence variation (e.g., distribution of SNPs by AMD status) in large-scale projects with high-quality data on genotypes and AMD phenotypes. For AMD-associated variants (those identified with nominal $p$-values ≤.05 from age-, sex-, and smoking-adjusted logistic regression models on additive, dominant, or recessive transmission models), we generated exact $p$-values with permutation tests. The permutation test is based on evaluation of empirical distributions constructed from significance testing on multiple datasets produced in an iterative process of swapping genotype and phenotype values within the sample. The swapping process destroys the phenotype–genotype relationship (and each new dataset is sampled under the null hypothesis), while preserving the patterns of coinheritance between SNPs. Empirically derived (exact) $p$-values permit inference free of the constraints of parametric methods; these include threats to validity of inference based on Hardy–Weinberg equilibrium (HWE), rare alleles, and small sample sizes. Permutation tests also provide a framework for correction for multiple testing. After association testing, we annotated our results for (1) replication at the level of the gene and SNP and (2) functional implication of AMD-MX-associated sequence variation (the consequence of nucleotide base substitution [leading to amino acid residue change] for nonsynonymous SNPs resident in exons, evaluation of variation in areas of conserved sequences, and localization of sequence variants to nonexonic regions with potential for posttranscriptional regulation).

### 6.3.2 Relationships of AMD with DNA Sequence Variation

Sections on our MX gene sets are organized into six functional groups. Each section starts with an overview of constituent genes—this information is followed by brief statements on function and localization to retinal cell types and laminae. The relationship of each gene or gene product with retinal MX status is then presented. This information is followed by a review of mechanistic studies on gene–disease relationships with a focus on potential influence of MXs on AMD-associated pathogenic processes. We then present pertinent findings on DNA sequence variations in the constituent genes and the consequence for AMD from the extant evidence base. This is followed by our novel results from four independent cohorts of elderly people participating in large-scale genotyping projects. In summary, our cohorts contained a total of 1555 people with AAMD (GA [retinal pigment epithelium [RPE] and photoreceptor degeneration] and/or NV AMD [pathologic angiogenesis of the choriocapillaris]) and 1170 of their peers who were both AMD-free and ≥65 years of age. Data from our 12-year prospective study (AREDS) were used to investigate age-, sex-, and smoking-adjusted relationships of AAMD with allelic variation in 1060 SNPs residing within genes encoding proteins that bind, cleave, or act in biosynthesis/metabolism of MXs. This discovery cohort was used to identify genes for analysis in a subsequent replication phase. We used age-, sex-, and smoking-adjusted meta-regression on 2710 target gene variants from the same MX gene set (we use the term "MX compendium" to represent this gene set) on three university hospital-based cohorts (1177 people with AAMD and 1024 AMD-free controls) in our replication study. Figure 6.1 shows the distribution of $-\log_{10}$ $p$-values in all sequence variants tested (lowest $p$-values appear at the top of the figure). Figure 6.2 contains findings from primate immunolocalization studies on the products of genes we analyzed.

Because the genotyping approach in genome-wide association (GWA) studies is nonexhaustive (does not involve fine genetic mapping of most sequence variants in a gene), lack of relationship in our study does not refute the putative importance of a MX-related gene in AMD pathogenesis. Our inferences are constrained by the density of SNP coverage in the gene chip feature set and thus should be considered

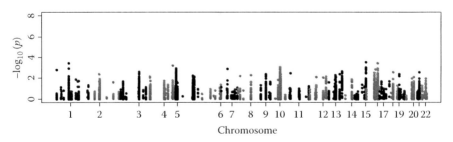

**FIGURE 6.1** Distribution of $p$-values for association of MX-related sequence variants with AMD.

as based on best available evidence. We apply data on linkage disequilibrium (LD, a measure of how often two SNPs or specific DNA sequences are coinherited) in people of European ancestry from the International HapMap Project (http://hapmap.ncbi.nlm.nih.gov) and the Ensembl Project (http://useast.ensembl.org/info/about/intro.html) to compare results across the gene chip feature sets used in our discovery and replication cohorts. In many instances, valid proxy SNPs for the discovery (AREDS) cohort exist in the replication cohort (viz. different SNPs may have been tested in different cohorts but these SNPs may be coinherited—essentially providing highly related findings). We comment on SNPs with LD measures (D' and $r^2$) indicating a high degree of coinheritance (those with $r^2$ values $\geq 0.85$).

The starting point for analysis of each sequence variant was evidence in the extant literature of a gene–MX relationship influencing MX status. In cases where gene–MX relationships had not been tested, we considered published evidence on gene–carotenoid ($\alpha$- and $\beta$-carotenes, $\beta$-cryptoxanthin, and lycopene) relationships because these molecules share bonding frameworks and functional groups with MXs. Whenever possible, we provide commentary on emerging research opportunities, knowledge gaps, and tools that offer promise for meaningful investigation and inference in the field. Our efforts are aimed primarily at informing applied clinical and mechanistic research in the planning and implementation phases. Details on our methods appear in Section 6.5. An overview of our findings exists in Table 6.1. Shaded cells represent an MX-associated gene–AMD relationship. The table is organized by cohort. In instances where replication existed at the level of the gene, shaded cells are aligned within a row. Replication at the level of the SNP or proxy replication with a SNP in nearly complete LD ($r^2$ value ~ 1) is discussed in the text.

### 6.3.2.1 Group 1: Binding, Transport, and Accretion of Macular Xanthophylls

In 1997, Bone et al. (1997) stated that "The ability to increase the amount of macular pigment by dietary supplementation with lutein has been demonstrated. Such a strategy may become recognized as an effective means of reducing the risk, and/or progression, of AMD in some individuals. Understanding the mechanism of pigment accumulation and possible transformation in the macula is vitally important to the development of such a therapy." We examined SNPs in seven genes encoding putative MX binding or transport proteins identified from a comprehensive review on the subject by Li et al. (Neale et al. 2010). We start with two genes of the cluster determinant 36 superfamily, *CD36* and *SCARB1*. Products of these genes are involved in binding of MXs within the gastrointestinal tract, choriocapillaris, and RPE. Next, we examine the roles of genes encoding putative MX transport proteins and their components or activators; these include apolipoprotein E (*APOE*) and *ABCA1*. We then present results on DNA variants in genes encoding two high-affinity retinal receptors principally responsible for binding and sequestration of MXs in the macula; these are the lutein-specific StAR-related lipid transfer domain containing 3 (*STARD3*) and zeaxanthin-specific glutathione S-transferase pi 1 (*GSTP1*). We also take an agnostic approach to investigate members of the tubulin family—these highly expressed

## TABLE 6.1
## Overview of the Results for Advanced AMD-Associated Genes Encoding Factors Influenced by or Influencing Macular Xanthophyll Activity, Uptake, Transport, and Association with Processes Implicated in AMD Pathogenesis

| Analysis Group and MX-Related Gene | Gene Symbol | Evidence (Reference) Gene-MX | Gene-AMD | Present Study MMAP | AAMD | AREDS |
|---|---|---|---|---|---|---|
| *1. Uptake, Transport, and Accretion of MXs* | | | | | | |
| Transport in intestine, choroid, RPE | CD36 | Borel et al. (2011), During et al. (2008), Neale et al. (2010) | Kondo et al. (2009) | | | |
| Transport in intestine, choroid, RPE | SCARB1 | Borel et al. (2011), During et al. (2008) | Zerbib et al. (2009) | ● | ● | |
| Accretion/transport in OPL | Tubulin genes[a] | Bernstein et al. (1997), Crabtree et al. (2001), Yemelyanov et al. (2001) | Machalinska et al. (2011) | ● | ● | |
| Transport in circulation | APOE | Fernandez-Robredo et al. (2009), Loane et al. (2010a, 2010b) | Baird et al. (2004), Francis et al. (2009), Klaver et al. (1998a), Souied et al. (1998), Tikellis et al. (2007), Wong et al. (2006), Zareparsi et al. (2004, 2005) | | | ● |
| Transport/uptake on HDL-c | ABCA1 | Connor et al. (2007) | Chen et al. (2010), Fauser et al. (2011), Neale et al. (2010), Yu et al. (2011a, 2011b), Zareparsi et al. (2005) | ● | | ● |
| | APOA1 | Connor et al. (2007)[b] | Colak et al. (2011) | | | |
| Specific zeaxanthin binding in retina | GSTP1 | Bhosale et al. (2004), Neale et al. (2010) | Guven et al. (2011) | | | |

# Molecular Genetics, AMD, and MXs

| Function | Gene | References | | |
|---|---|---|---|---|
| Specific lutein binding in retina | STARD3 | Bhosale et al. (2009), Li et al. (2011) | | |
| 2. Cleavage of MXs | | | | |
| 15, 15′ (symmetric) cleavage | BCMO1 | Borel et al. (2011) | | ● |
| 9′,10′ (eccentric) cleavage | BCO2 | Amengual et al. (2011), Mein et al. (2011) | | |
| Synthesis/degradation of MX-rich HDL-c | LPL | Chen et al. (2010), Fauser et al. (2011) | ● | |
| 3. Antioxidant Potential DNA Preservation | | | | |
| Structural integrity of DNA | POLB | Horie et al. (2010) | | |
| | POLL | Horie et al. (2010) | | |
| 4. Influence on Adaptive Cellular Response | | | | |
| Adhesion complex (AC) signaling | AC Genes[a] | Demmig-Adams and Adams (2002), Zhang et al. (1991)[b] | | ● |
| Proliferation/apoptosis | CDK2 | Mein et al. (2008)[b] | | |
| | CCND1 | Mein et al. (2008)[b] | | |

*(Continued)*

## TABLE 6.1 (Continued)
### Overview of the Results for Advanced AMD-Associated Genes Encoding Factors Influenced by or Influencing Macular Xanthophyll Activity, Uptake, Transport, and Association with Processes Implicated in AMD Pathogenesis

| Analysis Group and MX-Related Gene | Gene Symbol | Evidence (Reference) Gene-MX | Gene-AMD | Present Study AAMD MMAP | Present Study AAMD AREDS |
|---|---|---|---|---|---|
| 5. *Interference with Growth Factors (GFs)* | GF Genes[a] | Mein et al. (2008)[b] | | ● | ● |
| 6. *Disease Associated with Low MX in Retina* | | | | | |
| Sjögren–Larsson syndrome | ALDH3A2 | van der Veen et al. (2010) | — | ● | |
| Stargardt's disease 1/cone rod dystrophy 3 | ABCA4 | Aleman et al. (2007) | Allikmets et al. (1997, 1999) Brion et al. (2010) | ● | ● |

AAMD, advanced age-related macular degeneration; MX, macular xanthophyll; NV AMD, neovascular age-related macular degeneration; RPE, retinal pigmented epithelium.

*Notes:* Results for MMAP cohort are based on a meta-analysis on Michigan, Mayo, University of Pennsylvania (MMAP) cohorts. Cells with filled circles (●: far right columns) indicate association of a sequence variant with AMD at $p \leq .05$. Results are from age-, sex-, and smoking-adjusted logistic regression analyses. Three models were run for each SNP: additive, dominant, and recessive. Full names for gene symbols exist at http://www.ncbi.nlm.nih.gov/gene.

[a] Results for individual genes are discussed in the text.

[b] Indirect relationship (e.g., MXs not tested, but carotenoids sharing some of the same functional groups were tested).

# Molecular Genetics, AMD, and MXs

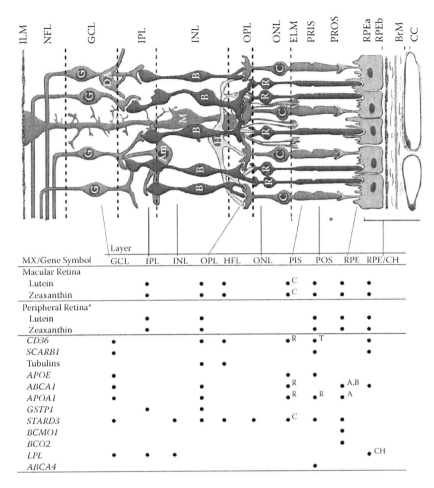

FIGURE 6.2 Retinal localization of macular xanthophylls (MXs) and proteins involved in uptake, transport, accretion, and cleavage of MXs in primates. Data for macular xanthophyll (MX) localization exist in Bernstein et al. (2001), Snodderly et al. (1984), and Sommerburg et al. (1999). Immunohistochemical work in primate retina reported by Tserentsoodol et al. (2006), Li et al. (Neale et al. 2010), Anderson et al. (2001), Bhosale et al. (2004), Lindqvist et al. (2005), and Casaroli-Marano et al. (1996) provided information on cellular and tissue localization of a number of proteins encoded by genes in our MX compendium. RPEa, apical area of RPE cell; Am, amacrine cell; RPEb, basal area of RPE cell; B, bipolar cell; BM, Bruch's membrane; C, cone photoreceptors; CC, retinal choroid layer; DA, displaced amacrine cell; ELM, external limiting membrane; G, ganglion cell; GCL, ganglion cell layer; H, horizontal cell; ILM, inner limiting membrane; INL, inner nuclear layer; IPL, inner plexiform layer (interneurons); M, Müller cell; NFL, nerve fiber layer; ONL, outer nuclear layer; OPL, outer plexiform layer; PRIS, photoreceptor inner segments; PROS, photoreceptor outer segments; R, rod photoreceptors; RPEa, retinal pigmented epithelium apical area; RPEb, retinal pigmented epithelium basal area. *Retinal layers concentrated with MXs. Schematic was created by D. Fisher and reproduced with permission from Zheng, W., et al., *PLoS One, 7*, e37926, 2012. *Peripheral retina represents extramacular regions at ≥2.5mm eccentricity from the fovea. Full names for genes represented by symbols exist at http://www.ncbi.nlm.nih.gov/gene.

proteins show potentially large MX storage capacity in the retina, yet demonstrate nonspecific and substantially weaker binding properties than STARD3 and GSTP1.

#### 6.3.2.1.1 Scavenger Receptor Class B Type I

The scavenger receptor class B type I (SR-BI), a cell surface glycoprotein of the CD36 superfamily with high affinity to HDL and localized to the apical surface of the enterocyte, is thought to act in nonspecific MX absorption at the intestinal brush border (Li et al. 2010). SR-BI is encoded by the *SCARB1* gene (12q24.31). *SCARB1* is expressed in primary human RPE cells (Duncan et al. 2002). SR-BI mRNA was detected in total human neural retina using RT-PCR (Tserentsoodol et al. 2006); immunohistochemical localization of the protein in monkey retina demonstrated concentration of the protein in retinal ganglion cells, outer segments of photoreceptor rods and cones, and the choriocapillaris (the vascular interface to the RPE and neural retina). Evidence of high SR-BI expression in primate RPE is equivocal. Primary human cultures showed expression of this protein (Duncan et al. 2002), while the immunolocalization study discussed in Section 6.3.1 did not (Tserentsoodol et al. 2006). During et al. (2008) used a specific SR-BI antibody blocking technique and small interfering RNA on a differentiated human RPE-derived cell line (ARPE-19) to demonstrate that zeaxanthin uptake can be driven by an SR-BI-dependent process. These authors discuss a gene–MX–disease link, observing that a mutation in *ninaD*, an insect gene with high sequence identity to *SCARB1* and *CD36*, is associated with reduced carotenoid uptake, reduced zeaxanthin deposition, and blindness in *Drosophila* (Giovannucci and Stephenson 1999; Kiefer et al. 2002; Voolstra et al. 2006). Provost et al. (2009) examined the retinal ultrastructure of SR-BI$^{-/-}$ and wild-type (WT) mice fed a diet rich in compounds carried on HDL. Relative to the WT mice, the SR-BI$^{-/-}$ animals showed increased lipid inclusions and disorganization of photoreceptor outer segments and the outer nuclear layer (ONL). Bruch's membrane (BM), a permeable five-layer structure of basement membranes, collagen, and elastin existing between the choroid and RPE, was thickened in the SR-BI$^{-/-}$ mice and showed sparse sub-RPE deposits. Altered flow of essential compounds across BM (due to accumulation of oxidized lipid–protein complexes) may lead to progression to AAMD. In this study, the choroid manifested abnormal distribution of collagen fibers and a vacuolization associated with local inflammation in the subretinal space (and linked to the infiltration of macrophages); SR-BI$^{-/-}$ mice did not exhibit abnormal choroidal neovascularization. The team also reported induction of vascular endothelial growth factor (VEGF) expression in the ONL of the SR-BI$^{-/-}$ mice. An exonic sequence variant in *SCARB1* (rs5888) linked to lower SR-BI expression (Constantineau et al. 2010) has been associated with AAMD in a large-scale genotyping project in French-based and United States–based cohorts (Zerbib et al. 2009). The gene chip used for analysis of our discovery cohort contained six SNPs resident in *SCARB1*. Among these, we observed relationships of one common intronic variant (rs989892, $p = .010$) in LD (D′ = 0.98, $r^2 = 0.82$) with rs5888. Of eight *SCARB1*-resident SNPs tested in our replication panel, two were associated with AAMD: rs701106 ($p = .032$) and rs838878 ($p = .007$); the latter of these is in complete LD with a SNP (rs838884) proximal to the 3′ untranslated region (UTR) of *SCARB1*. Neither SNP in the replication sample is in LD with rs989892. Our findings on *SCARB1* are promising because they replicate the rs5888-AMD finding through proxy.

### 6.3.2.1.2 Cluster Determinant 36

CD36 is a major glycoprotein that acts as a primary antiangiogenic receptor of thrombospondin-1. CD36 also binds collagen, anionic phospholipids, oxidized LDL in macrophages, and long chain fatty acids. The protein is involved in phagocytosis of photoreceptor outer segments, transport of oxidized LDL particles, internalization of HDL, and may act in the transport and/or as a regulator of fatty acid transport. CD36 is encoded by the *C36* gene (7q11.2); the protein is localized in the primate retina within RPE (in a punctuate pattern, with some suggestion of basolateral localization), the tips of rod outer segments, the rod inner segments, the choriocapillaris, the outer plexiform layer (retinal ganglion cells and Müller cells), and in the ganglion cell layer (Tserentsoodol et al. 2006). During et al. (2008) did not observe MX transport actions of CD36 in their work on a fully differentiated ARPE-19 system; here a specific antibody against CD36 did not prevent accumulation of zeaxanthin after addition of the MX to the medium. Borel et al. (2011) examined the relationship of five sequence variants in *CD36* with MPOD response to a 6-month regimen of daily supplementation with a formula containing 10 mg lutein esters in a 30-person French cohort. Variation at one *CD36* locus (rs1761667) was associated with variation in macular status of MXs; people homozygous for the major allele (G) had significantly higher MPOD than those carrying the minor allele (A).

$CD36^{-/-}$ mice manifest an age-related (progressive) choroidal involution usually involving a 100%–300% increase in the avascular area within the choriocapillaris (Houssier et al. 2008). This progressive choroidal degeneration is accompanied by reduced expression of cyclooxygenase-2 (COX-2) and VEGF (Houssier et al. 2008). CD36 activating antibody stimulates COX-2 expression in RPE cell cultures, while CD36 deficiency was associated with inhibition of COX-2 and subsequent lack of VEGF response to outer segment or antibody stimulation *in vitro* (Houssier et al. 2008). Picard et al. (2010) used $CD36^{-/-}$ and $CD36^{+/+}$ mice to demonstrate an age-related CD36-dependent process of deposition and clearance of subretinal deposits that bear similarity to those typically seen in AMD; in this study, $CD36^{-/-}$ animals developed basal laminar deposits. In a mouse model showing similar pathology ($ApoE^{-/-}$) to the $CD36^{-/-}$ model, administration of a CD36 agonist inhibited the formation of subretinal deposits. CD36 may be linked to AMD-like retinal pathology in spontaneous hypertensive rats (SHRs) as well. These animals develop retinal and choroidal degeneration independent of hypertension (Funk and Rohen 1985). CD36 mutations exist in some SHR strains (Aitman et al. 1999). Kondo et al. (2009) examined the allelic frequency of 19 SNPs resident in CD36 for association with NV AMD in a Japanese cohort of 109 people with NV AMD and 182 unrelated controls. There was a 50% lower likelihood of having NV AMD among carriers of the minor allele for two intronic sequence variants in the gene (rs3173798 and rs3211883). The feature sets of our discovery and replication gene chips contained five and four sequence variants of the *CD36* gene, respectively; none of these SNPs were in LD with the AMD-associated CD36 SNPs published by Kondo. We did not observe CD36–AMD relationships in any of the SNPs tested.

### 6.3.2.1.3 Tubulins

Tublins are constituent globular proteins present in retinal microtubules. Microtubules are key components of the cytoskeleton, acting to impart a structural stability to the cell—they are also involved in intracellular transport (reviewed in Bernstein et al. 1997). Tubulin is abundant within the outer plexiform layer of the retina (including the Henle fiber layer in the macula) (Neale et al. 2010). In 1997, Bernstein et al. used a bovine retinal model to demonstrate that tubulin is a major (albeit nonspecific; Neale et al. 2010) soluble MX-binding protein. Four years later, this group showed that MXs copurified with tubulin extracted from human macula (Yemelyanov et al. 2001). Crabtree et al. (2001) suggested the existence of a β-tubulin paclitaxel (Taxol)–binding site in microtubules of rhesus monkey. β-tubulin may be a receptor suited for "high-capacity deposition" of MXs in the retina. There is reason to consider influence of tubulin–MX links in retinal function. In a project identifying GSTP1 as a high specificity and affinity ligand for zeaxanthin in the macula, Bhosale et al. demonstrated that the MX–tubulin complex produces a spectral (bathochromic) shift that differs from that measured in MX alone. These findings have been used to suggest that MX–tubulin interactions may have some functional significance in the retina (Bhosale et al. 2004).

To our knowledge, there is no published work examining the relationships of AMD with DNA sequence variants in tubulin genes expressed in the human retina. As such, we applied an agnostic (data-driven) approach to examine allelic frequencies in sequence variants of 26 tubulin genes for their association with AAMD. rs3813739, a common sequence variant proximal to the 5′UTR of the gene encoding the TUBGCP3 (13q34) was associated with AAMD in discovery ($p = .010$) and replication cohorts ($p = .046$). In the discovery cohort, we observed AAMD relationships with sequence variants in the tubulin β 3 (*TUBB3*, 16q24.3) and tubulin delta 1 (*TUBD1*, 17q23.1) genes. The *TUBB3* SNPs (rs4395073 and rs4558416) were proximal to the 3′UTR. The *TUBD1* SNPs included an exonic variant yielding a synonymous substitution (rs2250526 and Tyr287Tyr). Machalinska et al. (2011) report significant intracellular overexpression of TUBB3 in peripheral blood nuclear cells derived from NV AMD patients, and it would thus be important to examine the relationship of AMD and MX binding with other SNPs in proximity to rs2250526. There were no relationships of TUBB3 and TUBD1 with AMD in the replication cohort. There is currently no known relationship of TUBGC3 with MXs or AMD.

### 6.3.2.1.4 Apolipoprotein E

APOE is a major apolipoprotein acting in lipid metabolism (as a ligand to lipoprotein receptors) and response to injury within the central nervous system (reviewed in Klaver et al. 1998a and Zareparsi et al. 2004). APOE is necessary for the normal catabolism of triglyceride-rich components of chylomicrons, very low-density lipoprotein (VLDL), and HDL (Loane et al. 2010a). SNPs in the *APOE* gene (19q13.2) are associated with increased plasma concentrations of cholesterol and triglycerides. DNA sequence variants in *APOE* have been a focus of work on AMD, as APOE is a component of drusen (retinal lipid deposits first evident in early AMD). In the retina, APOE is expressed primarily in astrocytes and Müller cells; it has been localized to

BM, the RPE, and the photoreceptor outer segment layer (reviewed in Tserentsoodol et al. 2006). MXs are transported in serum on APOE in HDL and LDL (Loane et al. 2010a, 2010b).

Six-month-old apoE$^{-/-}$ C57BL/6 mice that were fed a standard lab chow (9605/8, Harlan Teklad TRM) showed a 50% lower concentration of lutein in the neural retina, relative to WT animals of the same inbred strain that were fed the same diet. Zeaxanthin levels were unchanged (Fernandez-Robredo et al. 2009). Retinal ultrastructure in the apoE$^{-/-}$ mice (with low lutein status) was characterized by AMD-associated lesions, including severe basal laminar deposits/vacuolization and thickening of BM. Choroid–RPE homogenates of these animals showed 42% higher VEGF levels than those in WT mice, as analyzed by Western blots. Lee et al. report abnormal lipid accumulation in the RPE and BM in a transgenic mouse expressing the human *APOε2* risk variant. Compared to WT mice, the *APOε2* transgenic animals exhibited an overexpression of VEGF and basic fibroblast growth factor (bFGF). Dysregulation of VEGF and bFGF are central events in diseases characterized by pathologic retinal angiogenesis.

DNA variants in *APOE* are associated with MPOD concentrations. Loane et al. (2010a) report people carrying at least one *APOε4* allele have higher MPOD values across the macula than noncarriers. The variation in MPOD associated with APOE allelic variants (Loane et al. 2010a) provides a reasonable basis to examine the interplay of factors influencing the HDL lipoprotein profile in the context of processes influencing retinal MX accretion and AMD risk. The APOE–AMD relationships reported by Souied et al. (1998) and Klaver et al. (1998a) were among the first reported genetic associations in AMD. Findings have been replicated numerous times (Baird et al. 2004; Francis et al. 2009; Tikellis et al. 2007; Wong et al. 2006; Zareparsi et al. 2004, 2005). An age-, sex-, and smoking-adjusted pooled analysis of 15 studies ($n = 21,160$) confirmed the protective association of the *APOε4* alleles and risk for the *APOε2* alleles on late AMD (McKay et al. 2011). We did not have sufficient coverage of *APOE*-resident sequence variants on our gene chips to test associations in either of our cohorts. Our replication cohort gene chip panel carried rs405509, a functional (promoter) polymorphism in *APOE* influencing gene expression. We did not see a relationship of AMD with the distribution of allelic variants in this SNP. Our findings on rs405509 as a single marker (versus an essential contribution as a member of an AMD-associated haplogroup defined by multiple SNPs) are similar to that of the pooled analysis by McKay et al. (2011).

### 6.3.2.1.5 ATP-Binding Cassette, Subfamily A, Member 1

The membrane-associated ATP-binding cassette (ABC), subfamily A, member 1 (ABCA1) protein, encoded by the *ABCA1* gene (9q31.1), is a member of the superfamily of ABC transporters. The protein acts as a major lipoprotein transporter and works with apolipoprotein A1 (APOA1) in the process of cholesterol and phospholipid metabolism as an efflux pump from tissue to nascent HDL. HDL is a main transport lipoprotein for MXs in extrahepatic tissue (Connor et al. 2007). *ABCA1* controls the intracellular transport and secretion of APOE and APOA1 (Tserentsoodol et al. 2006). A mutation mapped to *ABCA1* exists in people with Tangier disease (OMIM #205400), an autosomal recessive disorder characterized by extremely reduced levels of plasma HDL leading to tissue accumulation of cholesterol esters. Immunolocalization of ABCA1 in primate retina indicates that highest concentrations of the protein are found in the macular ganglion cell layer, MX-rich outer plexiform layer, and RPE (Tserentsoodol et al. 2006). Within the polarized RPE cell, both basal (side of choroidal apposition) and apical (side of photoreceptor apposition) aspects showed specific ABCA1 staining.

Connor et al. (2007) applied an avian model expressing a sex-linked recessive mutation in *ABCA1* (the Wisconsin hypoalpha mutant [WHAM] chick) to demonstrate the critical role of HDL-mediated MX transport to the retina. The WHAM chick exhibits similar concentrations of VLDL and LDL to WT Leghorn chicks, but shows a 90% reduction in HDL cholesterol. Hepatic accumulation of MXs in the WHAM chick was not appreciably different from that in the WT birds; however, differences are seen in plasma, heart, adipose, and retina. Repletion of all tissues except retina was attained with a 1-month lutein-rich feeding regimen. With the lutein-rich diet, the absolute concentration of retinal MXs remained 15-fold lower than those in the WT chicks. The WHAM mutation acts by inhibiting APOA1-mediated efflux of hydrophobic molecules from peripheral tissue to HDL. APOA1 is rapidly degraded when it is low in lipid content.

Lakkaraju et al. (2007) used immortalized human and bovine primary RPE cell cultures to demonstrate that activation of *ABCA1* by the peroxisome proliferator-activated receptor γ (PPARG) agonist pioglitazone and liver X receptor (LXR) agonist TO901317 is effective in hydrolyzing A2E, a quaternary amine and retinoid by-product of the visual cycle responsible for pathologic accumulation of free and esterified cholesterol in RPE cells. Blue light exposure in RPE cells induces A2E to generate singlet oxygen that has the capacity to damage DNA and lead to apoptotic cell death (Sparrow and Boulton 2005); MXs act as a filters for blue light and block A2E photooxidation (Kim et al. 2006). The implications of these findings for MX-AMD research may not be readily apparent. Pioglitazone acts on PPARG that forms a bioactive heterodimer with the retinoid X receptor (RXR). RXR heterodimerizes with the retinoid A receptors (RARs). RXR heterodimerizes with PPARG and the LXR—this process activates *ABCA1* leading to A2E hydrolysis. Lutein is a ligand to the RARs and may thus influence the activation of the RAR–RXR–PPARG–LXR complex (Matsumoto et al. 2007). This putative relationship ties the ABCA1–MX relationship to a process implicated in AMD pathogenesis and raises the possibility that lutein (or a lutein metabolite) may activate ABCA1 to influence A2E catabolism. Work by Masumoto et al. (2007) indicates that β-cryptoxanthin (a monohydroxy xanthophyll with similar bonding structure and functional groups to MXs) was effective in activating ABCA1. Duncan et al. (2009) report that exposure of human RPE cell cultures to glyburide, a nonspecific *ABCA1* inhibitor (also inhibiting *SCARB1*), prevents HDL-stimulated basal transport of photoreceptor-derived lipids implicated in AMD pathogenesis. These findings are intriguing, considering the MX-binding capacity of SCARB1, the role of *ABCA1* in MX transport, and putative activation of *ABCA1* by MX through RARA and RARG.

Sequence variants in ABCA1 have been associated with AMD in large-scale genotyping projects (Chen et al. 2010; Fauser et al. 2011; Neale et al. 2010; Yu et al. 2011a, 2011b). We identified AAMD-associated allelic variants in *ABCA1*. The gene chip used on our discovery cohort allowed testing of 16 sequence variants in the *ABCA1* gene. The panel for the replication cohort chip contained 42 *ABCA1* SNPs—7 of which showed associations with AAMD. An intronic SNP in the discovery panel (rs2575876) in proximity to an exon was associated with AAMD ($p \leq .005$) and was in moderate LD ($D' = 1.0$, $r^2 = 0.58$) with an AMD-associated variant in the replication cohort (rs2575875); *p*-values for AAMD and NV AMD relationships are .007 and .010, respectively. An AAMD-associated exonic sequence variant yielding a nonsynonymous amino acid substitution (rs2230808, Lys1587Arg) existed ($p \leq .038$), but was not in LD with (rs2575876).

### 6.3.2.1.6 Apolipoprotein A1

APOA1, a secreted soluble lipoprotein and the major protein constituent in HDL, is encoded by the *APOA1* gene (11q23-q24). APOA1 forms a complex with ABCA1 (reviewed in Section 6.3.2.1.5) to optimize reverse cholesterol transport by HDL. Immunolocalization of APOA1 in the primate retina demonstrated labeling in the ganglion cell layer, outer plexiform layer, photoreceptor inner segments, rod outer segments, and the apical side of RPE cells (the area in which photoreceptor outer segments are embedded and proximal to the interphotoreceptor matrix) (Tserentsoodol et al. 2006). The basis for examining APOA1 is the protein–protein interaction it shares with ABCA1. Colak et al. (2011) examined serum APOA1 levels in 79 people with AMD and 84 aged-matched controls; differences in concentrations were negligible. To our knowledge, there are no extant works examining the distribution of allelic variants in APOA1 and how they may be related to AMD. There was insufficient coverage of APOA1 sequence variants on gene chips used in both discovery and replication cohorts to test relationships with our endpoints.

### 6.3.2.1.7 Glutathione S-Transferase Pi 1

GSTP1 protein is a soluble phase II detoxification enzyme encoded by the *GSTP1* gene (11q13). GSTP1 catalyzes conjugation of reduced glutathione with a number of hydrophobic and electrophilic compounds, acts as a *cis-trans* isomerase for retinoic acid, a nitric oxide scavenger protein, and a regulator of reactive oxygen- and nitrogen-induced S-glutathionylation (reviewed in Neale et al. 2010 and Bhosale et al. 2004). In addition to these actions, GSTP1 serves as a specific (saturable), high (submicromolar)-affinity zeaxanthin- and *meso*-zeaxanthin-binding protein in the primate retina (Bhosale et al. 2004; Neale et al. 2010). GSTP1 has been immunolocalized to MX-rich inner and outer plexiform layers (Bhosale et al. 2004). Carotenoid-protein binding may alter the spectral absorption properties of the carotenoid—in the case that the interaction is of physiologic significance, one would expect the absorption properties to mimic those of macular pigment measured in live primates. Bhosale et al. (2004) demonstrated that the MX–GSTP1 complex shows an absorption spectrum different than that of free zeaxanthin or *meso*-zeaxanthin and similar to that measured *in vivo* within primates. Oz et al. (2006) and Guven et al. (2011) compared allelic distributions of *GSTP1* variants in people with AAMD to those in people without AMD; neither study reported AMD-associated SNPs in this gene. Bhosale et al. (2004) reviewed putative functions of the MX–GSTP1 complex, suggesting that it may act in (1) macular membrane stabilization within proximity of the fovea; (2) MX metabolism/catabolism; and (3) increases in antioxidant potency beyond that of either molecule alone (Bhosale and Bernstein 2005).

The gene chip used on our discovery cohort contained two variants in resident in *GSTP1* (rs4147581 and rs4891); the gene chip used in the replication cohort contained three (rs1138272, rs6591256, and rs1695). Allelic frequencies of these SNPs did not vary significantly with AAMD status. Because the GSTP1 protein is a key molecule driving highly specific zeaxanthin binding, we applied a network analysis approach using the Ingenuity® Knowledge Base to examine the possible influence of AMD-associated sequence variants in genes whose products interact GSTP1. The protein encoded by the Fanconi anemia complementation group C (*FANCC*, 9q22.3) gene increases catalytic activity of GSTP1 during apoptosis by preserving the structure of bonds in GSTP1 that normally sustain disruptions in disulfide structure with exposure to oxidizing agents (Cumming et al. 2001). Direct interaction of FANCC and GSTP1 has been demonstrated in a model using the *in vitro* co-immunoprecipitation paradigm (Reuter et al. 2003). To our knowledge, there are no published works identifying sequence variants that alter FANCC–GSTP1 interactions, and such studies would add value to inference on FANCC–GSTP1–AMD relationships. We (SanGiovanni and Neuringer 2012) observed a number of AMD-associated variants in FANCC both in discovery and in replication cohorts. One such SNP in FANCC (rs356666, $p \leq .004$) in the discovery cohort was in nearly complete LD ($D' = 0.96$–$1.0$, $r^2 = .92$) with AAMD-associated SNPs in the replication panel (rs356677, $p \leq .001$ and rs356669, $p \leq .0005$). There are currently no publically accessible data with which to examine the coinheritance of these variants with others in regulatory or protein-coding regions of FANCC.

*6.3.2.1.8 StAR-Related Lipid Transfer Domain Containing 3*

STARD3 is a protein involved in lipid trafficking and encoded by the *STARD3* gene (17q11-q12). The STARD3 product has been immunolocalized to the ganglion cell, inner nuclear, outer plexiform (Henle fibers in the macula), and outer nuclear layers—cone inner segments and photoreceptor outer segments are also labeled (Li et al. 2011). A nonspecific binding signal existed for RPE cells. The Bernstein lab has recently identified the STARD3 protein as a specific, high-affinity lutein-binding protein in the primate retina (Li et al. 2011; Neale et al. 2010). Li et al. (2011) report that the MX–STARD3 complex shows an absorption spectrum different than that of free lutein and similar to that measured *in vivo* on macular pigment within primates. To our knowledge, there are no extant reports applying targeted mutagenesis or analysis of DNA sequence variation in this gene to examine its role in AMD pathogenesis. The gene chip used on our discovery sample contained two SNPs resident in *STARD3* (the coding variant rs1877031, Arg117Gln, and the intronic variant rs881844). The gene chip used on the replication sample contained three intronic *STARD3* SNPs (rs2271308, rs931992, and rs9892427). AMD status did not vary with the allelic frequency of these SNPs.

### 6.3.3 Group 2: Cleavage of Macular Xanthophylls

We identified human genes encoding MX cleavage enzymes from a review covering an emerging field on the activity of a highly conserved family of non-heme iron monooxygenases (Amengual et al. 2011; Mein et al. 2011) that target the hydrocarbon backbone of carotenes and xanthophylls. These genes encode β-carotene-15, 15′-monooxygenase 1 (BCMO1), and β-carotene-9′, 10′-oxygenase 2 (BCO2). BCMO1 is active in cytosol and BCO2 is active in mitochondria. Amengual et al. (2011) reviewed the structural chemistry of carotenoid oxygenases, discussing homology in a centrally active binding component containing a coordinated ferrous iron (accessed through a large kinked tunnel). These authors cite homology of RPE65 (the third member in the human carotenoid monooxygnase family) with an insect protein (*ninaB*) exhibiting bacterial apocarotenoid-oxygenase activity. An ionone ring configuration binds this component in an orientation-dependent manner within ninaB and BCMO1. In addition to BCMO1 and BC02, we examined lipoprotein lipase (LPL, 8p22). The protein product of LPL has the capacity to cleave MXs from resident lipoproteins and carries a functional sequence variant associated with altered levels of MXs in serum.

### 6.3.3.1 β-Carotene-15, 15′-Monooxygenase 1

BCMO1 is encoded by the *BCMO1* gene (16q23.2). This enzyme symmetrically cleaves provitamin A carotenoids (β-carotene, α-carotene, and β-cryptoxanthin) at the 15 and 15′ double bond of the constituent polyene chain; BCMO1 is essential for catalyzing the first reaction in the production of vitamin A (carotene → 2 retinaldehyde). The active binding component in BCMO1 is specific for a nonsubstituted β-ionone ring, thus permitting the major dietary carotenes to act as substrates, but not the MXs. Although not frequently discussed, the RPE–choroid contains appreciable quantities of β-carotene (Bernstein et al. 2001). In the primate retina, the BCMO1 protein is concentrated in the RPE–choroid (Lindqvist et al. 2005). A meta-analysis on GWA data from three large-scale genotyping projects ($n = \sim 3900$) identified an allelic variant in the proximity of *BCMO1* (rs6564851) associated with higher circulating levels of β-carotene, lower circulating levels of MXs, and no alteration in plasma retinol (Ferrucci et al. 2009). The authors suggest that this sequence variant results in a reduced BCMO1 activity. Because the molecular structure of MXs does not permit interactions with BCMO1, the altered circulating levels are likely the result of some unknown receptor-mediated process of competitive uptake and transport between provitamin A carotenoids and xanthophylls.

*BCMO1* variants are associated with altered concentrations of MXs in the human retina. In a 6-month placebo-controlled clinical trial with a lutein-rich supplement in 29 healthy male French participants (mean age: 49.7 ± 0.8 years), participants homozygous for the major allele of the exonic *BCMO1* sequence variant A379V (TT, rs7501331) had higher MPOD than those carrying the minor allele (Borel et al. 2011). A379V is a functional variant: *in vitro* models in TC7 Caco-2 cells using a targeted mutagenesis of A379V with R267S (rs12934922) showed that carriers of the 379V and 379V + 267S had reduced catalytic activity in the protein (amounting to ~60% reduction in the double mutant). When 28 female volunteers were challenged with a high dose of β-carotene (120 mg in a high-fat meal), those participants carrying both mutant alleles were less efficient in their conversion of β-carotene to retinyl palmitate than those with the WT alleles (Leung et al. 2009). The authors of the MPOD study acknowledge that the gene–MX relationship is likely to be indirect, because the requisite BCMO1-binding motifs are not present on lutein or zeaxanthin. They propose that the sequence variants in the gene encoding this enzyme alter the status of vitamin A, vitamin A precursors, and then their products to activate factors and processes influencing MX transport, binding, and storage. The single BCO2 SNP tested in this study did not yield an association with MPOD. To our knowledge, there are no extant works reporting tests of *BCMO1* variants for their relationship with AMD. We were able to test the association of AMD with five sequence variants (rs12449108, rs6564856, rs6564860, rs7205123, and rs7500996) in the *BCMO1* gene from the gene chip used in the discovery cohort and six SNPs (rs3803651, rs4889294, rs6564862, rs6564863, rs7192178, and rs8046134) on the gene chip used in the replication cohort. None of these SNPs are in LD with rs7501331 (the variant associated with altered MPOD). We did not observe any AMD–BCMO1 relationship in variants examined.

### 6.3.3.2 β-Carotene-9′, 10′-Oxygenase 2

BCO2 is encoded by the *BCO2* gene (11q22.3-q23.1). This enzyme asymmetrically cleaves carotenoids between the 9′ and 10′ carbons of the constituent polyene chain. Activity on the 9 and 10 bond has also been observed. Cleavage of MXs by BCO2 yields apocarotenals with hydroxyl ionones whose acid derivatives may have the capacity to act as signaling molecules in adaptive cellular response and transcription (Mein et al. 2011). The BCO2 protein is expressed in the primate RPE (Lindqvist et al. 2005). The binding pocket of BCO2 accepts both β- and ε-3OH-ionone ring structures and thus binds MXs (Amengual et al. 2011). Mein et al. used *in vitro* baculovirus-generated recombinant ferret BCO2 to demonstrate cleavage activity toward lutein and zeaxanthin and showed conversion of one MX metabolite (3-OH-β-apo-10′-carotenal) to 3-OH-β-apo-10′-carotenoic acid. Sequence variants in the *BCO2* gene are associated with accumulation of lutein in adipose tissue of chicken (Castaneda et al. 2005; Eriksson et al. 2008) and sheep (Vage and Boman). 3-OH-β-apo-10′-carotenoic acid and other apocarotenoids (e.g., apo-lycopenic acid) sharing some of the same functional groups as MX metabolites may act as signaling molecules to regulate cell proliferation and apoptosis through stimulation of cellular differentiation (Winum et al. 1997), induction of phase II enzymes through activation of the NFE2L2 transcription factor (Lian and Wang 2008), inhibition of cell growth (Lian et al. 2007; Prakash et al. 2004; Tibaduiza et al. 2002), trans-activation of nuclear receptors (Lian et al. 2007), or inhibition of nuclear receptor activation (Ziouzenkova et al. 2007).

Amengual et al. applied tests on HepG2 human hepatocyte and COS7 monkey kidney cell model systems and adult BCO2$^{-/-}$ and BCO2$^{+/-}$ mice on 8-week vitamin A–deficient, MX-supplemented, AIN-93G-diet-based feeding regimens to examine the physiological function of BCO2 (50 mg/kg lutein in BCO2$^{-/-}$ mice and 50 mg/kg zeaxanthin in BCO2$^{+/-}$ mice) (Amengual et al. 2011). BCO2 was localized to mitochondria in the kidney cell line and *in vivo*. In BCO2$^{-/-}$ mice and BCO2$^{+/-}$ mice, metabolism or secretion of MX-derived apocarotenoid cleavage products was impaired in blood, heart, and liver. To our knowledge, there are no published reports of work examining the relationship of sequence variants in *BCO2* with AMD. We did not observe any consistent AMD-BCO2 or NFE2L2 sequence variant relationships in our cohorts.

### 6.3.3.3 Lipoprotein Lipase

LPL is a water-soluble enzyme, encoded by the *LPL* gene (8p22), that hydrolyzes triglycerides in lipoproteins and enables cellular uptake of chylomicron remnants,

# Molecular Genetics, AMD, and MXs

cholesterol-dense lipoproteins, and free fatty acids. The enzyme is encoded by the *LPL* gene (8p22) and requires the apolipoprotein C2 as a cofactor in these processes. In the human eye, the LPL protein is localized within the inner retinal layers (nerve fiber, ganglion cell, inner plexiform, and inner nuclear layers) and the choroid (Casaroli-Marano et al. 1996). Within the choroid, it is most likely to be attached to the luminal epithelium. Carriers of an exonic sequence variant X447 (rs328) in the *LPL* gene show a 20% reduction in serum MXs, relative to carriers of the S447S alleles (Herbeth et al. 2007). The authors of this work suggest that this stop polymorphism alters shedding of MXs from their surface positions on chylomicrons during lypolysis. A sequence variant in *LPL* (rs12678919) known to influence HDL-c (the amount of cholesterol associated with ApoA-1/HDL) has been reported in association with advanced (Chen et al. 2010), but not early AMD (Fauser et al. 2011). We had access to genotypes on nine *LPL* SNPs in our discovery cohort (rs1059611, rs248, rs255, rs301, rs316, rs327, rs330, rs331, and rs4922115) and five SNPs in our replication cohort (rs10099160, rs1534649, rs264, rs270, and rs316). Of these, only rs10099160 showed association with AAMD ($p \leq .002$). This SNP was neither in LD with the others yielding null results nor with the AMD-associated variant reported previously (rs12678919) by Chen et al. (2010).

### 6.3.4 GROUP 3: ANTIOXIDANT POTENTIAL OF MXs IN DNA PRESERVATION AND REPAIR

MXs have been implicated in repair of endogenous oxidant-induced DNA base damage and strand interruptions\breaks that disrupt homeostatic mechanisms in the cell (Serpeloni et al.). Lutein protected Chinese hamster ovary cells against mutation and chromosomal damage as assessed by the Ames test and the chromosomal aberration (Wang et al. 2006). In postmenopausal women, lutein supplementation with 12 mg/day for a 2-month period decreased endogenous and $H_2O_2$-induced DNA damage (Zhao et al. 2006). Inhibition of DNA polymerases (POLs) has been associated with an anti-inflammatory response in model systems (Horie et al. 2010). The influence of MXs on POLs involved in DNA base excision repair for single-strand damage was studied by Horie et al. (2010). In a tested panel of mammalian genes encoding these proteins, MXs blocked POL (DNA-directed) beta (POLB, 8p11.2) and POL (DNA-directed) lambda (POLL, 10q23) activity *in vitro*. POLB and POLL act through short-patch base excision repair. We examined sequence variation in *POLB* and *POLL* genes for their association with AAMD, testing five variants and one variant in *POLB* and three variants in each cohort for *POLL* in the discovery and replication cohorts, respectively. In no instance did allelic frequency of *POLs* vary significantly with AAMD.

### 6.3.5 GROUP 4: INFLUENCE ON ADAPTIVE CELLULAR RESPONSE

Connexins, Gap Junction Intercellular Signaling and the Adhesion Complex. Demig-Adams and Adams (2002) discuss the role of dietary carotenoids in regulation of connexin expression. Connexins are membrane proteins serving as key structural and signaling constituents of gap junctions (Dbouk et al. 2009), the small molecule channeling complexes that permit time- and voltage-dependent electrical and metabolic coupling between cells. Gap junctions are active in maintaining homeostasis within the RPE, photoreceptors, and retinal ganglion cells (Cook and Becker 1995; Hornstein et al. 2005; Vaney 1991, 2002). With adherens and tight junctions, gap junctions form a system known as the adhesion complex. The adhesion complex is involved in sensing of soluble mediators (e.g., cytokines, chemokines, and growth factors) and cellular response through alteration of cellular scaffolding and initiation/integration of signaling cascades. The adhesion complex acts in the differentiation and redifferentiation of polarized cells regulating barrier function (Dbouk et al. 2009). RPE cells are one such polarized cell type (Kojima et al. 2008). Connexins interact with many adhesion complex proteins (reviewed in Dbouk et al. 2009); these include (1) caveolins; (2) adherens junction proteins—cadherins and catenins; (3) tight junction proteins—zonulins; (4) constituents of the cytoskeleton—microtubules and actin; (5) actin-binding proteins—α-spectrin and drebrin; and (6) phosphatases and kinases involved in adhesion complex construction, function, and disassembly.

A number of connexins are expressed in the human RPE and neural retina (reviewed in Hoang et al. 2010). Connexin 43 (*GJA1* gene, 6q21-q23.2) and connexin 46 (*GJA3* gene, 13q12.11) have been identified as essential proteins in gap junction intercellular signaling within the RPE (Malfait et al. 2001; Udawatte et al. 2008). An immunolocalization study limited to study of connexin 36 (*GJD2* gene, 15q14), connexin 45 (*GJC1* gene, 17q21.31), connexin 59 (*GJA9* gene, 1p34), and connexin 62 (*GJA10* gene, 6q15-q16) showed a strong signal in the inner plexiform and ganglion cell layers for GJD2. Staining was strongest in the outer plexiform layer for GJC1 (Sohl et al.). There was weak staining for GJD2 in the outer plexiform layer and for GJC1 in the inner plexiform and ganglion cell layers. Carbenoxolone, a gap junction inhibitor, abolished both electrical and metabolic tracer coupling in macaque photoreceptors (Hornstein et al. 2005). Work in model systems demonstrated that the gap junction blocker heptanol is capable of disrupting electrophysiologic properties of retinal ganglion cells (Ye et al. 2008) Although these cells do not sustain the bulk of damage in AMD, they are important in sending signals from the retina to the brain.

β-carotene, β-cryptoxanthin, and canthaxanthin enhance gap junction communication in model systems (Stahl et al. 1997). There are few reports on the influence of MXs on gap junctions. Lutein efficiently stimulated gap junction intercellular communication in the C3H/10T1/2 mouse embryo fibroblast line (Zhang et al. 1991). The research team reporting these findings attributed the effect to an increased expression of the GJA1 protein that was associated with binding of the 6-carbon ring structure of the MX ionone ring (Basu et al. 2001). In addition to a role in gap junction intercellular signaling, the GJA1 protein acts as an essential factor in the regulation of redifferentiation in RPE cells (an important process in wound healing in this tissue) through a cyclic adenosine monophosphate–dependent (gap junction–independent) process (Kojima et al. 2008). The GJA1 protein also shows direct interactions with debrin, zonulin-1, β-catenin, and caveolin-1 (reviewed in Kojima et al. 2008).

The evidence presented in this section provides a basis for considering a means by which MX exposure may lead to connexin regulation, and connexin regulation may alter adhesion complex function. Adhesion complex function may dysregulate barrier function. Barrier function is implicated in AMD pathogenesis. We are unaware of work in the retina examining the direct effect of MXs on adhesion complex proteins. This topic provides an excellent research opportunity. To the extent that connexins interact with these proteins, any AMD-associated sequence variants yielding structural changes in an assembly complex constituent would offer promise as a candidate for gene–MX–AMD investigations.

### 6.3.6 GROUP 5: INTERFERENCE OF CAROTENOIDS WITH GROWTH FACTORS

MXs may have the capacity to influence activity of growth factors involved in cell cycle and survival. Mein et al. (2008) review mechanistic evidence implicating protective role of lycopene in the inhibition of insulin-like growth factor (IGF) signaling system in cancer prevention. IGFs are mitogens acting on IGF receptors (IGFRs) and regulated in circulation by IGF-binding proteins (IGFBPs). IGF1 (*IGF1* gene, 12q23.2) binding at IGF1R (*IGF1R* gene, 15q26.3) activates the Akt/PI3K and Ras/Raf/MAPK signaling pathways (reviewed in Mein et al. 2008). Akt/PI3K and Ras/Raf/MAPK pathways are implicated in regulation of cell survival (Mein et al. 2008). The IGFs *IGF1* and *IGF1R* are involved in pathologic retinal angiogenesis (Smith 2005; Whitmire et al.) and stimulate differentiation and proliferation of RPE and neural retinal cells (Lambooij et al. 2003). In humans, IGF1 and IGF1R proteins colocalize throughout the neural retina and RPE. In the mouse, the IGF1R homolog protein localizes to photoreceptors and vessels, whereas the IGF1 homolog protein is found throughout the retina (Lofqvist et al. 2009). Because MXs share structural elements and functional groups with lycopene, we examined association of AMD with genes encoding IGFs (*IGF1* and *IGF2*), their receptors (*IGF1R* and *IGF2R*), and their binding proteins (IGFBP1–IGFBP6). Two AMD-associated *IGF1R* variants (rs1815009 and rs2684788) in the discovery cohort exist within the UTR of the gene. These SNPs were not in LD with AMD-associated SNPs in the replication sample. *IGFBP1* carried an exonic variant (rs4619, Ile253Met) that was associated with AAMD in both cohorts. To our knowledge, there are no published reports investigating the influence of MXs in the IGF system in the retina. Our findings provide reasonable basis for planning mechanistic studies on the topic.

### 6.3.7 GROUP 6: DISEASES ASSOCIATED WITH LOW MX IN THE RETINA

We examined AMD relationships in two genes associated both with inherited retinal dystrophies and with reduced MPOD. These genes are (1) aldehyde dehydrogenase 3 family member A2, carrying Sjögren–Larsson syndrome (SLS)–related DNA variants and (2) ABC subfamily A member 4, carrying Stargardt disease 1 (STGD1)–related and cone rod dystrophy 3 (CORD3)–related variants.

#### 6.3.7.1 Sjögren–Larsson Syndrome (Aldehyde Dehydrogenase 3 Family Member A2)

SLS (OMIM: #270200) is an autosomal recessive disorder manifesting a cystoid foveal atrophy (Fuijkschot et al. 2008; Willemsen et al. 2000) with a crystalline juvenile macular dystrophy in the retinal ganglion cell and inner plexiform layers (van der

Veen et al. 2010). People with SLS have defects in the aldehyde dehydrogenase 3 family member A2 protein, an enzyme encoded by the *ALDH3A2* gene (OMIM: *609523, 17p11.2). The ALDH3A2 protein plays a major role in detoxification of aldehydes generated by alcohol metabolism and peroxidation of lipid polyene chains, acting in this capacity by oxidizing long-chain aliphatic aldehydes to fatty acid. People with SLS often exhibit pallor in the fovea caused by a lack of macular pigment. To our knowledge, there are no published findings on the distribution of the ALDH3A2 protein in the retina. van der Veen et al. (2010) used fundus autofluorescence imaging and reflectance spectroscopy in a cross-sectional observational case study to examine distribution profiles and quantity of MXs in 14 people with genetic confirmation of SLS. The authors report nearly complete absence of MXs in the central retina and conclude that SLS is the first known disease with a genetically caused deficiency in MXs. We had access to genotypes on five sequence variants of *ALDH3A2* in the discovery cohort and four sequence variants in the replication cohort. The AMD-related SNPs in the discovery cohort (rs1004490, $p \leq .04$; rs1004491, $p \leq .04$; and rs2072332, $p \leq .04$) were in complete LD with one SNP in the replication cohort (rs962801, $p \leq .03$). These 4 SNPs are proximal to exons 4 and 5—in some cases, less than 100 base pairs from coding regions. It is interesting to note that this same region in the *ALDH3A2* gene carries at least nine missense mutations related to SLS; these are designated sequentially as rs72547562-to-rs72547570 (Rizzo and Carney 2005). Neither of the gene chips used in our study nor the HapMap cohort contain data on these SLS variants. It would be informative to examine either the LD or (preferably) the direct relationships of SLS-related SNPs with AMD and MPOD concentration. If relationships exist, these SNPs may guide clinical practice toward identification of genetic factors driving retinal response to MX supplementation in people at risk for AAMD.

### 6.3.7.2 Stargardt Disease (ATP-Binding Cassette Subfamily A Member 4)

Stargardt disease 1 (OMIM: #248200) and cone rod dystrophy 3 (OMIM#: 604116) are autosomal recessive disorders manifesting photoreceptor degenerations with macular involvement. People with STGD1 and CORD3 have defects in the ABC subfamily A member 4 protein. This protein, encoded by the *ABCA4* gene (OMIM: *601691, 1p22.1) and localized to outer segments (Molday et al. 2000), acts in removal of all-*trans*-retinal from photoreceptors by transporting A2-phosphatidylethanolamine (PEA), a retinoid adduct formed by all-*trans*-retinal and PEA. People with STGD1 show homozygous or compound heterozygous *ABCA4* variants at risk loci. People with CORD3 carry homozygous null mutations. Aleman et al. (2007) used an observational design on 14 people with STGD1, 3 people with CORD3, and 29 people without ocular disease to examine the distribution of MPOD in people with *ABCA4* variants. The study also included a 6-month open-label trial with lutein (20 mg/day with a high-fat meal) on 11 patients and 8 controls to determine the influence of dietary intake on retinal MX response. Participant's serum MX concentrations were reduced, and MPOD was marked reduced in some cases. Four patients did not show a significantly increased MPOD with supplementation. Responders were more likely to be female, have lower serum MXs and baseline MPOD, and have greater retinal thickness at baseline.

AMD-associated variants in *ABCA4* variants have been reported over the past decade (Allikmets et al. 1997; Bernstein et al. 2002; Brion et al. 2010; De La Paz et al. 1999; Shroyer et al. 1999). The gene chip used in our discovery cohort allowed us to investigate 12 *ABCA4* variants; the chip used in the replication cohort allowed us to investigate 33. Two variants (rs2297634 and rs544830) in our discovery cohort and four variants in our replication cohort (rs1889548, rs3818778, rs472908, and rs560426) showed relationships with AAMD. None of these SNPs are in LD or in LD with variants leading to nonsynonymous substitutions in amino acids (peptide shifts in protein structure). The STGD1 and CORD3 participants in the Aleman et al. (2007) MPOD study carried the following *ABCA4* sequence variants: rs1801269, rs1800553, rs1801581, rs61750120, rs61753033, rs76157638, and rs80309162. None of these existed in the AMD exploration or replication gene chip panel. As none of our tested SNPs were in LD with the STGD1- and CORD3-associated SNPs, we were unable to make inferences on ABCA4–MPOD–AMD relationships. The relationship between the AMD, STGD1/CORD3, and MPOD-related sequence variants should be a focus of future research, as findings may provide a basis for understanding variations in retinal response to MX supplementation in people at risk for AMD.

## 6.4 CONCLUSION

Emerging technologies in molecular genetics and genomics permit the development of an MX–AMD gene compendium for use in mechanistic and translational research. The compendium may be applied to interrogate biologic actions of MXs in biochemical pathways and networks enriched with AMD-associated DNA variants. To the extent that MXs and/or their metabolites are causally related to AMD, "molecular phenotypes" may effectively guide systems-based preventive interventions. To make meaningful inferences from such efforts, multidisciplinary teams should work to examine systems of enzymes, transporters, ligands, receptors influencing or influenced by MX uptake, transport, retention, and repair. These groups should also focus their efforts toward determining how MXs and MX metabolites

- interact with proteins and lipids in retinal membranes
- modulate oxidative stress and redox balance
- interact with molecules in signal transduction cascades

## 6.5 METHOD

Data used for genetic analyses in this report were obtained from the NEI Age-Related Eye Disease Database (NEI-AREDS) and the NEI Study of Age-Related Macular Degeneration (NEI-AMD) Database at the U.S. National Center for Biotechnology Information (NCBI) database of Genotypes and Phenotypes (dbGaP). Details on NEI-AREDS exist at http://www.ncbi.nlm.nih.gov/projects/gap/cgi-bin/study.cgi?study_id=phs000001.v2.p1. Details on NEI-AMD exist at http://www.ncbi.nlm.nih.gov/projects/gap/cgi-bin/study.cgi?study_id=phs000182.v2.p1. NEI-AREDS was a collaborative of scientists and clinicians at the AREDS Project Office NEI Clinical Trials Branch, the AREDS Coordinating Center, the AREDS Photographic Reading Center, the NEI Office of the Director, and the AREDS clinical sites. NEI-AMD is a collaborative of researchers from the University of Michigan, Mayo Clinic, University of Pennsylvania, and the AREDS group including National Eye Institute intramural investigators. NEI-AREDS and NEI-AMD researchers collected clinical data and DNA from people affected with AMD and their elderly AMD-free peers. The NEI was the primary sponsor of this effort. Institutional review boards at each NEI-AREDS and NEI-AMD study site reviewed and approved the study protocols. Each participant provided written informed consent in accordance with the *Declaration of Helsinki*.

### 6.5.1 Subjects and Study Design

#### 6.5.1.1 AREDS Cohort

Details on AREDS and the demographics of NEI-AREDS participants involved in the NEI-AREDS GWA study design exist at http://www.ncbi.nlm.nih.gov/projects/gap/cgi-bin/study.cgi?study_id=phs000182.v2.p1. In brief, AREDS was a 12-year multicenter natural history study (with a 5-year phase III clinical trial) on 4757 elderly U.S. residents designed to assess the clinical course of, and risk factors for, and the development and progression of AMD by collecting data on possible risk factors, measuring changes in visual acuity, photographically documenting changes in macula status, and assessing self-reported visual function. Eleven retinal specialty clinics enrolled participants aged 55–80 years from November 1992 through January 1998 and followed them until April 2001. A natural history study extending to December 2005 was implemented in April 2001. Our analytic sample contained 378 people with AAMD (GA and/or NV AMD) and 146 of their AMD-free peers.

#### 6.5.1.2 NEI-AMD Cohorts

Details on the NEI-AMD GWA study exist http://www.ncbi.nlm.nih.gov/projects/gap/cgi-bin/study.cgi?study_id=phs000182.v2.p1. In brief, three independent cohorts from the University of Michigan in Ann Arbor, the University of Pennsylvania in Philadelphia, and the Mayo Clinic in Rochester, Minnesota, contributed data to this GWA study. Our analytic sample contained, respectively, 675, 227, and 275 people with AAMD from the University of Michigan, University of Pennsylvania, and the Mayo Clinic. There were 512, 198, and 314 AMD-free people, aged ≥65 years in these respective sites.

### 6.5.2 Outcome Ascertainment

We restricted our AMD-free comparison cohort to people aged ≥65 years in both NEI-AREDS and NEI-AMD. The likelihood of having AMD increases twofold to sixfold after the age of 75 years and it was therefore essential to select our oldest AMD-free participants to reduce the chances of including false negatives in analyses (that would otherwise result from nonrandom misclassification in the youngest members of the control group).

#### 6.5.2.1 AREDS Cohort

AREDS Report 1 contains information on outcome ascertainment in the NEI-AREDS cohort (AREDS Research Group 1999). People classified with AAMD manifested the following characteristics: (1) presence in either eye of GA or NV AMD defined as photocoagulation or other treatment for choroidal neovascularization (based on clinical center reports) and (2) photographic documentation of any of the following: non-drusenoid RPE detachment, serous or hemorrhagic retinal detachment, hemorrhage under the retina or RPE, and/or subretinal fibrosis either at baseline or during the course of the study. Our AMD-free NEI-AREDS controls have the following three distinguishing characteristics:

- Phenotype was determined annually over a 12-year period (AREDS) or across an 8-year period with a standardized protocol by multiple professional graders who were masked to phenotypic information from previous years. Adjudication with a standardized protocol occurred when discrepancies emerged.
- The criteria for AMD-free classification (<5 drusen of ≤63 μm in both eyes for the entire follow-up period) is stringent relative to those applied in previous association studies for AMD.
- The age of the AREDS AMD-free group is in the range in which AMD prevalence increases approximately three times (from ~4% in those aged 74–79 years to ~12% in those ≥80 years) in population-based studies.

### 6.5.2.2 NEI-AMD Cohorts

People with AMD had GA and/or NV AMD. Experienced graders (ophthalmologists) classified outcomes according to AMD diagnosis in the worse eye; NV AMD was coded as most severe outcome—presence of large drusen was coded as the least severe. Here we report outcomes for the NV AMD and GA + NV AMD. Our AMD-free comparison group was composed of people ≥65 years of age who had no large or intermediate drusen in either eye; these participants received examinations and gradings by the study ophthalmologists. If small drusen or pigment changes were present in the AMD-free group, they were neither bilateral nor extensive. No participant exhibited history or evidence of (1) retinal insult rendering the fundus ungradable; (2) severe macular disease or vision loss onset before 40 years of age; or (3) diagnosis of juvenile macular or retinal degeneration, macular damage resulting from ocular trauma, retinal detachment, high myopia, chorioretinal infection, or inflammatory disease, or choroidal dystrophy. The AMD-free comparison cohort was restricted to people ≥65 years of age. These people had no more than five hard drusen or small drusen and pigment changes in one eye only.

### 6.5.3 GENOTYPING

All NEI-AREDS and NEI-AMD specimens were genotyped with DNA microarrays at the Johns Hopkins University Center for Inherited Disease Research (CIDR, Baltimore, MD, USA). The NEI-AREDS cohort was genotyped with AFFYMETRIX Mapping50K_Hind240 (SNP batch IDs at http://www.ncbi.nlm.nih.gov/SNP/snp_viewBatch.cgi?sbid=33750), AFFYMETRIX Mapping50K_Xba240 (SNP batch IDs at http://www.ncbi.nlm.nih.gov/SNP/snp_viewBatch.cgi?sbid=33751), and ILLUMINA ILMN_Human-1 (SNP batch IDs at http://www.ncbi.nlm.nih.gov/

SNP/snp_viewBatch.cgi?sbid=33668). The NEI-AMD cohort was genotyped using ILLUMINA HumanCNV370v1 (SNP batch IDs at http://www.ncbi.nlm.nih.gov/SNP/snp_viewBatch.cgi?sbid=1047132).

The feature set on the 100K chip used in the discovery cohort allowed us to test 1060 SNPs from 275 genes in the NEI-AREDS cohort. The feature set on the 300K chip used in the NEI-AMD cohort allowed us to test 2710 variants.

### 6.5.4 Statistical Analyses

All sequence variants analyzed for AAMD relationships passed process quality and analytic filters for missingness (<5%), minor allele frequency (>1%), and HWE ($p \leq 1 \times 10^{-6}$ in the AMD-free group). We used Plink (version 1.07, http://pngu.mgh.harvard.edu/purcell/plink/) and SAS (version 9.1, Cary, NC) software for these purposes. The analytic plan was implemented as follows:

1. We examined the public and proprietary literature to identify 225 genes with putative relationships of MXs.
2. Positional coordinates of these genes were obtained from public access databases and used to filter ($\pm$ 2000 base pairs) all high-quality SNPs from the genome-wide microarray data in our 12-year prospective discovery cohort (NEI-AREDS) and our three replication cohorts (NEI-AMD).
3. We examined the allelic distributions of these SNPs in people with AMD (relative to the AMD-free comparison group) with age-, sex-, and smoking-adjusted logistic regression analyses using additive, dominant, and recessive models in each cohort. The NEI-AMD cohort was examined for replication with NEI-AREDS findings after combined effects of the NEI-AMD cohorts were estimated with meta-analytic techniques under constraints of sample heterogeneity (using Cochrane's Q for evaluation). $p$-values and odds ratios were chosen from random effects models in instances when significant heterogeneity was demonstrated.
4. We computed empirical $p$-values on AMD-associated variants resident in genes replicating in the NEI-AMD cohort using a max(T) permutation procedure set to 10,000 iterations.

### ACKNOWLEDGMENTS

The data used for the genetic analyses were obtained from the NEI-AREDS Database and the NEI-AMD Database found at http://www.ncbi.nlm.nih.gov/projects/gap/cgi-bin/study.cgi?study_id=phs000001.v2.p1 and http://www.ncbi.nlm.nih.gov/projects/gap/cgi-bin/study.cgi?study_id=phs000182.v2.p1 respectively. Funding support for NEI-AREDS and NEI-AMD was provided by the National Eye Institute. We thank NEI-AREDS and NEI-AMD participants and the NEI-AREDS and NEI-AMD Research Groups for their valuable contributions to this research. Dr. Neuringer was supported by NIH grant P51OD011092 and The Foundation Fighting Blindness.

## REFERENCES

Aitman, T. J., Glazier, A. M., Wallace, C. A., et al. (1999). Identification of Cd36 (Fat) as an insulin-resistance gene causing defective fatty acid and glucose metabolism in hypertensive rats. *Nat Genet* **21**(1), 76–83.

Aleman, T. S., Cideciyan, A. V., Windsor, E. A., et al. (2007). Macular pigment and lutein supplementation in ABCA4-associated retinal degenerations. *Invest Ophthalmol Vis Sci* **48**(3), 1319–29.

Allikmets, R., Seddon, J. M., Bernstein, P. S., et al. (1999). Evaluation of the Best disease gene in patients with age-related macular degeneration and other maculopathies. *Hum Genet* **104**(6), 449–53.

Allikmets, R., Shroyer, N. F., Singh, N., et al. (1997). Mutation of the Stargardt disease gene (ABCR) in age-related macular degeneration. *Science* **277**(5333), 1805–7.

Amengual, J., Lobo, G. P., Golczak, M., et al. (2011). A mitochondrial enzyme degrades carotenoids and protects against oxidative stress. *FASEB J* **25**(3), 948–59.

Anderson, D. H., Ozaki, S., Nealon, M., et al. (2001). Local cellular sources of apolipoprotein E in the human retina and retinal pigmented epithelium: implications for the process of drusen formation. *Am J Ophthalmol* **131**(6), 767–81.

AREDS Research Group. (1999). The Age-Related Eye Disease Study (AREDS): design implications. AREDS report no. 1. *Control Clin Trials* **20**(6), 573–600.

Baird, P. N., Guida, E., Chu, D. T., et al. (2004). The epsilon2 and epsilon4 alleles of the apolipoprotein gene are associated with age-related macular degeneration. *Invest Ophthalmol Vis Sci* **45**(5), 1311–5.

Basu, H. N., Del Vecchio, A. J., Flider, F., et al. (2001). Nutritional and potentential disease prevention properties of carotenoids. *J Am Oil Chem Soc* **78**, 665–75.

Beatty, S., Boulton, M., Henson, D., et al. (1999). Macular pigment and age related macular degeneration. *Br J Ophthalmol* **83**(7), 867–77.

Beatty, S., Koh, H., Phil, M., et al. (2000). The role of oxidative stress in the pathogenesis of age-related macular degeneration. *Surv Ophthalmol* **45**(2), 115–34.

Bernstein, P. S., Balashov, N. A., Tsong, E. D., et al. (1997). Retinal tubulin binds macular carotenoids. *Invest Ophthalmol Vis Sci* **38**(1), 167–75.

Bernstein, P. S., Delori, F. C., Richer, S., et al. (2010). The value of measurement of macular carotenoid pigment optical densities and distributions in age-related macular degeneration and other retinal disorders. *Vision Res* **50**(7), 716–28.

Bernstein, P. S., Khachik, F., Carvalho, L. S., et al. (2001). Identification and quantitation of carotenoids and their metabolites in the tissues of the human eye. *Exp Eye Res* **72**(3), 215–23.

Bernstein, P. S., Leppert, M., Singh, N., et al. (2002). Genotype-phenotype analysis of ABCR variants in macular degeneration probands and siblings. *Invest Ophthalmol Vis Sci* **43**(2), 466–73.

Bhosale, P., and Bernstein, P. S. (2005). Synergistic effects of zeaxanthin and its binding protein in the prevention of lipid membrane oxidation. *Biochim Biophys Acta* **1740**(2), 116–21.

Bhosale, P., Larson, A. J., Frederick, J. M., et al. (2004). Identification and characterization of a Pi isoform of glutathione S-transferase (GSTP1) as a zeaxanthin-binding protein in the macula of the human eye. *J Biol Chem* **279**(47), 49447–54.

Bhosale, P., Li, B., Sharifzadeh, M., et al. (2009). Purification and partial characterization of a lutein-binding protein from human retina. *Biochemistry* **48**(22), 4798–807.

Bone, R. A., Landrum, J. T., Friedes, L. M., et al. (1997). Distribution of lutein and zeaxanthin stereoisomers in the human retina. *Exp Eye Res* **64**(2), 211–8.

Bone, R. A., Landrum, J. T., Guerra, L. H., et al. (2003). Lutein and zeaxanthin dietary supplements raise macular pigment density and serum concentrations of these carotenoids in humans. *J Nutr* **133**(4), 992–8.

Bone, R. A., Landrum, J. T., and Tarsis, S. L. (1985). Preliminary identification of the human macular pigment. *Vision Res* **25**(11), 1531–5.

Borel, P., de Edelenyi, F. S., Vincent-Baudry, S., et al. (2011). Genetic variants in BCMO1 and CD36 are associated with plasma lutein concentrations and macular pigment optical density in humans. *Ann Med* **43**(1), 47–59.

Brion, M., Sanchez-Salorio, M., Corton, M., et al. (2010). Genetic association study of age-related macular degeneration in the Spanish population. *Acta Ophthalmol* **89**(1), e12–22.

Casaroli-Marano, R. P., Peinado-Onsurbe, J., Reina, M., et al. (1996). Lipoprotein lipase in highly vascularized structures of the eye. *J Lipid Res* **37**(5), 1037–44.

Castaneda, M. P., Hirschler, E. M., and Sams, A. R. (2005). Skin pigmentation evaluation in broilers fed natural and synthetic pigments. *Poult Sci* **84**(1), 143–7.

Chen, W., Stambolian, D., Edwards, A. O., et al. (2010). Genetic variants near TIMP3 and high-density lipoprotein-associated loci influence susceptibility to age-related macular degeneration. *Proc Natl Acad Sci USA* **107**(16), 7401–6.

Cho, E., Hankinson, S. E., Rosner, B., et al. (2008). Prospective study of lutein/zeaxanthin intake and risk of age-related macular degeneration. *Am J Clin Nutr* **87**(6), 1837–43.

Chucair, A. J., Rotstein, N. P., Sangiovanni, J. P., et al. (2007). Lutein and zeaxanthin protect photoreceptors from apoptosis induced by oxidative stress: relation with docosahexaenoic acid. *Invest Ophthalmol Vis Sci* **48**(11), 5168–77.

Colak, E., Kosanovic-Jakovic, N., Zoric, L., et al. (2011). The association of lipoprotein parameters and C-reactive protein in patients with age-related macular degeneration. *Ophthalmic Res* **46**(3), 125–32.

Congdon, N., O'Colmain, B., Klaver, C. C. W., et al. (2004). Causes and prevalence of visual impairment among adults in the United States. *Arch Ophthalmol* **122**(4), 477–85.

Connor, W. E., Duell, P. B., Kean, R., et al. (2007). The prime role of HDL to transport lutein into the retina: evidence from HDL-deficient WHAM chicks having a mutant ABCA1 transporter. *Invest Ophthalmol Vis Sci* **48**(9), 4226–31.

Constantineau, J., Greason, E., West, M., et al. (2010). A synonymous variant in scavenger receptor, class B, type I gene is associated with lower SR-BI protein expression and function. *Atherosclerosis* **210**(1), 177–82.

Cook, J. E., and Becker, D. L. (1995). Gap junctions in the vertebrate retina. *Microsc Res Tech* **31**(5), 408–19.

Crabtree, D. V., Ojima, I., Geng, X., et al. (2001). Tubulins in the primate retina: evidence that xanthophylls may be endogenous ligands for the paclitaxel-binding site. *Bioorg Med Chem* **9**(8), 1967–76.

Cumming, R. C., Lightfoot, J., Beard, K., et al. (2001). Fanconi anemia group C protein prevents apoptosis in hematopoietic cells through redox regulation of GSTP1. *Nat Med* **7**(7), 814–20.

Dbouk, H. A., Mroue, R. M., El-Sabban, M. E., et al. (2009). Connexins: a myriad of functions extending beyond assembly of gap junction channels. *Cell Commun Signal* 7:4, doi: 10.1186/1478-811X-7-4.

De La Paz, M. A., Guy, V. K., Abou-Donia, S., et al. (1999). Analysis of the Stargardt disease gene (ABCR) in age-related macular degeneration. *Ophthalmology* **106**(8), 1531–6.

Delcourt, C., Carriere, I., Delage, M., et al. (2006). Plasma lutein and zeaxanthin and other carotenoids as modifiable risk factors for age-related maculopathy and cataract: the POLA Study. *Invest Ophthalmol Vis Sci* **47**(6), 2329–35.

Demmig-Adams, B., and Adams, W. W., 3rd. (2002). Antioxidants in photosynthesis and human nutrition. *Science* **298**(5601), 2149–53.

Duncan, K. G., Bailey, K. R., Kane, J. P., et al. (2002). Human retinal pigment epithelial cells express scavenger receptors BI and BII. *Biochem Biophys Res Commun* **292**(4), 1017–22.

Duncan, K. G., Hosseini, K., Bailey, K. R., et al. (2009). Expression of reverse cholesterol transport proteins ATP-binding cassette A1 (ABCA1) and scavenger receptor BI (SR-BI) in the retina and retinal pigment epithelium. *Br J Ophthalmol* **93**(8), 1116–20.

During, A., Doraiswamy, S., and Harrison, E. H. (2008). Xanthophylls are preferentially taken up compared with beta-carotene by retinal cells via a SRBI-dependent mechanism. *J Lipid Res* **49**(8), 1715–24.

Eriksson, J., Larson, G., Gunnarsson, U., et al. (2008). Identification of the yellow skin gene reveals a hybrid origin of the domestic chicken. *PLoS Genet* **4**(2), e1000010.

Fauser, S., Smailhodzic, D., Caramoy, A., et al. (2011). Evaluation of serum lipid concentrations and genetic variants at high-density lipoprotein metabolism loci and TIMP3 in age-related macular degeneration. *Invest Ophthalmol Vis Sci* **52**(8), 5525–8.

Fernandez-Robredo, P., Recalde, S., Arnaiz, G., et al. (2009). Effect of zeaxanthin and antioxidant supplementation on vascular endothelial growth factor (VEGF) expression in apolipoprotein-E deficient mice. *Curr Eye Res* **34**(7), 543–52.

Ferrucci, L., Perry, J. R., Matteini, A., et al. (2009). Common variation in the beta-carotene 15,15'-monooxygenase 1 gene affects circulating levels of carotenoids: a genome-wide association study. *Am J Hum Genet* **84**(2), 123–33.

Francis, P. J., Hamon, S. C., Ott, J., et al. (2009). Polymorphisms in C2, CFB and C3 are associated with progression to advanced age related macular degeneration associated with visual loss. *J Med Genet* **46**(5), 300–7.

Friedman, D. S., O'Colmain, B. J., Munoz, B., et al. (2004). Prevalence of age-related macular degeneration in the United States. *Arch Ophthalmol* **122**(4), 564–72.

Fuijkschot, J., Cruysberg, J. R., Willemsen, M. A., et al. (2008). Subclinical changes in the juvenile crystalline macular dystrophy in Sjogren-Larsson syndrome detected by optical coherence tomography. *Ophthalmology* **115**(5), 870–5.

Funk, R., and Rohen, J. W. (1985). Comparative morphological studies on blood vessels in eyes of normotensive and spontaneously hypertensive rats. *Exp Eye Res* **40**(2), 191–203.

Gale, C. R., Hall, N. F., Phillips, D. I., et al. (2003). Lutein and zeaxanthin status and risk of age-related macular degeneration. *Invest Ophthalmol Vis Sci* **44**(6), 2461–5.

Giovannucci, D. R., and Stephenson, R. S. (1999). Identification and distribution of dietary precursors of the *Drosophila* visual pigment chromophore: analysis of carotenoids in wild type and ninaD mutants by HPLC. *Vision Res* **39**(2), 219–29.

Guven, M., Gorgun, E., Unal, M., et al. (2011). Glutathione S-transferase M1, GSTT1 and GSTP1 genetic polymorphisms and the risk of age-related macular degeneration. *Ophthalmic Res* **46**(1), 31–7.

Hammond, B. R., Jr., Johnson, E. J., Russell, R. M., et al. (1997a). Dietary modification of human macular pigment density. *Invest Ophthalmol Vis Sci* **38**(9), 1795–801.

Hammond, B. R., Jr., Wooten, B. R., and Snodderly, D. M. (1997b). Individual variations in the spatial profile of human macular pigment. *J Opt Soc Am A Opt Image Sci Vis* **14**(6), 1187–96.

Handelman, G. J., Dratz, E. A., Reay, C. C., et al. (1988). Carotenoids in the human macula and whole retina. *Invest Ophthalmol Vis Sci* **29**(6), 850–5.

Handelman, G. J., Snodderly, D. M., Krinsky, N. I., et al. (1991). Biological control of primate macular pigment. Biochemical and densitometric studies. *Invest Ophthalmol Vis Sci* **32**(2), 257–67.

Heiba, I. M., Elston, R. C., Klein, B. E., et al. (1994). Sibling correlations and segregation analysis of age-related maculopathy: the Beaver Dam Eye Study. *Genet Epidemiol* **11**(1), 51–67.

Herbeth, B., Gueguen, S., Leroy, P., et al. (2007). The lipoprotein lipase serine 447 stop polymorphism is associated with altered serum carotenoid concentrations in the Stanislas Family Study. *J Am Coll Nutr* **26**(6), 655–62.

Ho, L., van Leeuwen, R., Witteman, J. C., et al. (2011). Reducing the genetic risk of age-related macular degeneration with dietary antioxidants, zinc, and omega-3 fatty acids: the Rotterdam study. *Arch Ophthalmol* **129**(6), 758–66.

Hoang, Q. V., Qian, H., and Ripps, H. (2010). Functional analysis of hemichannels and gap-junctional channels formed by connexins 43 and 46. *Mol Vis* **16**, 1343–52.

Horie, S., Okuda, C., Yamashita, T., et al. (2010). Purified canola lutein selectively inhibits specific isoforms of mammalian DNA polymerases and reduces inflammatory response. *Lipids* **45**(8), 713–21.

Hornstein, E. P., Verweij, J., Li, P. H., et al. (2005). Gap-junctional coupling and absolute sensitivity of photoreceptors in macaque retina. *J Neurosci* **25**(48), 11201–9.

Houssier, M., Raoul, W., Lavalette, S., et al. (2008). CD36 deficiency leads to choroidal involution via COX2 down-regulation in rodents. *PLoS Med* **5**(2), e39.

Johnson, E. J., Neuringer, M., Russell, R. M., et al. (2005). Nutritional manipulation of primate retinas, III: effects of lutein or zeaxanthin supplementation on adipose tissue and retina of xanthophyll-free monkeys. *Invest Ophthalmol Vis Sci* **46**(2), 692–702.

Kiefer, C., Sumser, E., Wernet, M. F., et al. (2002). A class B scavenger receptor mediates the cellular uptake of carotenoids in *Drosophila*. *Proc Natl Acad Sci USA* **99**(16), 10581–6.

Kim, S.R., Nakanishi, K., Itagaki, Y., Sparrow, J. R. (2006). Photooxidation of A2-PE, a photoreceptor outer segment fluorophore, and protection by lutein and zeaxanthin. *Exp Eye Res.*, 82(5):828–39.

Klaver, C. C., Kliffen, M., van Duijn, C. M., et al. (1998a). Genetic association of apolipoprotein E with age-related macular degeneration. *Am J Hum Genet* **63**(1), 200–6.

Klaver, C. C., Wolfs, R. C., Assink, J. J., et al. (1998b). Genetic risk of age-related maculopathy. Population-based familial aggregation study. *Arch Ophthalmol* **116**(12), 1646–51.

Klein, M. L., Mauldin, W. M., and Stoumbos, V. D. (1994). Heredity and age-related macular degeneration. Observations in monozygotic twins. *Arch Ophthalmol* **112**(7), 932–7.

Klein, R., Chou, C. F., Klein, B. E., et al. (2011). Prevalence of age-related macular degeneration in the US population. *Arch Ophthalmol* **129**(1), 75–80.

Kojima, A., Nakahama, K., Ohno-Matsui, K., et al. (2008). Connexin 43 contributes to differentiation of retinal pigment epithelial cells via cyclic AMP signaling. *Biochem Biophys Res Commun* **366**(2), 532–8.

Kondo, N., Honda, S., Kuno, S., et al. (2009). Positive association of common variants in CD36 with neovascular age-related macular degeneration. *Aging (Albany NY)* **1**(2), 266–74.

Krinsky, N. I., Landrum, J. T., and Bone, R. A. (2003). Biologic mechanisms of the protective role of lutein and zeaxanthin in the eye. *Annu Rev Nutr* **23**, 171–201.

Lakkaraju, A., Finnemann, S. C., and Rodriguez-Boulan, E. (2007). The lipofuscin fluorophore A2E perturbs cholesterol metabolism in retinal pigment epithelial cells. *Proc Natl Acad Sci USA* **104**(26), 11026–31.

Lambooij, A. C., van Wely, K. H., Lindenbergh-Kortleve, D. J., et al. (2003). Insulin-like growth factor-I and its receptor in neovascular age-related macular degeneration. *Invest Ophthalmol Vis Sci* **44**(5), 2192–8.

Leung, I. Y., Sandstrom, M. M., Zucker, C. L., et al. (2004). Nutritional manipulation of primate retinas, II: effects of age, n-3 fatty acids, lutein, and zeaxanthin on retinal pigment epithelium. *Invest Ophthalmol Vis Sci* **45**(9), 3244–56.

Leung, I. Y., Sandstrom, M. M., Zucker, C. L., et al. (2005). Nutritional manipulation of primate retinas. IV. Effects of n-3 fatty acids, lutein, and zeaxanthin on S-cones and rods in the foveal region. *Exp Eye Res* **81**(5), 513–29.

Leung, W. C., Hessel, S., Meplan, C., et al. (2009). Two common single nucleotide polymorphisms in the gene encoding beta-carotene 15,15'-monoxygenase alter beta-carotene metabolism in female volunteers. *FASEB J* **23**(4), 1041–53.

Li, B., Ahmed, F., and Bernstein, P. S. (2010). Studies on the singlet oxygen scavenging mechanism of human macular pigment. *Arch Biochem Biophys* **504**(1), 56–60.

Li, B., Vachali, P., Frederick, J. M., et al. (2011). Identification of StARD3 as a lutein-binding protein in the macula of the primate retina. *Biochemistry* **50**(13), 2541–9.

Lian, F., Smith, D. E., Ernst, H., et al. (2007). Apo-10'-lycopenoic acid inhibits lung cancer cell growth *in vitro*, and suppresses lung tumorigenesis in the A/J mouse model *in vivo*. *Carcinogenesis* **28**(7), 1567–74.

Lian, F., and Wang, X. D. (2008). Enzymatic metabolites of lycopene induce Nrf2-mediated expression of phase II detoxifying/antioxidant enzymes in human bronchial epithelial cells. *Int J Cancer* **123**(6), 1262–8.

Liew, S. H., Gilbert, C. E., Spector, T. D., et al. (2005). Heritability of macular pigment: a twin study. *Invest Ophthalmol Vis Sci* **46**(12), 4430–6.

Lindqvist, A., He, Y. G., and Andersson, S. (2005). Cell type-specific expression of beta-carotene 9',10'-monooxygenase in human tissues. *J Histochem Cytochem* **53**(11), 1403–12.

Loane, E., McKay, G. J., Nolan, J. M., et al. (2010a). Apolipoprotein E genotype is associated with macular pigment optical density. *Invest Ophthalmol Vis Sci* **51**(5), 2636–43.

Loane, E., Nolan, J. M., and Beatty, S. (2010b). The respective relationships between lipoprotein profile, macular pigment optical density, and serum concentrations of lutein and zeaxanthin. *Invest Ophthalmol Vis Sci* **51**(11), 5897–905.

Loane, E., Nolan, J. M., McKay, G. J., et al. (2011). The association between macular pigment optical density and CFH, ARMS2, C2/BF, and C3 genotype. *Exp Eye Res* **93**(5), 592–8.

Lofqvist, C., Willett, K. L., Aspegren, O., et al. (2009). Quantification and localization of the IGF/insulin system expression in retinal blood vessels and neurons during oxygen-induced retinopathy in mice. *Invest Ophthalmol Vis Sci* **50**(4), 1831–7.

Machalinska, A., Klos, P., Safranow, K., et al. (2011). Neural stem/progenitor cells circulating in peripheral blood of patients with neovascular form of AMD: a novel view on pathophysiology. *Graefes Arch Clin Exp Ophthalmol* **249**(12), 1785–94.

Maguire, M. G., Ying, G. S., McCannel, C. A., et al. (2009). Statin use and the incidence of advanced age-related macular degeneration in the Complications of Age-related Macular Degeneration Prevention Trial. *Ophthalmology* **116**(12), 2381–5.

Malfait, M., Gomez, P., van Veen, T. A., et al. (2001). Effects of hyperglycemia and protein kinase C on connexin43 expression in cultured rat retinal pigment epithelial cells. *J Membr Biol* **181**(1), 31–40.

Matsumoto, A., Mizukami, H., Mizuno, S., et al. (2007). Beta-cryptoxanthin, a novel natural RAR ligand, induces ATP-binding cassette transporters in macrophages. *Biochem Pharmacol* **74**(2), 256–64.

McKay, G. J., Patterson, C. C., Chakravarthy, U., et al. (2011). Evidence of association of APOE with age-related macular degeneration: a pooled analysis of 15 studies. *Hum Mutat* **32**(12), 1407–16.

Mein, J. R., Dolnikowski, G. G., Ernst, H., et al. (2011). Enzymatic formation of apo-carotenoids from the xanthophyll carotenoids lutein, zeaxanthin and beta-cryptoxanthin by ferret carotene-9',10'-monooxygenase. *Arch Biochem Biophys* **506**(1), 109–21.

Mein, J. R., Lian, F., and Wang, X. D. (2008). Biological activity of lycopene metabolites: implications for cancer prevention. *Nutr Rev* **66**(12), 667–83.

Molday, L. L., Rabin, A. R., and Molday, R. S. (2000). ABCR expression in foveal cone photoreceptors and its role in Stargardt macular dystrophy. *Nat Genet* **25**(3), 257–8.

Neale, B. M., Fagerness, J., Reynolds, R., et al. (2010). Genome-wide association study of advanced age-related macular degeneration identifies a role of the hepatic lipase gene (LIPC). *Proc Natl Acad Sci USA* **107**(16), 7395–400.

Nussbaum, J. J., Pruett, R. C., and Delori, F. C. (1981). Historic perspectives. Macular yellow pigment. The first 200 years. *Retina* **1**(4), 296–310.

Oz, O., Aras Ates, N., Tamer, L., et al. (2006). Glutathione S-transferase M1, T1, and P1 gene polymorphism in exudative age-related macular degeneration: a preliminary report. *Eur J Ophthalmol* **16**(1), 105–10.

Peponis, V., Chalkiadakis, S. E., Bonovas, S., et al. (2010). The controversy over the association between statins use and progression of age-related macular degeneration: a mini review. *Clin Ophthalmol* **4**, 865–9.

Picard, E., Houssier, M., Bujold, K., et al. (2010). CD36 plays an important role in the clearance of oxLDL and associated age-dependent sub-retinal deposits. *Aging (Albany NY)* **2**(12), 981–9.

Prakash, P., Liu, C., Hu, K. Q., et al. (2004). Beta-carotene and beta-apo-14'-carotenoic acid prevent the reduction of retinoic acid receptor beta in benzo[a]pyrene-treated normal human bronchial epithelial cells. *J Nutr* **134**(3), 667–73.

Provost, A. C., Vede, L., Bigot, K., et al. (2009). Morphologic and electroretinographic phenotype of SR-BI knockout mice after a long-term atherogenic diet. *Invest Ophthalmol Vis Sci* **50**(8), 3931–42.

Rapp, L. M., Maple, S. S., and Choi, J. H. (2000). Lutein and zeaxanthin concentrations in rod outer segment membranes from perifoveal and peripheral human retina. *Invest Ophthalmol Vis Sci* **41**(5), 1200–9.

Reuter, T. Y., Medhurst, A. L., Waisfisz, Q., et al. (2003). Yeast two-hybrid screens imply involvement of Fanconi anemia proteins in transcription regulation, cell signaling, oxidative metabolism, and cellular transport. *Exp Cell Res* **289**(2), 211–21.

Rizzo, W. B., and Carney, G. (2005). Sjogren-Larsson syndrome: diversity of mutations and polymorphisms in the fatty aldehyde dehydrogenase gene (ALDH3A2). *Hum Mutat* **26**(1), 1–10.

SanGiovanni, J. P., Chew, E. Y., Clemons, T. E., et al. (2007). The relationship of dietary carotenoid and vitamin A, E, and C intake with age-related macular degeneration in a case-control study: AREDS Report No. 22. *Arch Ophthalmol* **125**(9), 1225–32.

SanGiovanni, J. P., and Neuringer, M. (2012). The putative role of lutein and zeaxanthin as protective agents against age-related macular degeneration: promise of molecular genetics for guiding mechanistic and translational research in the field. *Am J Clin Nutr* **96**(5), 1223S–33S.

Seddon, J. M., Ajani, U. A., and Mitchell, B. D. (1997). Familial aggregation of age-related maculopathy. *Am J Ophthalmol* **123**(2), 199–206.

Seddon, J. M., Ajani, U. A., Sperduto, R. D., et al. (1994). Dietary carotenoids, vitamins A, C, and E, and advanced age-related macular degeneration. Eye Disease Case-Control Study Group. *JAMA* **272**(18), 1413–20.

Seddon, J. M., Cote, J., Page, W. F., et al. (2005). The US twin study of age-related macular degeneration: relative roles of genetic and environmental influences. *Arch Ophthalmol* **123**(3), 321–7.

Serpeloni, J. M., Grotto, D., Mercadante, A. Z., et al. (2010). Lutein improves antioxidant defense *in vivo* and protects against DNA damage and chromosome instability induced by cisplatin. *Arch Toxicol* **84**(10), 811–22.

Shroyer, N. F., Lewis, R. A., Allikmets, R., et al. (1999). The rod photoreceptor ATP-binding cassette transporter gene, ABCR, and retinal disease: from monogenic to multifactorial. *Vision Res* **39**(15), 2537–44.

Smith, L. E. (2005). IGF-1 and retinopathy of prematurity in the preterm infant. *Biol Neonate* **88**(3), 237–44.

Snellen, E. L., Verbeek, A. L., Van Den Hoogen, G. W., et al. (2002). Neovascular age-related macular degeneration and its relationship to antioxidant intake. *Acta Ophthalmol Scand* **80**(4), 368–71.

Snodderly, D. M. (1995). Evidence for protection against age-related macular degeneration by carotenoids and antioxidant vitamins. *Am J Clin Nutr* **62**(6 Suppl), 1448S–61S.

Snodderly, D. M., Auran, J. D., and Delori, F. C. (1984). The macular pigment. II. Spatial distribution in primate retinas. *Invest Ophthalmol Vis Sci* **25**(6), 674–85.

Sohl, G., Joussen, A., Kociok, N., et al. (2010). Expression of connexin genes in the human retina. *BMC Ophthalmol* **10**, 27.

Sommerburg, O. G., Siems, W. G., Hurst, J. S., et al. (1999). Lutein and zeaxanthin are associated with photoreceptors in the human retina. *Curr Eye Res* **19**(6), 491–5.

Souied, E. H., Benlian, P., Amouyel, P., et al. (1998). The epsilon4 allele of the apolipoprotein E gene as a potential protective factor for exudative age-related macular degeneration. *Am J Ophthalmol* **125**(3), 353–9.

Sparrow, J. R., and Boulton, M. (2005). RPE lipofuscin and its role in retinal pathobiology. *Exp Eye Res* **80**(5), 595–606.

Stahl, W., Nicolai, S., Briviba, K., et al. (1997). Biological activities of natural and synthetic carotenoids: induction of gap junctional communication and singlet oxygen quenching. *Carcinogenesis* **18**(1), 89–92.

Sunness, J. S., Bressler, N. M., Tian, Y., et al. (1999). Measuring geographic atrophy in advanced age-related macular degeneration. *Invest Ophthalmo Vis Sci* **40**(8), 1761–9.

Swaroop, A., Chew, E. Y., Rickman, C. B., et al. (2009). Unraveling a multifactorial late-onset disease: from genetic susceptibility to disease mechanisms for age-related macular degeneration. *Annu Rev Genomics Hum Genet* **10**, 19–43.

Tan, J. S., Wang, J. J., Flood, V., et al. (2008). Dietary antioxidants and the long-term incidence of age-related macular degeneration: the Blue Mountains Eye Study. *Ophthalmology* **115**(2), 334–41.

The Eye Disease Case-Control Study Group. (1992). Risk factors for neovascular age-related macular degeneration. *Arch Ophthalmol* **110**(12), 1701–8.

Tibaduiza, E. C., Fleet, J. C., Russell, R. M., et al. (2002). Excentric cleavage products of beta-carotene inhibit estrogen receptor positive and negative breast tumor cell growth *in vitro* and inhibit activator protein-1-mediated transcriptional activation. *J Nutr* **132**(6), 1368–75.

Tikellis, G., Sun, C., Gorin, M. B., et al. (2007). Apolipoprotein e gene and age-related maculopathy in older individuals: the cardiovascular health study. *Arch Ophthalmol* **125**(1), 68–73.

Tserentsoodol, N., Gordiyenko, N. V., Pascual, I., et al. (2006). Intraretinal lipid transport is dependent on high density lipoprotein-like particles and class B scavenger receptors. *Mol Vis* **12**, 1319–33.

Udawatte, C., Qian, H., Mangini, N. J., et al. (2008). Taurine suppresses the spread of cell death in electrically coupled RPE cells. *Mol Vis* **14**, 1940–50.

Vage, D. I., and Boman, I. A. (2010). A nonsense mutation in the beta-carotene oxygenase 2 (BCO2) gene is tightly associated with accumulation of carotenoids in adipose tissue in sheep (Ovis aries). *BMC Genet* **11**, 10.

van der Veen, R. L., Fuijkschot, J., Willemsen, M. A., et al. (2010). Patients with Sjogren-Larsson syndrome lack macular pigment. *Ophthalmology* **117**(5), 966–71.

Vaney, D. I. (1991). Many diverse types of retinal neurons show tracer coupling when injected with biocytin or Neurobiotin. *Neurosci Lett* **125**(2), 187–90.

Vaney, D. I. (2002). Retinal neurons: cell types and coupled networks. *Prog Brain Res* **136**, 239–54.

Voolstra, O., Kiefer, C., Hoehne, M., et al. (2006). The *Drosophila* class B scavenger receptor NinaD-I is a cell surface receptor mediating carotenoid transport for visual chromophore synthesis. *Biochemistry* **45**(45), 13429–37.

Wang, M., Tsao, R., Zhang, S., et al. (2006). Antioxidant activity, mutagenicity/anti-mutagenicity, and clastogenicity/anti-clastogenicity of lutein from marigold flowers. *Food Chem Toxicol* **44**(9), 1522–9.

Wenzel, A. J., Sheehan, J. P., Burke, J. D., et al. (2007). Dietary intake and serum concentrations of lutein and zeaxanthin, but not macular pigment optical density, are related in spouses. *Nutr Res* **27**(8), 462–9.

Whitehead, A. J., Mares, J. A., and Danis, R. P. (2006). Macular pigment: a review of current knowledge. *Arch Ophthalmol* **124**(7), 1038–45.

Whitmire, W., Al-Gayyar, M. M., Abdelsaid, M., et al. (2011). Alteration of growth factors and neuronal death in diabetic retinopathy: what we have learned so far. *Mol Vis* **17**, 300–8.

Willemsen, M. A., Cruysberg, J. R., Rotteveel, J. J., et al. (2000). Juvenile macular dystrophy associated with deficient activity of fatty aldehyde dehydrogenase in Sjogren-Larsson syndrome. *Am J Ophthalmol* **130**(6), 782–9.

Winum, J. Y., Kamal, M., Defacque, H., et al. (1997). Synthesis and biological activities of higher homologues of retinoic acid. *Farmaco* **52**(1), 39–42.

Wong, T. Y., Shankar, A., Klein, R., et al. (2006). Apolipoprotein E gene and early age-related maculopathy: the Atherosclerosis Risk in Communities Study. *Ophthalmology* **113**(2), 255–9.

Ye, J. H., Ryu, S. B., Kim, K. H., et al. (2008). Functional connectivity map of retinal ganglion cells for retinal prosthesis. *Korean J Physiol Pharmacol* **12**(6), 307–14.

Yemelyanov, A. Y., Katz, N. B., and Bernstein, P. S. (2001). Ligand-binding characterization of xanthophyll carotenoids to solubilized membrane proteins derived from human retina. *Exp Eye Res* **72**(4), 381–92.

Yu, Y., Bhangale, T. R., Fagerness, J., et al. (2011a). Common variants near FRK/COL10A1 and VEGFA are associated with advanced age-related macular degeneration. *Hum Mol Genet* **20**(18), 3699–709.

Yu, Y., Reynolds, R., Fagerness, J., et al. (2011b). Association of variants in the LIPC and ABCA1 genes with intermediate and large drusen and advanced age-related macular degeneration. *Invest Ophthalmol Vis Sci* **52**(7), 4663–70.

Zareparsi, S., Buraczynska, M., Branham, K. E., et al. (2005). Toll-like receptor 4 variant D299G is associated with susceptibility to age-related macular degeneration. *Hum Mol Genet* **14**(11), 1449–55.

Zareparsi, S., Reddick, A. C., Branham, K. E., et al. (2004). Association of apolipoprotein E alleles with susceptibility to age-related macular degeneration in a large cohort from a single center. *Invest Ophthalmol Vis Sci* **45**(5), 1306–10.

Zeimer, M., Hense, H. W., Heimes, B., et al. (2009). The macular pigment: short- and intermediate-term changes of macular pigment optical density following supplementation with lutein and zeaxanthin and co-antioxidants. The LUNA Study. *Ophthalmologe* **106**(1), 29–36.

Zerbib, J., Seddon, J. M., Richard, F., et al. (2009). rs5888 variant of SCARB1 gene is a possible susceptibility factor for age-related macular degeneration. *PLoS One* **4**(10), e7341.

Zhang, L. X., Cooney, R. V., and Bertram, J. S. (1991). Carotenoids enhance gap junctional communication and inhibit lipid peroxidation in C3H/10T1/2 cells: relationship to their cancer chemopreventive action. *Carcinogenesis* **12**(11), 2109–14.

Zhao, X., Aldini, G., Johnson, E. J., et al. (2006). Modification of lymphocyte DNA damage by carotenoid supplementation in postmenopausal women. *Am J Clin Nutr* **83**(1), 163–9.

Zheng, W., Reem, R. E., Omarova, S., et al. (2012). Spatial distribution of the pathways of cholesterol homeostasis in human retina. *PLoS One* **7**(5), e37926.

Ziouzenkova, O., Orasanu, G., Sukhova, G., et al. (2007). Asymmetric cleavage of beta-carotene yields a transcriptional repressor of retinoid X receptor and peroxisome proliferator-activated receptor responses. *Mol Endocrinol* **21**(1), 77–88.

# 7 A Review of Recent Data on the Bioavailability of Lutein and Zeaxanthin

*Mareike Beck and Wolfgang Schalch*

## CONTENTS

7.1 Introduction ............................................................................................. 129
7.2 Experiments in Rhesus Monkeys ........................................................... 130
    7.2.1 Xanthophyll-Free Status of Animals ........................................... 130
    7.2.2 Plasma and Macular Pigment Optical Density Responses to Supplementation with Xanthophylls .......................................... 131
    7.2.3 Vulnerability to Blue Light .......................................................... 131
    7.2.4 Conclusions .................................................................................. 134
7.3 Comparison of Plasma Responses to Two Preparations Containing Either Esterified or Nonesterified Lutein ............................................... 134
    7.3.1 Esterified Versus Nonesterified ("Free") Lutein ........................... 134
    7.3.2 Plasma Lutein Responses ............................................................ 135
    7.3.3 Conclusions .................................................................................. 137
7.4 Comparison of Plasma Responses to Nonesterified Lutein Formulated by Two Different Processes ................................................ 138
    7.4.1 Starch Versus Alginate Matrix ..................................................... 138
    7.4.2 Design of Study ........................................................................... 139
    7.4.3 Results ........................................................................................... 140
    7.4.4 Conclusions .................................................................................. 141
References ....................................................................................................... 143

## 7.1 INTRODUCTION

Humans cannot biosynthesize carotenoids and are dependent on diet or dietary supplementation for a continuous provision of lutein and zeaxanthin, the two key xanthophylls of the human and nonhuman primate retina. To be enriched in the macula to exert their protective function against photooxidative damage and to contribute to lowering the risk of age-related macular degeneration (AMD), lutein and zeaxanthin must be released from the food matrix or from nutritional supplements, absorbed in the gastrointestinal tract and reach the systemic circulation to be transported into the retina. In other words, they must be bioavailable at the target organ. General bioavailability of xanthophylls, however, is not the topic of this chapter since it has been reviewed numerous times (more recent reviews being Thurnham 2007; Loane, Nolan et al. 2008; Maiani, Castón et al. 2009).

Rather, this chapter first highlights a study in rhesus monkeys that demonstrated how xanthophyll-free monkeys upon supplementation with lutein or zeaxanthin readily accumulated xanthophylls in plasma and retina and how replenishing their maculae with lutein and zeaxanthin restored protection of the macula from blue light damage. This protection had been lost because these animals had been raised on a semipurified, carotenoid-free diet since conception.

Secondly, we look into the results of two studies that compared plasma responses of healthy volunteers to lutein when supplemented with different commercially available lutein preparations. The first study investigated the differences in bioavailability between esterified and nonesterified lutein. The second study compared plasma lutein responses after supplementation with nonesterified lutein produced using two different formulation technologies.

Ingested xanthophylls can fulfill their protective function in the macula only if they are transported through the blood to their ultimate target organ, the retina. The evidence available to date suggests that the quantitative response of xanthophyll plasma concentrations to supplementation can be considered a surrogate biomarker for their augmentation in the retina (Bone and Landrum 2010; Loane, Nolan et al. 2010).

## 7.2 EXPERIMENTS IN RHESUS MONKEYS

### 7.2.1 Xanthophyll-Free Status of Animals

Investigations of the bioavailability of substances normally occurring in the food chain such as lutein and zeaxanthin are often hampered by incomplete or missing information on the plasma baseline concentrations and baseline fluctuation of the nutrients in question during the investigation. The use of animals depleted of specific nutrients circumvents these problems.

The Oregon National Primate Research Center raised rhesus monkeys from conception on a semipurified diet free of carotenoids. Consequently, these animals did not have any measurable concentrations of lutein and zeaxanthin in plasma or tissues including the macula. This provided two unique opportunities: first, the possibility to evaluate consequences of a life-long xanthophyll deficiency. Secondly, by supplementing these animals with *pure* lutein or *pure* zeaxanthin, it was possible to study the uptake of these xanthophylls into plasma and retina as well as their metabolism, particularly in the retina.

Monkeys are an important animal model for studies relevant to AMD, because they are the only animals that have a macula that is similar to that in humans and can develop drusen (Hope, Dawson et al. 1992) as well as AMD (Monaco and Wormington 1990; Borel, Grolier et al. 1996). In this section, we review a publication that describes the effects of xanthophyll depletion and repletion on the vulnerability to blue light damage of the retina of these monkeys (Barker, Snodderly et al. 2011). In addition to the other experiments done with this precious monkey colony (Neuringer, Sandstrom et al. 2004; Leung, Sandstrom et al. 2004; Johnson, Neuringer et al. 2005; Leung, Sandstrom et al. 2005; Albert, Hoeller et al. 2008), this blue-light experiment is a particularly relevant work in the context of the role of lutein and zeaxanthin in blue-light damage, known to be one of the important etiological factors for AMD (Algvere, Marshall et al. 2006).

### 7.2.2 PLASMA AND MACULAR PIGMENT OPTICAL DENSITY RESPONSES TO SUPPLEMENTATION WITH XANTHOPHYLLS

The "xanthophyll-free" rhesus monkeys were supplemented for 22–56 weeks with lutein or zeaxanthin at daily doses of 2.2 mg/kg body weight, equivalent to average doses of 20 mg/monkey/day. Concurrent with supplementation, the xanthophylls were avidly taken up and mean plasma concentrations increased from 0 to up to 1.2 µM during the first 4 weeks of supplementation (Figure 7.1a). During continued supplementation, mean plasma levels equilibrated at plateau values of around 0.8 µM, similarly for lutein and zeaxanthin. Macular pigment optical density (MPOD) increased (Figure 7.1b) as well but slower, reaching plateau levels only after 16 weeks, again similar for lutein and zeaxanthin. For further details, see Neuringer, Sandstrom et al. (2004).

### 7.2.3 VULNERABILITY TO BLUE LIGHT

The vulnerability of the retina to blue light was tested by controlled exposures to blue laser light at discrete small locations. Specifically, this was done by recording the energy delivered and measuring the area of the lesions that developed at the exposed retinal spots. There is a linear relationship between the exposure energy and the size of the developing lesion in the retina. For an exposed spot, the slope of the resulting line can be interpreted as reflecting its vulnerability to blue light, with larger slopes indicating higher vulnerabilities (Figure 7.2a and b).

Because of the physiological characteristics of the distribution of macular pigment in the retina, the density of the macular pigment is decreasing with increasing distance from the center of the macula. The area in the very center where macular pigment density is highest is called the fovea. More peripheral locations, with eccentricities between approximately 6° and 8°, belong to the parafovea, where MPOD normally is negligible.

Before and after supplementation, blue light vulnerability was compared between these two locations (i.e., fovea vs. parafovea). Statistical evaluation was done using the Generalized Estimating Equations (Zeger and Liang 1986).

Because of the complete absence of macular pigment, the vulnerability to blue light of xanthophyll-free monkeys was virtually identical in the fovea and in the parafovea, as shown in Figure 7.2a. After supplementation, this situation changed drastically with differences in vulnerability between fovea and parafovea becoming statistically significant, demonstrating lower susceptibility to damage in the fovea, where macular pigment had accumulated during supplementation. Figure 7.2b shows the effects supplementation had upon vulnerability in the fovea. As can been seen, the slope of the xanthophyll-free monkeys was significantly lowered by supplementation. Interestingly, the protection of the fovea resulting from this supplementation statistically was indistinguishable from that of rhesus monkeys raised on normal monkey chow and with normal yellow macular pigmentation. In other words, protection of

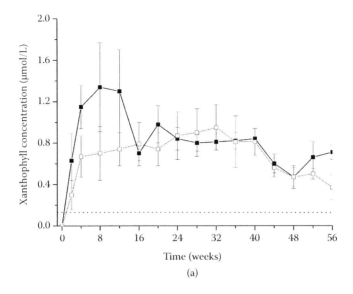

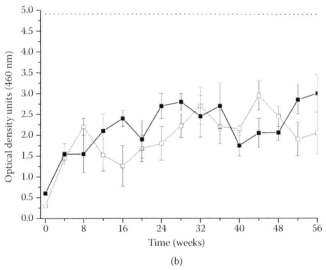

**FIGURE 7.1** (a) Time course of xanthophyll plasma concentrations (mean ± SEM) of originally xanthophyll-free monkeys supplemented with lutein (filled squares, solid line) or zeaxanthin (open squares, dotted line) over a period of 56 weeks. Overall and at any time point, differences between the supplementation groups are not statistically significant. The dashed straight line indicates the average plasma xanthophyll concentration of control (chow-fed) monkeys (approximately 0.1 µmol/L). (b) Time course of MPOD (mean ± SEM) of originally xanthophyll-free monkeys supplemented with lutein (filled squares, solid line) or zeaxanthin (open squares, dotted line) over a period of 56 weeks. Overall and at any time point, differences between the supplementation groups are not statistically significant. The dashed straight line indicates the average MPOD level of control (chow-fed) monkeys (approximately 4.9 optical density units).

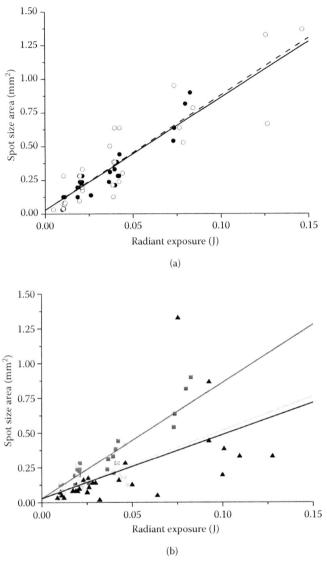

**FIGURE 7.2** (a) Blue light lesion area and corresponding radiation exposure energy in the *fovea* (*solid line*) and the *parafovea* (*dashed line*) for the 12 animals of this experiment *before* supplementation. The plot illustrates how increasing radiant exposure energies result in larger spot size lesion areas. Each data point represents the area of a single lesion (as measured on the fundus photographs) and its corresponding energy. Regression lines were computed by GEE (Generalized Estimating Equations) analysis. Note that the slope of the lines are indistinguishable from each other, because irradiation was done before supplementation and macular pigment is absent in the fovea. (b) Blue light lesion area and radiation exposure energy in the *fovea* of the same animals as in Figure 7.2a. Note the relatively steeper slope in the xanthophyll-free monkeys before supplementation (squares) and the relatively shallower slope in the normal (triangles) and supplemented (circles) monkeys.

the fovea was restored by supplementation to levels typical for normal monkeys. For more details, see Barker, Snodderly et al. (2011).

### 7.2.4 Conclusions

This is the first *in vivo* study to directly document that supplementation with xanthophylls can restore the macula's protection from blue light damage, when it had been lost, for example, as a consequence of xanthophyll depletion. Blue light damage to the retina is one of the important factors in the etiology of AMD and evidence is accumulating that lifetime exposure to blue light is related to increased risk of AMD, particularly in subjects with low levels of xanthophylls in plasma (Fletcher, Bentham et al. 2008).

## 7.3 COMPARISON OF PLASMA RESPONSES TO TWO PREPARATIONS CONTAINING EITHER ESTERIFIED OR NONESTERIFIED LUTEIN

### 7.3.1 Esterified Versus Nonesterified ("Free") Lutein

In foods of plant origin, lutein may occur as "free," nonesterified lutein (e.g., in green leafy or nonleafy vegetables) or as lutein esters (in fruits) (Rodriguez-Amaya 2010). In the petals of flowers like marigold, the main commercial source of lutein, lutein is almost exclusively present as lutein esters and only traces of free lutein are detectable. The hydroxyl groups of lutein are acylated with saturated fatty acids, mainly palmitic, myristic, and stearic acids forming mainly diesters, but monoesters are also found. The diesters can be symmetrical or nonsymetrical, containing different fatty acids. Thus, marigold lutein is a mixture of different lutein ester molecules (Breithaupt, Wirt et al. 2002), which is sold as such by some manufacturers, while others, in additional production steps, first cleave the fatty acid esters by saponification (i.e., treatment with alkali and heat) and then remove the fatty acids, finally providing relatively pure nonesterified or "free" lutein.

Absorption of lutein requires its release from the food or supplement matrix (emulsification), its incorporation into mixed micelles, and its uptake into the intestinal mucosa. Compared to unesterified lutein, lutein esters are less hydrophilic, and after emulsification tend to be located in the core of lipid droplets, whereas the more polar nonesterified lutein is thought to be preferentially located at their surface (Borel, Grolier et al. 1996). This may facilitate the transfer of free lutein into mixed micelles and accordingly its absorption as compared to lutein esters.

Despite the fact that by ingestion of fruits or vegetables a mixture of esterifed and free lutein is taken up, in human plasma, and in the retina, only free and no esterified lutein is found. This indicates that nonesterified lutein can be absorbed without prior processing, while lutein esters have to be cleaved in the gastrointestinal tract (by, e.g., bile salt–dependent lipase and/or cholesterol esterase) before absorption (Chitchumroonchokchai and Failla 2006).

This ester hydrolysis is thought to be mediated by lipases (bile salt–dependent lipase and/or cholesterol esterase), enzymes specialized to cleave fatty acid esters.

These differences in characteristics and processing of the lutein molecules during absorption made it interesting to compare the relative bioavailability of lutein esters and nonesterified (free) lutein. In this study, this was done by the comparison of two commercially available products as described in a recently published article (Norkus, Norkus et al. 2010).

### 7.3.2 Plasma Lutein Responses

In a randomized, parallel group study design, 36 volunteers/group received daily lutein doses of either 12.2 mg free lutein or 27 mg of lutein ester (determined as being equivalent to 13.5 mg free lutein) over 4 weeks. Fasting blood was obtained at baseline and following 7, 14, 21, and 28 days of supplementation. Supplements were consumed each day with a standard breakfast consisting of a portion of dry, ready-to-eat cereal and 8 oz of 2% milk.

Plasma samples were analyzed for lutein concentration by high-performance liquid chromatography (HPLC), and data were dose-normalized to compensate for the slightly different lutein doses (12.2 and 13.5 mg lutein equivalents per day from free lutein and lutein esters, respectively).

Changes in plasma lutein concentrations were larger in subjects supplemented with free lutein than in subjects who ingested lutein esters (see Figure 7.3). From day 21 onward, differences were statistically significant ($p = .0012$) and remained so until after 28 days of supplementation ($p = .0011$). The area under the concentration time curve ($AUC_{day\ 0-28}$) was more than 16% higher ($p < .02$) for free lutein (1558 day × nmol/(L × mg dose)) than for esterified lutein (1334 day × nmol/[L × mg dose]).

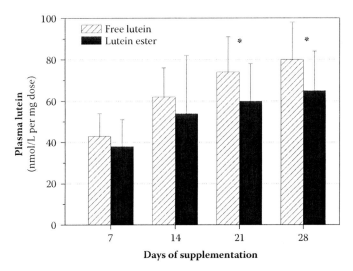

**FIGURE 7.3** Dose normalized increase in plasma lutein concentrations from baseline (mean + SD, $n = 36$) during 28 days of supplementation with free lutein (striated bars) or lutein esters (dark bars). Daily doses were 12.5 mg free lutein or 27 mg lutein esters (equivalent to 13.5 mg free lutein). The asterisks indicate statistical significance of difference ($p < .002$).

In multiple regression analyses, the baseline plasma lutein level and the form of lutein molecule ingested (free or esters) were found to significantly influence the lutein plasma response: both a higher initial plasma lutein concentration and ingesting the nonesterified lutein supplement predicted a greater plasma increase. In contrast, subject age, gender, body mass index (BMI), and plasma lipids had no effect.

Another interesting result of the study relates to the question of whether esterified lutein is present in plasma of human subjects who have ingested lutein esters. This question was investigated by HPLC analyses of unsaponified plasma samples at the end of the 1-month supplementation period, after the subjects from the lutein ester group had ingested cumulative doses of lutein ester of approximately 750 mg. As can be seen in Figure 7.4, no lutein esters but only free lutein could be detected in these plasma samples. This was independent from whether the subjects had ingested free lutein equivalents of 12.2 or 13.5 mg/day from free or esterified lutein, respectively. This strongly supports the evidence that only free and not esterified lutein is ultimately transported by the blood to the retina. It also suggests that free lutein, if generated from lutein ester or ingested as such, is not (re)esterified by metabolic transformations in the body.

In contrast to our results Bowen, Herbst-Espinosa et al. (2002) observed higher bioavailability of lutein esters in a crossover kinetic study with single doses of either free or esterified lutein, but the observed differences were not statistically significant. It is, however, difficult to compare these two studies because of their fundamentally different approaches.

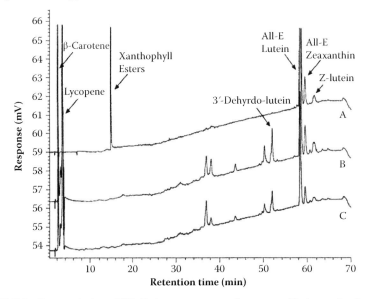

**FIGURE 7.4** Reversed-phase HPLC chromatograms of nonsaponified samples from capsules (A) or plasma (B and C). The chromatogram labeled A is of a mixture of free and esterified lutein to illustrate the retention time of xanthophyll esters in this chromatographic system. Chromatograms labeled B and C refer to plasma samples of subjects taking free, nonesterified (B) or esterified (C) lutein, respectively. Samples were taken at the end of the 28-week supplementation period.

The Bowen et al. results suggested that ester hydrolysis in the gut may not have been a limiting factor for absorption, at least not in the relatively young human subjects of their study. The age of their volunteers ranged between 22 and 35 years, while in our study the range was 23–52 years.

At present, limited data are available on the effect of age on plasma increases of fat-soluble compounds during supplementation. While our study results suggest that age was not a factor in plasma lutein increases, a negative correlation between age and plasma lutein levels after lutein ester supplementation has been reported by Chung, Rasmussen et al. (2004), who explained their findings by assuming a decline in lipase activity with age. The difference to our data may reflect the fact that Chung et al. had included even older subjects, with ages ranging up to 75 years.

Bone and Landrum (2010) reported smaller increases in lutein plasma concentrations with 20 mg/day of lutein esters in the older group as compared to a younger population tested. In this study, the efficacy of lutein esters to increase MPOD was investigated as well. The daily average increase of MPOD was at least twice as high in younger subjects than in older subjects, correlating well with the higher lutein plasma responses reached in the younger population. Similar differences in MPOD increases are also apparent when comparing the results of two other studies that both supplemented lutein for a duration of 6 months. Trieschmann, Beatty et al. (2007) and Zeimer, Hense et al. (2009) gave 12 mg/day of lutein to 100 subjects in esterified form and reported a 10%–15% increase in MPOD, while Stringham and Hammond (2008) supplemented 10 mg/day free lutein with an MPOD increase of 39% ($n = 40$).

Caimari, Oliver et al. (2008) and Van den Maagdenberg, Claeys et al. (2007) reported that in rats and boars, adipose and muscle lipase activities decrease with age. Other studies in experimental animals suggest that the exocrine pancreas, secreting digestive enzymes such as lipase and cholesterol esterase, possesses sufficient reserve to maintain normal digestive capacity throughout lifetime, but under conditions of dietary stress the aged pancreas may not retain the adaptive potential for quick and sufficient activity (Majumdar, Jaszewski et al. 1997). Whether or not aging affects carotenoid ester cleavage activity in humans remains to be resolved.

### 7.3.3 Conclusions

To date, our study is the largest study to examine the relative plasma response from free lutein versus lutein esters. The results support a statistically significant advantage of nonesterified lutein over the ester forms. Lutein plasma concentrations and AUCs were 10%–20% higher with the free form. In other words, higher plasma lutein steady-state levels were reached by ingestion of free lutein as opposed to lutein ester formulations.

On the basis of presently available evidence, it is difficult to decide whether there is an intrinsic difference in bioavailability between esterified and nonesterified lutein. Stringently, this could only be shown by a clinical study when applying identically formulated lutein ester and free lutein preparations, which are, however, difficult to prepare. The substantial heterogeneity of lutein esters, which are a mixture of many different molecules, adds particular complexity in this context. Furthermore, lutein can be present as *cis* or *trans* isomers. Other factors may have a similar or even higher importance in determining the bioavailability of lutein. Such factors include the amount of lutein consumed, the amount of simultaneous consumption of dietary fat, the characteristics of the matrix in which lutein is administered, not to forget individual factors such as age, health/nutritional status, and genetics, which all can influence the bioavailability of lutein. The age factor may be particularly important when considering xanthophyll supplementation of patients at risk of or with early AMD. Such patients are mainly elderly subjects, and if their capability to cleave lutein esters indeed would be reduced because of the age dependency of lipase activity, a prudent supplementation choice for these individuals would accordingly rather be free lutein. Whether the frequently documented decrease of MPOD with age (Nolan, Kenny et al. 2010) is also related to an age-related decline of lipase activity is unknown.

While age may thus be a factor influencing the individual bioavailability of xanthophylls, a considerably larger effect on lutein bioavailability had the final product formulation. Formulation of fat-soluble substances in general and xanthophylls in particular poses a number of challenges. The xanthophylls must be rendered dispersible in aqueous media and at the same time protected from oxidation. In Section 7.4, we focus on two different approaches to meet these challenges and on how these can have drastic influences on the bioavailability of xanthophylls.

## 7.4 COMPARISON OF PLASMA RESPONSES TO NONESTERIFIED LUTEIN FORMULATED BY TWO DIFFERENT PROCESSES

### 7.4.1 Starch versus Alginate Matrix

There are several approaches to deal with the two main challenges of the formulation of carotenoids, their insolubility in water and their sensitivity to oxidation. The Actilease technology developed by DSM Nutritional Products meets these challenges by embedding tiny microparticles of lutein in a protective hydrophilic matrix (starch). These characteristics facilitate dispersion in water and thus bioavailability. The embedding of lutein in the hydrocolloid ensures its protection from oxidation. One of the many alternative approaches, the "alginate" technology, uses a mixture of alginate, gum arabic, and pea starch as hydrocolloid. Alginic acid is a linear polysaccharide, the chains of which are noncovalently cross-linked by the addition of divalent cations (e.g., calcium). During this process of cross-link formation, the lutein can be considered to be "entrapped" in the matrix.

The investigated lutein products are two commercially available lutein preparations: Product A is manufactured by DSM (FloraGLO® Actilease® Lutein 5% CWS/S-TG beadlets) and product B is available from a competitor and contains lutein entrapped in a matrix of alginate and other ingredients. For both products, the source of lutein is extracts from marigold petals (*Tagetes erecta*) in which the lutein esters are cleaved through a saponification process. In addition to free lutein, both products also contain zeaxanthin from the marigolds. Relative to lutein, the zeaxanthin content was determined as 8.6% and 7.4% for products A and B, respectively.

While macroscopically and microscopically the two products appeared to be very similar, for example, in beadlet shape and size, they exhibited sharply contrasting differences in their dispersion characteristics (Figure 7.5). Product B was virtually

A Review of Recent Data on the Bioavailability of Lutein and Zeaxanthin 139

**FIGURE 7.5** (See color insert.) Top: Microscopic pictures of the two tested lutein products, A (left) and B (right). Bottom: Macroscopically evident differences in dispersion behavior: product A (left) disperses readily in artificial gastric juice (0.1 N HCl) while product B (right) does disperse very little if at all (pictures are taken within 1 minute after addition of 20 mg of the respective lutein product).

nondispersible, neither in water nor in artificial gastric juice (0.1 N HCl). Among other things, this latter characteristic precluded quantitative assessment of the particle size distribution of the lutein in product B. It accounts for excellent stability of B, even under acidic conditions, but may point to low bioavailability on the other hand. In contrast, product A readily and almost instantaneously dispersed in water. This latter product contains small lutein particles having an average diameter of around 0.2 μm.

Previous experimental investigations in rats (data not shown) have revealed unexpectedly large differences in lutein plasma responses to the administration of these two products. In these experiments, product A had a much better bioavailability than product B. The obvious question, therefore, was whether the results in rats could be confirmed in humans, because there are distinct differences in the absorption of fat-soluble micronutrients between rodents and humans.

Therefore, a comparative plasma kinetics study was undertaken with these two products. The results of this study have been recently been published (Evans, Beck et al. 2013).

### 7.4.2 Design of Study

A crossover design in which each subject is consecutively exposed to both products was selected for this comparison. Because of the relatively long half-life of lutein in plasma, the lutein was given in a single dose and not in a multiple dosing approach.

Following 2 days during which the lutein baseline concentrations in plasma were determined two times, individual plasma responses to a single dose of 20 mg of unesterified lutein from product A or B were recorded in a randomized, double-blind study involving 48 volunteers. The volunteers were healthy males or females with an average age of 39 years.

Plasma concentrations of lutein were determined by reversed-phase HPLC and regularly followed over 4 weeks for each crossover phase. The maximum lutein plasma concentration ($C_{max}$) and the time to reach this maximum concentration ($T_{max}$) were determined for each subject and for products A and B. The AUC was calculated using the linear trapezoidal rule. Paired t-tests were used to compare treatments with respect to outcome parameters such as AUC. Unpaired t-tests were used to compare the baseline parameters of the groups that received the treatment in different sequence, A–B versus B–A.

### 7.4.3 Results

Randomization resulted in two populations, 24 subjects who started with product A and 24 subjects who started with product B. In terms of age, BMI, and baseline lutein plasma concentrations, these two populations were well balanced.

In Figure 7.6, the time courses of average lutein and zeaxanthin plasma concentrations resulting from the ingestion of single doses of 20 mg of lutein from product A or B over 28 days are represented. For product A, rapid increases of plasma lutein concentration following ingestion of the 20-mg single dose are apparent, while for product B only a scant increase, if any, can be seen. A similar but quantitatively smaller response is observed with zeaxanthin, corresponding to its lower ingested dose.

Table 7.1 summarizes the main pharmacokinetic parameters, $C_{max}$, $T_{max}$, and $AUC_{(0-168\,h)}$ for lutein from A and B, their comparison, and statistics. Product A consistently exhibits significantly larger responses than product B: For $C_{max}$, the difference is almost 5-fold and for $AUC_{(0-168\,h)}$ it is over 10-fold with both differences being statistically significant ($p < .001$). No significant difference in time to reach the maximum ($T_{max}$) was observed.

While in Figure 7.6 average values per sampling time are compared, it is instructive to also look at the comparisons of individual AUC values as represented in Figure 7.7. From this figure it becomes apparent that the majority of subjects responded better to product A than to product B. But there are a few subjects who quantitatively show the same response to both products (at or close to the diagonal) and there is one subject who indicates a slightly higher, although still small absolutely, response to product B than to product A. The high interindividual variability in response to the two products is apparent from the distribution along the axes: the $AUC_{0-168\,h}$ range is 20–100 h × μmol/L for product A, and 10–60 h × μmol/L for product B. Low responders (in the lower left of the graph) usually also have a low lutein baseline concentration as can be observed in subject 52 (Figure 7.7), while good responders tend to have higher baseline concentrations. Since the absolute, not baseline-corrected $AUC_{0-168\,h}$ is shown in the graph, subjects with very high baseline concentration appear as extreme outliers (see subject 2 in Figure 7.7), although the average response may not be much smaller (see subject 48 in Figure 7.7).

As mentioned in Section 7.4.1, lutein from marigold also contains zeaxanthin. As expected, the zeaxanthin responses to product A or B are qualitatively very similar

# A Review of Recent Data on the Bioavailability of Lutein and Zeaxanthin

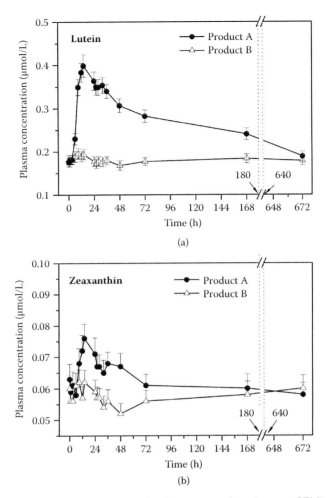

**FIGURE 7.6** (a) Time course of plasma lutein concentrations (mean ± SEM). A measure of 20.4 or 20.9 mg lutein, from product A (filled circles) or B (open triangles), respectively, was ingested in single doses at time point zero. Please note the break in the $x$-axis between 180 and 640 hours. (b) Time course of plasma zeaxanthin concentrations (mean ± SEM). A measure of 1.75 or 1.55 mg zeaxanthin, from product A (filled circles) or B (open triangles), respectively, was ingested in single doses at time point zero. Please note the break in the $x$-axis between 180 and 640 hours.

to those of lutein, but quantitatively smaller because of the smaller dose given, 1.7 mg of zeaxanthin for product A and 1.5 mg for product B, accounting for the approximately 7%–9% zeaxanthin content of the marigold lutein used.

### 7.4.4 Conclusions

The results of the study described in this section document the importance of the formulation technology to assure good bioavailability of a lutein product as measured by human plasma responses. In particular, the study emphasized how important a good

### TABLE 7.1
### Comparison of Kinetic Parameters for Lutein from Matrices A and B (Mean [SD], n = 46)

| | Maximum Concentration ($C_{max}$) | | Time to Reach $C_{max}$ ($T_{max}$) | Area Under the Curve (0–168 h) (AUC$_{(0-168 h)}$) | |
|---|---|---|---|---|---|
| | µmol/L | Increase from Baseline (%) | Hours | h × µmol/L | Increase from Baseline (%)[a] |
| Matrix A | 0.46 (0.17) | 177.0 (87.2) | 17.6 (7.7) | 48.0 (15.7) | 76.1 (48.9) |
| Matrix B | 0.23 (0.09) | 37.3 (26.0) | 22.2 (22.1) | 30.2 (10.6) | 5.9 (16.5) |
| Δ (A–B) | 0.23 | 139.6 | −4.6 | 17.8 | 70.2 |
| Paired t-test | $p < .001$ | $p < .001$ | $p = .22$ | $p < .001$ | $p < .001$ |

[a] A theoretical baseline AUC$_{0-168 h}$ was calculated using the baseline concentration.

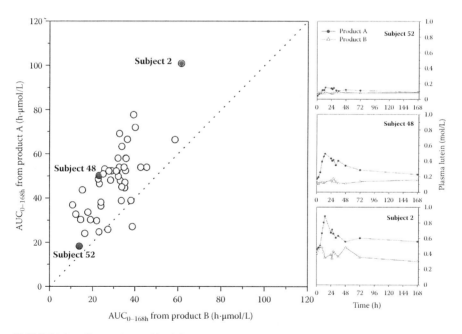

**FIGURE 7.7** Comparison of individual non-baseline-corrected AUCs (0–168 hours) resulting from product A (plotted on the ordinate) or product B (plotted on the abscissa). Points above the dotted line represent individuals who exhibit a larger response for product A than for product B. On the right-hand side, time courses of lutein plasma concentrations are shown for three selected individuals. Subject 2 has the largest response to both products. In contrast, subject 52 has a relatively small response to both A and B. Subject 48 was chosen as showing an "average" response to A and B.

dispersion of lutein is for its absorption into the body and indicated that superior lutein bioavailability was provided by the Actilease product from DSM (product A) as compared to the alginate formulation of lutein tested (product B). Numerous publications report that the response of plasma concentrations of xanthophylls toward supplementation are well correlated with retinal MPOD responses (e.g., Borel, de Edelenyi et al. 2011; Connolly, Beatty et al. 2010; Johnson, Chung et al. 2008), qualifying plasma responses as valid markers for retinal responses. Accordingly, it can be expected that supplementation with lutein from product B may not lead to substantial responses of MPOD. While, to our knowledge, nothing to the contrary has been reported for product B, it is known that supplementation with product A indeed leads to measurable MPOD responses (e.g., Stringham and Hammond 2008; Schalch, Cohn et al. 2007; and other publications). Consequently, a higher contribution to risk reduction of AMD can be expected for DSM's Actilease lutein.

## REFERENCES

Albert, G. I., U. Hoeller, et al. (2008). "Metabolism of lutein and zeaxanthin in rhesus monkeys: identification of (3R,6'R)- and (3R,6'S)-3'-dehydro-lutein as common metabolites and comparison to humans." *Comp Biochem Physiol B Biochem Mol Biol* **151**(1): 70–78.

Algvere, P. V., J. Marshall, et al. (2006). "Age-related maculopathy and the impact of blue light hazard." *Acta Ophthalmol Scand* **84**(1): 4–15.

Barker, F. M., 2nd, D. M. Snodderly, et al. (2011). "Nutritional manipulation of primate retinas, V: effects of lutein, zeaxanthin, and n-3 fatty acids on retinal sensitivity to blue-light-induced damage." *Invest Ophthalmol Vis Sci* **52**(7): 3934–3942.

Bone, R. A. and J. T. Landrum. (2010). "Dose-dependent response of serum lutein and macular pigment optical density to supplementation with lutein esters." *Arch Biochem Biophys* **504**(1): 50–55.

Borel, P., F. S. de Edelenyi, et al. (2011). "Genetic variants in BCMO1 and CD36 are associated with plasma lutein concentrations and macular pigment optical density in humans." *Ann Med* **43**(1): 47–59.

Borel, P., P. Grolier, et al. (1996). "Carotenoids in biological emulsions: solubility, surface-to-core distribution, and release from lipid droplets." *J Lipid Res* **37**(2): 250–261.

Bowen, P. E., S. M. Herbst-Espinosa, et al. (2002). "Esterification does not impair lutein bioavailability in humans." *J Nutr* **132**(12): 3668–3673.

Breithaupt, D. E., U. Wirt, et al. (2002). "Differentiation between lutein monoester regioisomers and detection of lutein diesters from marigold flowers (*Tagetes erecta* L.) and several fruits by liquid chromatography-mass spectrometry." *J Agric Food Chem* **50**(1): 66–70.

Caimari, A., P. Oliver, et al. (2008). "Impairment of nutritional regulation of adipose triglyceride lipase expression with age." *Int J Obes (Lond)* **32**(8): 1193–1200.

Chitchumroonchokchai, C. and M. L. Failla. (2006). "Hydrolysis of zeaxanthin esters by carboxyl ester lipase during digestion facilitates micellarization and uptake of the xanthophyll by Caco-2 human intestinal cells." *J Nutr* **136**(3): 588–594.

Chung, H. Y., H. M. Rasmussen, et al. (2004). "Lutein bioavailability is higher from lutein-enriched eggs than from supplements and spinach in men." *J Nutr* **134**(8): 1887–1893.

Connolly, E. E., S. Beatty, et al. (2010). "Augmentation of macular pigment following supplementation with all three macular carotenoids: an exploratory study." *Curr Eye Res* **35**(4): 335–351.

Evans, M., M. Beck, et al. (2013). "Effects of formulation on the bioavailability of lutein and zeaxanthin: A randomized, double-blind, cross-over, comparative, single-dose study in healthy subjects." *Eur J Nutr* **52**(4): 1381–1391.

Fletcher, A. E., G. C. Bentham, et al. (2008). "Sunlight exposure, antioxidants, and age-related macular degeneration." *Arch Ophthalmol* **126**(10): 1396–1403.

Hope, G. M., W. W. Dawson, et al. (1992). "A primate model for age-related macular drusen." *Br J Ophthalmol* **76**: 11–16.

Johnson, E. J., H. Y. Chung, et al. (2008). "The influence of supplemental lutein and docosahexaenoic acid on serum, lipoproteins, and macular pigmentation." *Am J Clin Nutr* **87**(5): 1521–1529.

Johnson, E. J., M. Neuringer, et al. (2005). "Nutritional manipulation of primate retinas, III: effects of lutein or zeaxanthin supplementation on adipose tissue and retina of xanthophyll-free monkeys." *Invest Ophthalmol Vis Sci* **46**(2): 692–702.

Leung, I. Y., M. M. Sandstrom, et al. (2004). "Nutritional manipulation of primate retinas, II: effects of age, n-3 fatty acids, lutein, and zeaxanthin on retinal pigment epithelium." *Invest Ophthalmol Vis Sci* **45**(9): 3244–3256.

Leung, I. Y., M. M. Sandstrom, et al. (2005). "Nutritional manipulation of primate retinas. IV. Effects of n-3 fatty acids, lutein, and zeaxanthin on S-cones and rods in the foveal region." *Exp Eye Res* **81**(5): 513–529.

Loane, E., J. M. Nolan, et al. (2008). "Transport and retinal capture of lutein and zeaxanthin with reference to age-related macular degeneration." *Surv Ophthalmol* **53**(1): 68–81.

Loane, E., J. M. Nolan, et al. (2010). "The respective relationships between lipoprotein profile, macular pigment optical density and serum concentrations of lutein and zeaxanthin." *Invest Ophthalmol Vis Sci* **51**(11): 5897–5905.

Maiani, G., M. J. P. Castón, et al. (2009). "Carotenoids: actual knowledge on food sources, intakes, stability and bioavailability and their protective role in humans." *Mol Nutr Food Res* **53**(Suppl. 2): 194–218.

Majumdar, A. P., R. Jaszewski, et al. (1997). "Effect of aging on the gastrointestinal tract and the pancreas." *Proc Soc Exp Biol Med* **215**(2): 134–144.

Monaco, W. A. and C. M. Wormington. (1990). "The rhesus monkey as an animal model for age-related maculopathy." *Optom Vis Sci* **67**: 532–537.

Neuringer, M., M. M. Sandstrom, et al. (2004). "Nutritional manipulation of primate retinas, I: effects of lutein or zeaxanthin supplements on serum and macular pigment in xanthophyll-free rhesus monkeys." *Invest Ophthalmol Vis Sci* **45**(9): 3234–3243.

Nolan, J. M., R. Kenny, et al. (2010). "Macular pigment optical density in an ageing Irish population: the Irish longitudinal study on ageing." *Ophthalmic Res* **44**(2): 131–139.

Norkus, E. P., K. L. Norkus, et al. (2010). "Serum lutein response is greater from free lutein than from esterified lutein during 4 weeks of supplementation in healthy adults." *J Am Coll Nutr* **29**(6): 575–585.

Rodriguez-Amaya, D. B. (2010). "Quantitative analysis, *in vitro* assessment of bioavailability and antioxidant activity of food carotenoids—a review." *J Food Compos Anal* **23**(7): 726–740.

Schalch, W., W. Cohn, et al. (2007). "Xanthophyll accumulation in the human retina during supplementation with lutein or zeaxanthin—the LUXEA (LUtein Xanthophyll Eye Accumulation) study." *Arch Biochem Biophys* **458**(2): 128–135.

Stringham, J. M. and B. R. Hammond. (2008). "Macular pigment and visual performance under glare conditions." *Optom Vis Sci* **85**(2): 82–88.

Thurnham, D. I. (2007). "Macular zeaxanthins and lutein—a review of dietary sources and bioavailability and some relationships with macular pigment optical density and age-related macular disease." *Nutr Res Rev* **20**(2): 163–179.

Trieschmann, M., S. Beatty, et al. (2007). "Changes in macular pigment optical density and serum concentrations of its constituent carotenoids following supplemental lutein and zeaxanthin: the LUNA Study." *Exp Eye Res* **84**(4): 718–728.

Van den Maagdenberg, K., E. Claeys, et al. (2007). "Effect of age, muscle type, and insulin-like growth factor-II genotype on muscle proteolytic and lipolytic enzyme activities in boars." *J Anim Sci* **85**(4): 952–960.

Zeger, S. L. and K.-Y. Liang. (1986). "Longitudinal data analysis for discrete and continuous outcomes." *Biometrics* **42**: 121–130.

Zeimer, M., H. W. Hense, et al. (2009). "[The macular pigment: short- and intermediate-term changes of macular pigment optical density following supplementation with lutein and zeaxanthin and co-antioxidants: the LUNA Study]." *Ophthalmologe* **106**(1): 29–36.

# 8 Multiple Influences of Xanthophylls within the Visual System

*Billy R. Hammond, Jr. and James G. Elliott*

## CONTENTS

8.1 Introduction ................................................................................................. 147
8.2 Functions within the Eye ............................................................................. 148
    8.2.1 Protection Hypothesis ...................................................................... 149
    8.2.2 Acuity Hypothesis ............................................................................ 153
    8.2.3 Glare Hypothesis .............................................................................. 155
    8.2.4 Visibility Hypothesis ........................................................................ 158
    8.2.5 Contrast Enhancement ..................................................................... 159
    8.2.6 Macular Pigment and Mesopic Acuity ............................................ 159
8.3 Function within the Retina and Brain ......................................................... 161
    8.3.1 Neural Efficiency Hypothesis .......................................................... 161
8.4 Multifunctionality of Lutein and Zexanthin ................................................ 164
References ............................................................................................................ 164

## 8.1 INTRODUCTION

Of the many carotenoids in the body, lutein (L) and zeaxanthin (Z) are unique in that they reach unusually high concentrations in the neural retina. The pigments can be found in millimolar concentrations in the central retina (the highest concentration of carotenoids in the body) and are detectable throughout the entire visual system (ocular fat, iridial tissue, choroid, visual and frontal cortex, etc.) (Bernstein et al. 2001; Craft et al. 2004; Vishwanathan et al. 2011). Retinal tissue accumulates both lutein and zeaxanthin, which are similar in chemical structure and properties: lutein ($3R,3'R,6'R$-lutein), zeaxanthin ($3R,3'R$-zeaxanthin). The pigments co-occur in many foods (largely, green leafy vegetables, colored fruits, and eggs) with an overall dietary ratio of about 5:1.

The very small difference in chemical structure (e.g., the placement of a single double bond) is, nonetheless, sufficient to create quite meaningful differences in retinal distribution: zeaxanthin is the dominant carotenoid in the center of the retina, whereas lutein, which also peaks in the central retina, is dominant throughout the rest of the retina but at much lower levels (Bone et al. [1988]; the macular ratio is reversed in infancy; Bone et al. [1997]) (see Figure 8.1). Data derived from model

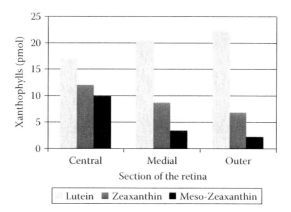

**FIGURE 8.1** Lutein, zeaxanthin, and meso-zeaxanthin content at central and eccentric locations in sections taken from 16 normal retinas. (From Landrum, J.T. et al., *Methods Enzymol.*, 299, 457–467, 1999.)

membrane systems suggest that this difference in spatial distribution may also reflect differences in function (Sujak et al. 1999). Both L and Z are linear molecules composed of a long, conjugated chain of double bonds with hydroxyl groups located on the terminal ionone ring end-groups. This electron-rich structure makes these pigments effective lipophilic antioxidants. L and Z also strongly absorb visible light (they appear yellow because they transmit long and mid wave [red and green] but absorb short wave or blue light). The absorbance peaks of Z are shifted hypsochromically by about 8 nm relative to L. The combined absorbance band of the macular pigment (MP), composed of both L and Z, is therefore broader than it would be where only a single carotenoid is present (encompassing a full third of the visible spectrum, approximately 400–500 nm). The Z and L concentrations peak sharply in the center of the fovea roughly following the spatial distribution of cone photoreceptors. The concentration of Z in the central fovea exceeds that of L by a factor of 2 or more (Bone et al. 1988; Handelman et al. 1988). This dense packing of the retinal photoreceptor mosaic as it approaches the foveola (one reason for our sharper acuity in the retinal center) is accompanied by associated increases in cellular and oxidative metabolism. Perhaps reflecting this functional need, Z is both more highly concentrated than L in the foveal center and a more effective antioxidant (Mortensen and Skibsted 1997) (see Figure 8.2). The retina has, apparently, evolved mechanisms for an elegant distribution of the pigments based on functional need.

## 8.2 FUNCTIONS WITHIN THE EYE

A number of hypotheses detail how the MPs might influence the visual system. In order of appearance, these hypotheses were named the Acuity hypothesis (posited in 1866 by Schultze), the Protection hypothesis (in 1982 by Kirschfeld; see also Malinow et al. 1980; Feeney-Burns et al. 1984, 1989; Haegerstrom-Portnoy 1988), the Visibility hypothesis (in 2002 by Wooten and Hammond, see the first mention by Walls and Judd 1933), and the Glare hypothesis (2007 by Stringham and Hammond, similarly

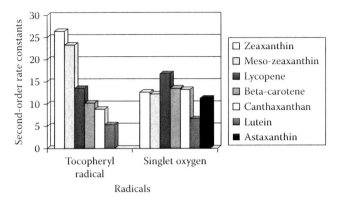

**FIGURE 8.2** Effect of different carotenoids on repair of α-tocopheryl radial cation and quenching of singlet oxygen. Lutein and zeaxanthin have quite different quenching capacity due to a difference of one double bond. Tocopheryl radical cation generated by pulse radiolysis; second-order rate constants $M^{-1}s^{-1}$ in hexane. Singlet oxygen generated by laser photo excitation; second-order rate constants $M^{-1}s^{-1}$ in benzene. (From Schalch, W. et al., *Nutritional and Environmental Influences on the Eye*, Boca Raton, FL, CRC Press, 215–250, 1999.)

mentioned by Walls and Judd 1933). A series of recent observations has prompted the newest hypothesis, the Neural Efficiency hypothesis (Renzi and Hammond 2010a; Zimmer and Hammond 2007). This hypothesis is based on the discovery that L is a dominant carotenoid in the brain (Craft et al. 2004) and the idea that the carotenoids may have direct effects on neural function (e.g., by influencing gap junction communication, as has been shown in model mitotic cells) (Stahl et al. 1997).

### 8.2.1 Protection Hypothesis

The Protection hypothesis is based on several fundamental premises. In the case of the retina and macular degeneration, damage manifests as a buildup of debris, both retinal/extracellular, and within the retinal pigment epithelium (RPE). This accumulation increases over the life span until it is sufficient to trigger a disease process characterized by a disruption in the interaction between the retina and RPE. This debris is largely oxidized lipids, and, hence, oxidative damage is thought to be a primary event in the pathogenesis of retinal disease. L and Z are lipophilic antioxidants that can quench radical reactions and, hence, retard this slow accumulation of damage. In the macula, L and Z reach a sufficient concentration to absorb a significant quantity of short-wave (blue) light. This filtering in the inner retina could reduce oxidative reactions in the outer retina.

There is, certainly, abundant evidence that L and Z are highly effective lipid antioxidants (Sujak et al. 1999). There is also evidence that they serve this function within the retina (Khachik et al. 1997). It is also clear that the retina suffers from oxidative damage (Krinsky et al. 2003). Similarly, there is convincing evidence that short-wave light in the visible region of the electromagnetic spectrum (400–500 nm) causes acute damage to the retina (the so-called "blue-light hazard") (Ham et al. 1978; van Norren and Schellekens 1990). Since MP attenuates such light, it follows that such attenuation is protective (see Barker et al. 2011). There is abundant experimental evidence that this is the case (Thomson et al. 2002) (Figure 8.3). If L and Z are important prophylactics, many individuals who have lower than normal macular levels of L and Z are at greater risk of oxidative damage. As shown in Figure 8.4, the average dietary intake of L and Z in the United States is under 2 mg (Johnson et al. 2010). However, several epidemiological studies indicate that protection is most evident at higher levels of intake (e.g., 5–6 mg/day) (see Figures 8.5 and 8.6).

What is the contradictory evidence? Although many epidemiological studies have linked protection against age-related macular degeneration (AMD) and cataract

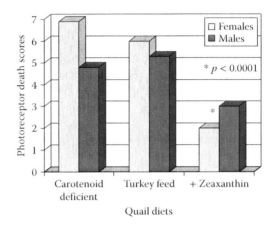

**FIGURE 8.3** Protective effect of zeaxanthin on photoreceptors in female and male Japanese quail, $n = 4$ per group 35 mg/kg diet zeaxanthin, 6 months exposed to light (1 hour on, 2 hours off), 27 hours. (From Thomson, L.R. et al., *Invest Ophthalmol Vis Sci.*, 43(11), 3538–3549, 2002.)

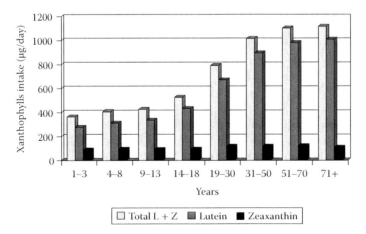

**FIGURE 8.4** Mean intakes of lutein, zeaxanthin, and total L + Z across age for males and females in Health and Examination Survey 2003–2004. (From Johnson, E. et al., *J Am Diet Assoc.*, 9, 1357–1362, 2010.)

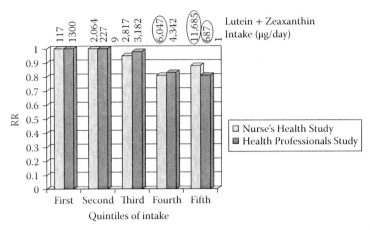

**FIGURE 8.5** Risk of AMD according to dietary intake of L and Z in Nurse's Health Study and Health Professionals Study. (From Cho, E. et al., *Am J Clin Nutr.*, 87(6), 1837–1843, 2008.)

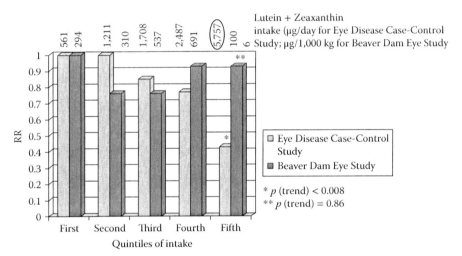

**FIGURE 8.6** Risk of AMD according to dietary intake of L and Z in Eye Disease Case-Control Study (Seddon et al. 1994) and Beaver Dam Eye Study. (From VandenLangenberg, G.M. et al., *Am J Epidemiol.*, 148, 204–214, 1998.)

(Karppi et al. 2012) to LZ intake, certainly not all have (Cho et al. 2008; Berrow et al. 2011; Dietzel et al. 2011). Some have further argued that AMD is primarily a genetic condition (explaining as much as 70% of the variance) (Seddon et al. 2005) and, hence, lifestyle alterations (a carotenoid-rich diet or avoiding excessive light exposure) might make little difference in altering risk for the disease. In contrast, a recent study found that even modest alterations in lifestyle (healthy diet, not smoking, physical activity) are associated with reducing risk of AMD by 71% (Mares et al. 2011). If lifestyle alterations can reduce risk, one approach could be increasing intake of L and Z containing foods (e.g., SanGiovanni and Neuringer 2012). Such an approach might be less fruitful if MP levels are determined genetically, as has been suggested by Liew et al. (2005) and Hammond (CJ) et al. (2012). Other data indicate, however, that the heritability of MP might be minimal (Hammond et al. 1995; Hammond and Renzi 2008), especially since MP density appears to be amenable to increase with supplementation of carotenoid-rich foods (Hammond et al. 1997) or purified L, Z, or MZ (Richer et al. 2004, 2011; Connolly et al. 2010).

An additional criticism is that most of the existing evidence is correlational and depends on assumptions that remain unproven (e.g., that acute exposure to high-intensity light in monkeys is analogous to chronic low-intensity exposure in humans) (Barker et al. 2011). Quantifying the effects of light exposure on late-stage disease has proven extremely difficult and is especially so since even light exposures during infancy may be critical (Hammond 2008, 2012). Age-related cataracts and macular degeneration are clearly multifactorial conditions with complex etiologies. Many lifestyle and genetic factors contribute to the incidence and severity of these diseases (Coleman 2011, Chang et al. 2011). Hence, although there appears to be a general consensus that L and Z protect ocular tissues, their relative importance, especially with respect to the disease process, is not known.

Another manifestation of the protective hypothesis does not pertain to protection in its most common connotation (i.e., prevention) per se, but rather the idea that supplementing L and Z might actually *treat* some aspects of retinal disease. The retina is comprised of neural tissue, which, like most other central nervous tissue, does not undergo mitosis. Consequently, once cell death has occurred, there is very little that can be done to *reverse* that damage, especially when considering advanced stages of the disease process (e.g., blindness in late AMD). Preventing degeneration is therefore the most likely benefit that could be expected from L and Z supplementation (preventive steps starting in infancy; Hammond, 2008, 2012). However, it is possible that supplementing L and Z even late in life but early in the disease process might be beneficial. For example, Margrain et al. (2004) argued that lipofuscin components, like A2E, are potent photosensitizers and that their accumulation in retinal tissue potentiates more lipofuscin accumulation. The end result of this cascade is an acceleration of the theoretical point where dysfunction of the RPE would initiate the disease process. Given the argument that increased lipofuscin accumulation in the elderly retina makes the elderly retina especially susceptible to blue light damage, L and Z supplementation may be of added benefit in this population (L and Z have been shown to prevent photooxidation of A2-PE/A2E, a toxic component of lipofuscin: Kim et al. 2006). This possibility becomes especially intriguing given the fact that phototoxicity of lipofuscin (after correction for dense anterior media, the peak phototoxicity is ~430 nm) closely matches the absorption spectrum of MP.

The idea that older retinas may benefit even more from supplementation of L and Z is not new. Other authors (Liang and Godley 2003; Wu et al. 2006) have argued that older retinas suffer increased levels of oxidative stress and therefore might benefit from extra antioxidant capacity. For example, photoreceptor loss (particularly rod loss) is a feature of older retinas and becomes profound in some disease states (e.g., retinitis pigmentosa, RP). It has been argued that increased rod loss could also increase oxygen tension by reducing oxygen consumption throughout the retina (Shen et al. 2005; Komeima et al. 2006). If increases in L and Z reduce photoreceptor loss (as shown in rodent models of RP, Komeima et al. 2006), then this increase in oxygen tension would also be reduced.

Izumi-Nagai et al. (2007) found, using rodents, that lutein could significantly suppress choroidal neovascularization. Choi et al. (2006) used high intraocular pressure to induce ischemia (known to promote cellular degeneration) in rat retinas. This manipulation results in numerous changes such as increases in the production of neuronal nitric oxide synthsase (nNOS), which generates nitric oxide, a promoter of oxidative stress (see Landrum 2013). Lutein reduced the production of nNOS in a dose-dependent manner. It also linearly decreased the expression of cyclooxygenase-2 (an inducible enzyme responsible for prostaglandin production during inflammatory response), which is a proinflammatory protein. Local inflammation has also been linked to AMD development. Klein et al. (2005), for example, have argued that a variant of the gene that encodes complement factor H (an inflammatory regulator) may account for as much as 50% of the AMD cases in the United States.

Isolating the *magnitude* of the effect of any single variable is difficult when considering the overall context of factors that together contribute to risk of retinal disease during the many decades of a person's life. Hence, a more fruitful line of inquiry might be to focus on the immediate as opposed to the very long-term influences of the MPs. For example, supplementation by L and Z, even if they have no impact on the biology of the disease per se, may influence the symptoms (i.e., visual dysfunction) of disease. This possibility is based on the idea that MP could improve visual performance through optical mechanisms.

### 8.2.2 Acuity Hypothesis

The original description of the spectral absorption characteristics of MP was made by Max Schultze in 1866. At that time, he theorized that MP might improve visual acuity in broadband illumination by filtering out short-wave (SW) energy before absorption by the photoreceptors. The surprising argument that one can see better with less light was based on the well-documented fact that not all wavelengths are equally focused on the retina. When the emmetropic eye is in focus for middle-wave light (as it typically would be given the photopic luminosity function and when exposed to natural sunlight), it will be myopic for short-wave light (i.e., it will be focused in front of the retina and therefore blurred) and slightly hyperopic for long-wave light (Gilmartin and Hogan 1985). This effect is known as longitudinal chromatic aberration.

For 460 nm light (the dominant wavelength of sunlight and peak absorption of MP), the magnitude of the focus error is approximately $-1.2$ D (Howarth and Bradley 1986). Given optimal focus at 550 nm, much of the SW region would be seriously out of focus. In addition to the focus problem, the wavelength dependency of the eye's focal length means that retinal image size is also inversely proportional to wavelength, that is, the shorter the wavelength, the larger the retinal image. This effect is known as lateral chromatic aberration. Thus, if a disc of white light is imaged on the fovea, a violet-blue penumbra will result (effectively making the visual image larger). Together, longitudinal and lateral chromatic aberrations are known simply as chromatic aberration. Clearly, both kinds of chromatic aberrations degrade the retinal image of any potential target (e.g., see Reading and Weale 1974). Schultze's

Acuity hypothesis predicts that retinal images are sharpened by MP's SW absorption and that visual acuity is consequently improved.

The Acuity hypothesis was tested by Engles et al. (2007) (also see Mclellan et al. 2002). MP was measured in 40 young healthy adults and the relation to acuity was evaluated. Two types of acuity were assessed: resolution acuity (RA) and vernier hyperacuity (HA) (i.e., the finest spatial discrimination achievable by the human visual system). These measures of spatial vision were made under different spectral conditions: an achromatic light condition (white light), which contained significant portions of SW light and therefore would be expected to be influenced by the optical filtering of MP, and a chromatic "yellow" light condition, which contained no SW light and therefore was unaffected by MP absorbance. No relation was found between MP and RA or HA in either the white or yellow conditions. In fact, the average RA and HA across conditions were highly similar, despite differences in MP screening. Average RA in the white and yellow conditions was 28 and 28.7 seconds, respectively. Average HA in the white and yellow conditions was 7.0 and 6.8, respectively. The data from that study clearly did not support the predictions of the Acuity hypothesis. Careful modeling of the results (see Engles et al. 2007) shows that it is very unlikely that MP *significantly* (although not significant perhaps in a real-world sense, some empirical data allow that MP might reduce chromatic aberration (CA) sufficiently to be detected in large samples, e.g., see Loughman et al. 2010) reduces the effects of chromatic aberration (especially since deleterious effects of chromatic aberration may simply be washed out by the many other ocular aberrations; Mclellan et al. 2002). Since publishing the original paper, the result has been extended to the entire contrast sensitivity function with similar, primarily null (with the exception of a small effect at the middle frequencies), results (Engles et al. 2008).

The finding that MP does not significantly improve acuity through an optical mechanism is important. Studies that do find that MP is related to acuity (e.g., like the Richer et al. (2004) or Olmedilla et al. (2003) studies; both double blind, placebo controlled) are probably finding effects that must be explainable based on biological, as opposed to optical, mechanisms. The Acuity hypothesis is still often cited as a primary function of the MPs in journal articles (Charbel Issa et al. 2009; Sasamoto et al. 2011; Zheng et al. 2012) and textbooks (Lim 2008; Schwartz, 2010).

The fact that MP does not improve spatial vision by reducing the effects of CA is consistent with phenomenological observation (people with low MP do not report violet penumbras on sunny days). Reading and Weale (1974) modeled the effects of MP when imaging a broad-band white disc on the fovea. As noted, when this is done and MP density is very low, CA will produce a violet penumbra surrounding the disc (again, based on their modeling). Reading and Weale noted that an average amount of MP is enough to effectively remove the halo caused by CA. However, one problem with this type of modeling is that the eye is considered as an independent entity. Vision, however, is not a passive process and the optics of the eye are not actually comparable to a camera when considering what we actually perceive. This is because the brain very actively corrects for deficiencies in input (e.g., filling in the scotoma caused by the optic nerve). There are many young subjects who have very little or no MP and have perfect Snellen Acuity. These subjects certainly do not report seeing blue fringes surrounding objects on a sunny day. The visual system actually has many means of decreasing the role of CA

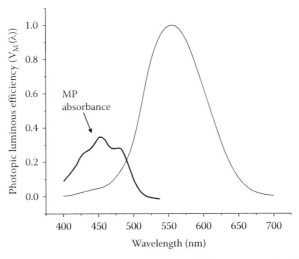

**FIGURE 8.7** Graph showing the human spectral sensitivity curve graphed relative to MP absorbance. (Spectral sensitivity created from data in Vos, J. J., *Color Research & Application*, 3: 125–128, 1978, Table 1.)

on spatial vision (reviewed by DeValois and DeValois 1990); these include the following: (1) Spectral sensitivity peaks in the center of the visible spectrum; we are relatively insensitive to short- and long-wave light where MP absorbs the most (see Figure 8.7). (2) Short-wave cones are numerically sparse. Consequently, mid- and long-wave cones contribute more to spatial vision tasks. Indeed, most spatial tasks are driven by the luminance channel that receives its input from the mid- and long-wave cones. (3) The young lens can accommodate to bring most images into focus (correct chromatic aberration when the light is dominantly short or long wave). (4) Brain mechanisms also exist for accommodating stable features of the environment (Rosinski et al. 1980). For example, Neitz et al. (2002) showed that wearing colored filters could shift unique yellow (see Dimmick and Hubbard 1939) by several nanometers. This shift persisted for 1–2 weeks after discontinuing use of the filters. Taken together, it is unlikely that MP improves vision significantly by reducing the deleterious effects of chromatic aberration.

### 8.2.3 Glare Hypothesis

As noted earlier, Walls and Judd (1933) and later Nussbaum et al. (1981) listed four effects one could generally expect based simply on the optics of yellow filters:

1. To increase visual acuity by reducing chromatic aberration
2. To promote comfort by the reduction of glare and "dazzle"
3. To enhance detail by the absorption of "blue haze"
4. To enhance "contrast"

The empirical results by Engles et al. (2007, 2008) make this first hypothesis unlikely (the Acuity hypothesis). In general, the data do not support the idea that MP significantly influences visual capabilities that are limited by refractive error (Neelam et al. 2006).

Many visual effects, however, are not based on errors in refraction. Rather, vision is often reduced due to other optical factors (intrinsic and extrinsic) that would be expected to be influenced by MP absorbance. For example, empirical data does support the idea that MP improves visual disability and discomfort due to glare and "dazzle" (#2). There is a strong inverse relationship ($r = -0.79$) between MP density and glare recovery time (Stringham and Hammond 2007, 2008) and Stringham et al. (2011). Subjects with the highest macular pigment optical density (MPOD) regained functional vision over twice as fast as the two subjects with the lowest MPOD (Stringham and Hammond 2007). Plotting the ratio of parafoveal to foveal recovery times across wavelength yields a difference spectrum that shows a peak at 460 nm which is consistent with the MP absorption spectrum. This similarity suggests that, like glare disability, filtering is the primary mechanism by which MP reduces photostress recovery time (e.g., it can reduce the bleaching of photopigment). Improvement in glare recovery has important practical implications for visual function when considering scenarios such as blinding headlights. The relation between MP and photostress recovery might also partially explain why certain groups have slower glare recovery. For example, Sandberg and Gaudio (1995) showed that photostress recovery is slower for patients with AMD. Beatty et al. (2001) and Bone et al. (2001) have showed that patients with AMD often have reduced levels of MP. Our results suggest the possibility that slower photostress recovery in patients might be at least partly due to reduced levels of MP for those groups.

In addition to producing substantial intraocular scatter, bright lights also cause considerable discomfort. A major complaint in many patients with AMD is visual discomfort as a result of exposure to even moderate lighting. This is termed "photophobia" or "discomfort glare" and refers to discomfort, or, in extreme cases, pain on exposure to sufficiently intense light. Stringham et al. (2003, 2004) showed that thresholds for photophobia responses (squinting of the eyes in reaction to an intense light) were much lower for lights of short wavelengths (those in the blue region of the visible spectrum), period compared to lights of middle (green) or long (red) wavelengths (Wenzel et al. 2006). In other words, it took much less light intensity to elicit an aversive response when the light was of a short wavelength. Interestingly, the action spectrum for photophobia (after correction for MP and ocular media absorption) was shown to approximate both the threshold retinal damage function for rhesus monkeys determined by Ham et al. (1978) and the action spectrum for aerobic photoreactivity of lipofuscin (a photosensitizer thought to generate singlet oxygen in the retina). It appears, therefore, that photophobia is likely a behavioral mechanism that is biased to protect biological tissue from potentially damaging energetic, short-wavelength light. With regard to MP level, subjects with higher levels of MP were shown to tolerate more short-wavelength light before the photophobia threshold was reached (as measured by electromyogram of the squint response, Stringham et al. 2003). A similar result was found in another study of photophobia in which thresholds to a broadband white light (containing short-wavelength energy) versus an orange light (containing no short-wavelength energy) were compared. Overall, the subjects were shown to be more sensitive to the broadband white light, but those subjects with higher levels of MP were able to tolerate higher levels of that light when viewed centrally (filtered by the MP) rather than viewed peripherally. Conversely, for orange light, the subjects were shown to be very similar in their photophobia sensitivity for central versus peripheral viewing conditions. From a functionality standpoint, these studies are consistent with MP increasing the bandwidth of comfortable visual operation via its action as a passive filter. For subjects with relatively high MP levels, a conservative estimate of this effect is roughly 0.5 log units (over three times the amount of broadband light intensity tolerated) compared with those with very little or no MP. Wenzel et al. (2006) supplemented four subjects with lutein esters (Xangold, 60 mg) for 12 weeks and found that increases in MP density led to proportional improvements in photophobia.

What can we conclude about MP and glare? Clearly, MP reduces glare disability and discomfort in young subjects. This effect is expected since MP is anterior to the cones and filters scattered light (see Figure 8.8). A few caveats, however, are worth noting. The first is that the spectral conditions of the stimulus are very important. MP will not reduce glare disability when the glare is not produced by light containing a significant proportion of short-wave light (see Loughman et al. 2010). MP filtering will also not reduce glare disability when the wavelength conditions between the target and surround are exactly the same. If MP absorbs light from both the target and surrounding in equal proportion, that ratio will stay the same irrespective of MP level. In such instances, high MP levels could reduce photostress and glare discomfort but it will not make a target more visible (i.e., improve glare disability). This same interpretation could be applied to other yellow filters (tinted intraocular lenses) and may explain why yellow filters improve visibility in some situations but not others. Most data on MP and glare come from studies of young subjects. Richer et al. (2004) and Olmedilla et al. (2003) found that glare recovery and disability was improved due to L supplementation in older subjects with cataracts and early retinal disease. Their measures of glare, however, were fairly gross and therefore this effect in the elderly needs to be confirmed.

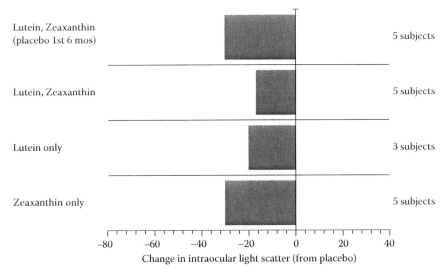

**FIGURE 8.8** Data from the LUXEA II trial (Kvansakul et al. 2006) showing the changes in intraocular scatter resulting from supplementation of L and Z.

## 8.2.4 Visibility Hypothesis

Glare effects related to MP are based on the idea that intense light entering the eye will scatter due to inhomogeneities in the ocular media anterior to the retina. This scattered light (intraocular) is then filtered by MP (proportional to the amount of MP and the spectral characteristics of the scattered light). Scattered light outside of the eye, like glare, also degrades vision, especially our ability to see far in the distance. Images of distant objects are focused by the cornea and lens on the retina (the proximal stimulus). In other words, both the scatter and the object of interest are relayed together to the retinal photoreceptors. The scatter superimposed on the image can veil the perception of the object. If there is minimal distal scatter, objects will appear more clearly. The term *visibility* refers to the clearness with which objects in the atmosphere stand out from their surroundings (Bennett 1930). Distant objects, however, are often obscured as a result of wavelength-dependent scattering in the atmosphere, which is greatest for short wavelengths. The Visibility hypothesis is based on the idea that MP, as a yellow filter, will filter out the blue-dominant atmospheric scatter that is transmitted with the image of an object onto the retinal surface. By filtering out the haze in the image, the target object becomes more visible (see Figure 8.9).

Extensive modeling based on known atmospheric physics and the equally well-known characteristics of the visual system has shown that this hypothesis is quantitatively feasible (Wooten and Hammond 2002). In fact, such modeling suggests a fairly significant visual effect for the MPs: an extension of visual range of about 30% when comparing individuals with very low to those with high MP density. Empirical data

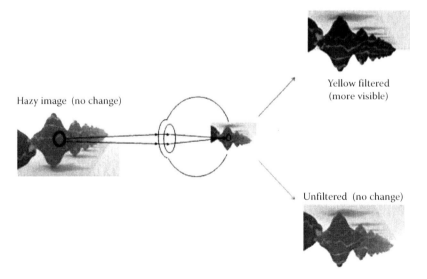

**FIGURE 8.9** (See color insert.) The example shows an achromatic test target and surround of nearly the same luminance. The central target becomes visible when either the target is just slightly brighter or the surrounding is just slightly darker. This very small change in brightness is enough to create a luminance edge (a phenomenon called brightness induction). These experiments are typically done with achromatic stimuli like those shown above but a similar effect would hold for colored stimuli.

using haze imposed over contrast gratings and simulated changes in MP (an artificial adjustable MP filter) found a similar magnitude (35%) of improvement when comparing low MP levels with that of high MP levels (Hammond et al. 2012).

### 8.2.5 Contrast Enhancement

A yellow target on a blue background is more visible (or more apparent) when the blue background is selectively filtered. At first, this effect seems obvious and trivial. It may also seem minor because it seems to be too specific, that is, it does not apply to many examples of everyday vision. Modeling the optics of vision outdoors, however, reveals that we are often looking at mid- to long-wave targets on predominantly blue backgrounds or surrounds (Wooten and Hammond 2002). This raises the possibility that yellow filters could improve chromatic contrast (e.g., by absorbing more of the background relative to a target) in a way that would benefit outdoor vision.

As noted earlier, Walls and Judd argued that yellow filters enhance contrast. Enhancing contrast is a very important aspect of spatial vision, particularly as they apply to edges. Edges provide the discontinuity in the image that allows objects to be discriminated from their backgrounds (the retina has been described as a "contrast engine") and the visual system is organized to accentuate edges (e.g., lateral inhibition in receptive fields). Edges have an exaggerated importance in many perceptual tasks. Retinex algorithms (a major theory of color vision), for example, emphasize the importance of color borders. Simple cells within the cortex are maximally sensitive to edges of a given orientation (Hubel and Wiesel 1968) and lateral inhibition within the retina accentuates discontinuities within our visual field. Anything that accentuates edges would be expected to improve spatial vision and the detection of objects against a background. Luminance differences are certainly one way an edge can be defined. Of course, in the real world, objects and backgrounds are rarely achromatic.

Consequently, other differences, such as wavelength composition (color), are used to define edges. This is one reason that colored filters can make objects appear more "crisp." Yellow filters, for instance, will make a yellow target with a blue surround more visible by selectively reducing the surround relative to the center (see Figure 8.10). This simple optical effect enhances the contrast between a mid- or long-wave target and a background with more short-wave light. Both Luria (1972) and Wolffsohn et al. (2000) have shown that the visibility of stimuli like these is improved when viewed through yellow lenses. We can also see such an effect when MP is measured directly: MP selectively absorbs the shortwave background making the target more visible. Renzi and Hammond (2010a) have shown that MP is related to chromatic contrast (a yellow target on a blue surround or background) in both young and older subjects. Using a minimum border technique, they showed that MPOD influences the perception of chromatic edges consistent with the absorbance spectrum. The effects of MP on chromatic contrast have recently been confirmed in a larger sample (Hammond et al. 2013).

### 8.2.6 Macular Pigment and Mesopic Acuity

It seems fairly obvious that MP will enhance contrast whenever the wavelength difference between an object and its surrounding/background is enhanced by selective

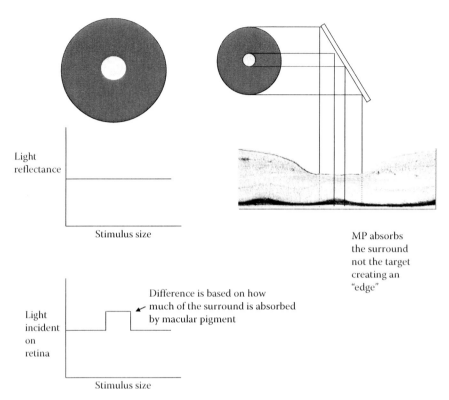

**FIGURE 8.10** **(See color insert.)** This example shows a blue surround (black in illustration) with a yellow (white in illustration) stimulus of equal luminance. Note that what defines the edge is based only on the wavelength or color difference. The luminance difference itself, however, will be exaggerated by differential absorption by macular pigment.

absorption by MP. When the luminance ratio between the target and a background is almost the same, this process could be enhanced due to the phenomenon of brightness induction. Kvansakul et al. (2006) suggested another situation where contrast sensitivity might be enhanced: mesopic vision levels. In the LUXEA II trial, L, Z, and L and Z were supplemented and shown to improve contrast acuity. This is graphed in Figure 8.11.

Pérez et al. (2003) reported a very similar effect of yellow filters on mesopic contrast acuity. Such results make sense. Humans have duplex vision. We have cones in our central retina that mediate color vision and fine acuity during the daytime when light levels are high. During mid-day, the photopigment of rods is effectively isomerized (bleached) and rods contribute little. At night, however, (low light levels), the photopigment in rods is regenerated and rods take over our visual function (we shift from photopic to scotopic vision). Because there are so many rods, ~90 million compared to ~5 million cones (rods are more sensitive and therefore more useful at times when little light is available). In moderate light levels, typical of twilight or dawn, cones and rods contribute strongly to our visual experience. This period is known as mesopic vision. Kvansakul et al. argued that, at such times, rods may decrease

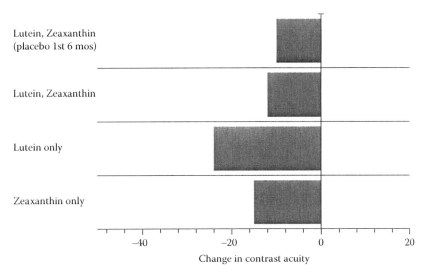

**FIGURE 8.11** Data from the LUXEA II trial (Kvansakul et al. 2006) showing the effects of supplementation on acuity measured under conditions where both cones and rods mediate the spatial task.

contrast sensitivity. Rods do, in fact, have poorer contrast sensitivity and temporal resolution compared with cones. Kvansakul et al. (2006) argued that high MP, by screening central rods, favors more cone-dominated mesopic vision that would confer superior contrast sensitivity. A visual effect of MP screening central rods is certainly plausible. Empirical evidence has shown that yellow filters can brighten the visual field (Kelly 1990) by increasing rod input to chromatic pathways. Yellow filters have also been shown to improve motion sensitivity, vergence eye movements, and reading performance (Ray et al. 2005) presumably due to influence on the rod-based magnocellular system. MP does screen central rods as shown in Figure 8.12.

## 8.3 FUNCTION WITHIN THE RETINA AND BRAIN

### 8.3.1 Neural Efficiency Hypothesis

The most recent hypothesis of L and Z function has been referred to as the Neural Efficiency hypothesis (Hammond and Wooten 2005; Zimmer and Hammond 2007; Renzi and hammond 2010a). This idea is based on physiological activity of L and Z postreceptorally and is supported by several observations: (1) L and Z are found throughout the visual pathway (e.g., brain areas like occipital and frontal lobes) (Craft et al. 2004; Vishwanathan et al. 2011) in amounts that vary significantly across subjects. (2) Ex vivo data have shown that L and Z directly influence cell-to-cell communication (e.g., enhancing gap junction communication, as has been shown in somatic cells) (Stahl et al. 1997). (3) Empirical results indicate that MP is related to temporal processing speeds (a visual measure known to be largely determined postreceptorally) (Hammond and Wooten 2005; Renzi and

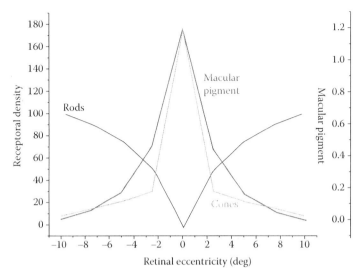

**FIGURE 8.12** (**See color insert.**) Data for the rod and cone densities were obtained from the original data by Osterberg (1935). The MP distribution was obtained from Werner et al. (2000) who measured MP density with HFP using a 12-degree reference. Note that for this example, MP is screening a significant number of rods in the central macula (around 10 degrees in diameter).

Hammond 2010a) and cognitive function (Johnson et al. 2008). Specifically, the Neural Efficiency hypothesis posits that increased levels of L and Z throughout the retina and brain (measures of retina reflecting levels within brain; Johnson, 2012) are related to improved neural function.

The definition of neural efficiency is, of course, meaningful. General good health is also related to improvements in neural activity along many dimensions (hence, good health is a major confounding factor for any direct cross-sectional comparison of LZ status and neural processing) (Gomez-Pinilla 2011). One effect that L and Z may have on the visual system and brain is increased neural processing speed (Hammond and Fletcher 2012; Renzi and Hammond 2010b). This assumes, of course, that a "faster" nervous system is beneficial (Salthouse 1996). Slowing does, in fact, appear to be a central event both in aging (Salthouse 2000) and degenerative conditions that extend from the eye (e.g., macular degeneration) (Phipps et al. 2004) to the brain (Alzheimer's disease; Curran and Wattis 1998). For example, the Stroop effect (interference caused when naming color words inked in contradictory colors) has often been studied as a measure of cognitive aging due to its marked decline with age and presumed reliance on a number of inhibitory processes. A large factor analysis of 20 studies by Verhaeghen and Meersman (1998), however, showed that the age-related reductions found for this complex static cognitive task was caused simply by the general slowing of the nervous system with age. It appears that many of the complex cognitive losses associated with aging (e.g., working memory) can be reduced to the simple factor of general slowing of neural response (often touted as the unifying principle behind the majority of age-related cognitive decline; Salthouse 2000; Park and Reuter-Lorenz 2009). Slowing has also been shown to be a central feature of numerous neurological deficits such as dyslexia (Breznitz and Meyler 2003), Parkinsons (Uc et al. 2005), schizophrenia (Knowles et al. 2010), and multiple sclerosis (Demaree et al. 1999). It is possible that LZ could improve neural function by simply increasing neural processing speed.

Undifferentiated increases in speed of processing, however, would not necessarily be beneficial. Visual perception is the result of various components of the visual image (color, form, motion, etc.) being processed and relayed to the brain along different channels. This segmented information is then recombined into a perceptual whole at relatively late stages of the visual process (the more frontal portions of the brain). When processing by the magnocellular and parvocellular system (the two main pathways from retina to cortex) are unbalanced, visual dyslexia results. Hence, processing speed must be increased in a highly coordinated manner for the overall efficiency of the entire system to be improved. There are means, however, by which carotenoids could help produce this outcome. One possibility, for example, is based on improving collective processing. One reason that processing speed is slowed in older adults is that more neural areas (e.g., more cortical volume) must be recruited to achieve the same result (an increasing lack of functional differentiation). Distributing processing in this manner takes more time and hence slows processing speeds (Park and Reuter-Lorenz 2009). Less neural real estate is necessary to solve a given problem in younger brains, probably because of the higher density of cells in any given neural region (Creasey and Rapoport 2004). If carotenoids induce neurons to connect laterally (as they presumably do through effects on connexin and gap junctions) (Stahl et al. 1997), they may facilitate the recruitment of adjoining areas during a processing task.

Other possibilities exist. For example, another hypothesis explaining cognitive decline and loss of executive skills with age is the selective age-related loss of fiber tracts within white matter causing, essentially, a cortical disconnection (see O'Sullivan et al. 2001). These white matter tracts form the essential pathways for much of the higher order reasoning that is facilitated by the neocortex. White matter (and myelin) has much higher lipid content than gray matter (O'Brien and Sampson 1965) and carotenoids in the brain associate more highly with white matter (Craft et al. 2004). It is possible that L and Z promote the preservation of white matter and the stabilization of informational tracts. This interpretation is consistent with the fact that L and Z are known to stabilize membranes (the xanthophylls are sometimes described as transmembrane rivets spanning lipid bilayers) and bind to microtubules (Bernstein et al. 1997) in the cytoskeleton. Another possibility is based on the observation that sensory function is strongly linked to a broad array of cognitive tasks in the elderly (Baltes and Lindenberger 1997). Sensory function does not appear to be linked to cognitive function at younger ages. If sensory ability does help to regulate intellectual performance at older ages, factors that improve sensory function (as has been argued for MP, Hammond et al. 1998) would also result in improved cognition.

At this stage, of course, any possible mechanism for how L and Z might improve neural efficiency remains purely speculative. Nonetheless, the first steps in testing the Neural Efficiency hypothesis are straightforward: L and Z can be supplemented and their effects on neural efficiency can be assessed. Neural efficiency itself can be measured using a battery of both direct (such as neuroimaging ortemporal processing) and indirect tests (such as psychomotor responses or careful cognitive testing,). Johnson et al. (2008) utilized this approach and found positive improvements in a number of cognitive indices tested in elderly women.

## 8.4 MULTIFUNCTIONALITY OF LUTEIN AND ZEXANTHIN

A general conclusion that can be drawn from the many hypotheses explaining MP function is that the macular carotenoids likely serve multiple functions within the central nervous system. In biology, a single molecular component often serves multiple purposes (e.g., the actions of dopamine within the basal ganglia [motor] and limbic [emotional] system). It is also plausible that the various functions of L and Z are more or less impactful at different stages of life, and in different geographic areas. For example, light protection effects of MP would likely be more important in high light-stress environments like the southwestern United States (e.g., Arizona) contrasted with the more northern latitudes (e.g., Wisconsin). Effects of light stress might also be more important in infancy when the crystalline lens is highly transparent (Hammond 2008, 2012). The importance of LZ as antioxidants might also be more significant when oxidative stress to the retina is high (such as during the maturational stages of infancy, Hammond 2008, 2012, or early in retinal disease). L and Z are concentrated in and around the fovea, which is the area that undergoes the largest maturational changes during the first year of life. Maturation of the retina and brain involves large-scale migration of neural cells, and pruning of synaptic circuitry (Hammond 2012). LZ are concentrated in the very regions undergoing these large-scale changes. Neural efficiency improvements would have great significance for the elderly as would optical improvements. For example, Richer et al. (2012), recently reported that increasing MPOD levels with supplementation was related to improved driving abilities (self-reported) in patients with early signs of macular degeneration. There is also evidence that photoreceptor loss tends to accelerate at older ages due to increases in inflammatory stress, and higher levels of photosensitizers (Margrain et al. 2004). Taken together, higher levels of protection might be particularly important both early and late in life. LZ effects on visual function and prophylaxis are likely significant across the entire life span.

## REFERENCES

Baltes, P.B., and Lindenberger, U. 1997. Emergence of a powerful connection between sensory and cognitive functions across the adult life span: a new window to the study of cognitive aging? *Psychol Aging.* 12(1), 12–21.

Barker, II, F.M., Snodderly, D.M., Johnson, E.J., et al. 2011. Nutritional manipulation of primate retinas, V: effects of lutein, zeaxanthin, and n-3 fatty acids on retinal sensitivity to blue-light–induced damage. *Invest Ophthalmol Vis Sci.* 52(7), 3934–3942.

Beatty, S., Murray, I.J., Henson, D.B., Carden, D., Koh, H., and Boulton, M.E. 2001. Macular pigment and risk for age-related macular degeneration in subjects from a Northern European population. *Invest Ophthalmol Vis Sci.* 42, 439–446.

Bernstein, P.S., Balashov, N.A., Tsong, E.D., and Rando, R.R. 1997. Retinal tubulin binds macular carotenoids. *Invest Ophthalmol Vis Sci.* 38, 167–175.

Bernstein, P.S., Khachik, F., Carvalho, L.S., Muir, G.J., Zhao, D.-Y., and Katz, N.B. 2001. Identification and quantification of carotenoids and their metabolites in the tissues of the human eye. *Exp Eye Res.* 72, 215–223.

Berrow, E.J., Bartlett, H.E., and Eperjesi, F. 2011. Do lutein, zeaxanthinandmacular pigment optical density differ with age or age-related maculopathy? *E-SPEN Eur E J Clin Nutr Metabol.* 6, e197–e201.

Bennett, M.G. 1930. The physical conditions controlling visibility through the atmosphere. *Q J Roy Met Soc.* 56, 1–29.

Bone, R.A., Landrum, J.T., Cao, Y., Howard, A.N., and Alvarez-Calderon, F. 2006. Macular pigment response to a supplement containing meso-zeaxanthin, lutein and zeaxanthin. *Nutr Metab.* 4, 12.

Bone, R.A., Landrum, J.T., Fernandez, L., and Tarsis, S.L. 1988. Analysis of the macular pigment by HPLC: retinal distribution and age study. *Invest Ophthalmol Vis Sci.* 29, 843–849.

Bone, R.A., Landrum, J.T., Friedes, L.M., et al. 1997. Distribution of lutein and zeaxanthin stereoisomers in the human retina. *Exp Eye Res.* 64, 211–218.

Bone, R.A., Landrum, J.T., Mayne, S.T., Gomez, C.M., Tibor, S.E., and Twaroska, E.E. 2001. Macular pigment in donor eyes with and without AMD: a case-control study. *Invest Ophthalmol Vis Sci.* 42, 235–240.

Breznitz, Z., and Meyler, A. 2003. Speed of lower-level auditory and visual processing as a basic factor in dyslexia: electrophysiological evidence. *Brain Lang.* 85, 166–184.

Chang, J.R., Koo, E., Agrón, E., et al. 2011. Risk factors associated with incident cataracts and cataract surgery in the Age-related Eye Disease Study (AREDS): AREDS report number 32. *Ophthalmology.* 118(11), 2113–2119.

Charbel Issa, P., van der Veen, R.L., Stijfs, A., Holz, F.G., Scholl, H.P., and Berendschot, T.T. 2009. Quantification of reduced macular pigment optical density in the central retina in macular telangiectasia type 2. *Exp Eye Res.* 89(1), 25–31.

Cho, E., Hankinson, S.E., Rosner, B., Willett, W.C., and Colditz, G.A. 2008. Prospective study of lutein/zeaxanthin intake and risk of age-related macular degeneration. *Am J Clin Nutr.* 87(6), 1837–1843.

Choi, J.S., Kim, D., Hong, Y., Mizuno, S., and Joo, C. 2006. Inhibition of nNOS and COX-2 expression by lutein in acute retinal ischemia. *Nutrition.* 22, 608–671.

Coleman, H.R. (2011). Modifiable risk factors of age-related macular degeneration. In *Age-related Macular Degeneration Diagnosis and Treatment* (pgs. 15–22). Springer, NY.

Connolly, E., Beatty, S., Thurnham, D.I., et al. 2010. Augmentation of macular pigment following supplementation with all three macular carotenoids: an exploratory study. *Curr Eye Res.* 35(4), 335–351.

Craft, N.E., Haitema, H.B., Garnett, K.M., Fitch, K.A., and Dorey, C.K. 2004. Carotenoid, tocopherol and retinol concentrations in the elderly human brain. *J Nutr Health Aging.* 8(3), 156–162.

Creasey, H., and Rapoport, S.I. 2004. The aging human brain. *Ann Neurol.* 17(1), 2–10.

Curran, S., and Wattis, J.P. 1998. Critical flicker fusion threshold: a useful research tool in patients with Alzheimer's disease. *Hum Psychopharmacol Clin Exp.* 13, 337–355.

Dartnall, H.J.A., and Thomson, L.C. 1949. Retinal oxygen supply and macular pigmentation. *Nature Lond.* 164, 876.

Davies, N.P., and Moreland, A.B. 2002. Color matching in diabetes: optical density of the crystalline lens and macular pigments. *Invest Ophthalmol Vis Sci.* 43, 281–284.

Demaree, H.A., DeLuca, J., Gaudino, E., and Diamond, B.J. 1999. Speed of information processing as a key deficit in multiple sclerosis. *J Neurol Neurosurg Psychiatry.* 67, 661–663.

DeValois, R.L., and DeValois, K.K. 1990. *Spatial Vision*, Oxford Psychology Series No. 14. New York, NY: Oxford University Press, 1–381.

Dietzel, M., Zeimer, M., Heimes, B., Claes, B., Pauleikhoff, D., and Hense, H.W. 2011. Determinants of macular pigment optical density and its relation to age-related maculopathy—results from the Muenster aging and retina study (MARS). *Invest Ophthalmol Vis Sci.* 52, 3452–3457.

Dimmick, F.L., and Hubbard, M.R. 1939. The spectral location of psychologically unique yellow, green, and blue. *Am J Psychol.* 52(2), 242–254.

Eye Disease Case-Control Study Group. 1992. Risk factors for neovascular age-related macular degeneration. *Arch Ophthalmol.* 110, 1701–1708.

Eye Disease Case-Control Study Group. 1993. Antioxidant status and neovascular age-related macular degeneration. *Arch Ophthalmol.* 111, 104–109.

Engles, M., Hammond, B.R., and Wooten, B.R. 2007. The relation between macular pigment and resolution acuity. *Invest Ophthalmol Vis Sci.* 48, 2922–2931.

Engles, M., Hammond, B.R., and Wooten, B.R. 2008. Evaluation of the acuity hypothesis of macular pigment using an assessment of the entire contrast sensitivity function. ARVO abstracts. *Invest Ophthalmol Vis Sci.* 4963.

Feeney-Burns, L., Hilderbrand, E.S., and Eldridge, S. 1984. Aging human RPE: morphometric analysis of macular, equatorial, and peripheral cells. *Invest Ophthalmol Vis Sci.* 25(2), 195–200.

Feeney-Burns, L., Neuringer, M., and Gao, C.-L. 1989. Macular pathology in monkeys fed semi-purified diets. *Prog Clin Biol Res.* 314, 601–622.

Gilmartin, B., and Hogan, R.E. 1985. The magnitude of longitudinal chromatic aberration of the human eye between 458 and 633 nm. *Vision Res.* 25(11), 1747–1753.

Gomez-Pinilla, F. 2011. Collaborative effects of diet and exercise on cognitive enhancement. *Nutr Health.* 20(3–4), 165–169.

Haegerstrom-Portnoy, G. 1988. Short-wavelength-sensitive-cone sensitivity with aging: a protective role for macular pigment? *J Opt Soc Am.* 5, 2140–2144.

Ham, W.T., Ruffolo, J.J., Mueller, H.A., Clarke, A.M., and Moon, M.E. 1978. Histologic analysis of photochemical lesions produced in rhesus retina by short wave-length light. *Invest Ophthalmol Vis Sci.* 17, 1029–1035.

Hammond Jr, B.R., Fuld, K., and Curran-Celentano, J. 1995. Macular pigment density in monozygotic twins. *Invest Ophthalmol Vis Sci.* 36(12), 2531–2541.

Hammond Jr, B.R., Johnson, E.J., Russell, R.M., et al. 1997. Dietary modification of human macular pigment density. *Invest Ophthalmol Vis Sci.* 38(9), 1795–1801.

Hammond, B.R., Wooten, B.R., and Snodderly, D.M. 1998. Preservation of visual sensitivity of older subjects: association with macular pigment density. *Invest Ophthalmol Vis Sci.* 39(2), 397–406.

Hammond, B.R., and Wooten, B.R. 2005. CFF Thresholds: relation to macular pigment optical density. *Ophthalmic Physiol Opt.* 25, 315–319.

Hammond, B.R. 2008. A possible role for dietary lutein and zeaxanthin in visual development. *Nutr Rev.* 66(12), 695–702.

Hammond, B.R., and Renzi, L. 2008. The characteristics and function of lutein and zeaxanthin within the human retina. In: *Phytochemicals: Aging and Health.* Meskin M. S., Bidlack W. and Randolph R.K. (ed). Boca Raton, FL: CRC Press, 89–106.

Hammond, B.R., Bernstein, B., and Dong, J. 2009. The effect of the Acrysof(R) natural lens on glare disability and photostress. *Am J Ophthalmol.* 148, 272–276.

Hammond, B.R., Renzi, L.M., Sachak, S., and Brint, S.F. 2010. A contralateral comparison of blue-filtering and non-blue-filtering intraocular lenses: glare disability, heterochromatic contrast threshold, and photostress recovery. *Clin Ophthalmol.* 4, 1465–1473.

Hammond, B.R. 2012. The dietary carotenoids lutein and zeaxanthin in pre-and-postnatal development. *Funct Food Rev.* 4(3), 130–137.

Hammond, B.R., and Fletcher, L.M. 2012. Influence of the dietary carotenoids lutein and zeaxanthin on visual performance: application to baseball. *Am J Clin Nutr.* 96(5), 1207S–1213S.

Hammond, B.R., Wooten, B.R., Engles, M., and Wong, J.C. 2012. The influence of filtering by the macular carotenoids on contrast sensitivity measured under simulated blue haze conditions. *Vision Res.* 63, 58–62.

Hammond, C.J., Liew, S.M., Van Kuijk, F.J., et al. 2012. The heritability of macular response to supplemental lutein and zeaxanthin: a classic twin study. *Invest Ophthalmol Vis Sci.* 53(8), 4963–4968.

Hammond, B.R., Fletcher, L.M., and Elliott, J.G. 2013. Glare disability, photostress recovery, and chromatic contrast: relation to macular pigment and serum lutein and zeaxanthin. *Invest Ophthalm Vis Sci* 54(1), 476–481.

Handelman, G.J., Dratz, E.A., Reay, C.C., and Van Kuijk, J.G. 1988. Carotenoids in the human macula and whole retina. *Invest Ophthalmol Vis Sci.* 29(6), 850–855.

Howarth, P.A., and Bradley, A. 1986. The longitudinal chromatic aberration of the human eye, and its correction. *Vis Res.* 26(2), 361–366.

Hubel, D.H., and Wiesel, T.N. 1968. Receptive fields and functional architecture of monkey striate cortex. *J Physiol.* 195(1), 215–243.

Izumi-Nagai, K., Nagai, N., Ohgami, K., et al. 2007. Macular pigment lutein is antiinflammatory in preventing choroidal neovascularization. *Arterioscler Thromb Vasc Biol.* 27(12), 2555–2562.

Johnson, E.J., McDonald, K., Caldarella, S.M., Chung, H.Y., Troen, A.M., and Snodderly, D.M. 2008. Cognitive findings of an exploratory trial of docosahexaenoic acid and lutein supplementation in older women. *Nutr Neurosci.* 11, 75–83.

Johnson, E., Maras, J., Rasmussen, H.M., and Tucker, K.L. 2010. Intake of lutein and zeaxanthin differ with age, sex, and ethnicity. *J Am Diet Assoc.* 9, 1357–1362.

Johnson, E.J. 2012. A possible role for lutein and zeaxanthin in cognitive function in the elderly. *Am J Clin Nutr.* 96(5), 1161S–1165S.

Karnaukhov, V.N. 1990. Carotenoids: recent progress, problems and prospects. *Comp Biochem Physiol B.* 95, 1–20.

Karppi, J., Laukkanen, J.A., and Kurl, S. 2012. Plasma lutein and zeaxanthin and the risk of age-related nuclear cataract among the elderly Finnish population. *Br J Nutr.* 108(1), 148–154.

Kelly, S.A. 1990. Effect of yellow-tinted lenses on brightness. *J Opt Soc Am A.* 7, 1905–1911.

Kim, S.R., Nakanishi, K., Itagaki, Y., and Sparrow, J.R. 2006. Photooxidation of A2-PE, a photoreceptor outer segment fluorophore, and protection by lutein and zeaxanthin. *Exp Eye Res.* 82(5), 828–839.

Khachik, F., Bernstein, P.S., and Garland, D.L. 1997. Identification of lutein and zeaxanthin oxidation products in human and monkey retinas. *Invest Ophthalmol Vis Sci.* 38(9), 1802–1811.

Kirschfeld, K. 1982. Carotenoid pigments: their possible role in protecting against photooxidation in eyes and photoreceptor cells. *Proc R Soc Lond, B, Biol Sci.* 216, 71–85.

Klein, R.J., Zeiss, C., Chew, E.Y., et al. 2005. Complement factor H polymorphism in age-related macular degeneration. *Science.* 308, 385–389.

Knowles, E.E, David, A.S, and Reichenberg, A. 2010. Processing-speed deficits in schizophrenia: re-examining the evidence. *Am J Psychiatry.* 167, 828–835.

Komeima, K., Rogers, B.S., Lu, L., and Campochiaro, P.A. 2006. Antioxidants reduce cone cell death in a model of retinitis pigmentosa. *PNAS.* 103(30), 11300–11305.

Krinsky, N.I., Landrum, J.T., and Bone, R.A. 2003. Biologic mechanisms of the protective role of lutein and zeaxanthin in the eye. *Annu Rev Nutr.* 23, 171–201.

Kvansakul, J., Rodriguez-Carmona, M., Edgar, D.F., et al. 2006. Supplementation with the carotenoids lutein or zeaxanthin improves human visual performance. *Ophthalmic Physiol Opt.* 26, 362–371.

Landrum, J.T., Bone, R.A., Moore, L.L., and Gomez, C.M. 1999. Analysis of zeaxanthin distribution within individual human retinas. *Methods Enzymol.* 299, 457–467.

Landrum, J.T. 2013. Reactive oxygen and nitrogen species in biological systems: reactions and regulation by carotenoids. In: *Carotenoids in Human Health*, Tanumihardjo, S. (ed). New York: Springer, 57–101.

Liang, F.Q., and Godley, B.F. 2003. Oxidative stress-induced mitochondrial DNA damage in human retinal pigment epithelial cells: a possible mechanism for RPE aging and age-related macular degeneration. *Exp Eye Res.* 76, 397–403.

Liew, S.H., Gilbert, C.E., Spector, T.D., et al. 2005. Heritability of macular pigment: a twin study. *Invest Ophthalmol Vis Sci.* 46, 4430–4436.

Lim, J.I. 2008. *Age-Related Macular Degeneration*, 2nd edition. Informa Healthcare, New York, NY, 1–366.

Lima, V.C., Rosen, R., Maia, M., et al. 2010. Macular pigment optical density measured by dual wavelength autofluorescence imaging in diabetic and non-diabetic patients: a comparative study. *Invest Ophthalmol Vis Sci.* 51, 5840–5845.

Loughman, J., Davison, P., Akkalli, M., Nolan, J., and Beatty, S. 2010. Visual performance and macular pigment. *J Optom.* 3(2), 73–89.

Luria, S.M. 1972. Vision with chromatic filters. *Am J Opt Arch Amer Acad Opt.* 10, 818–829.

Mainster, M.A. 2006. Voilet and blue light blocking intraocular lenses: photoprotection versus photoreception. *Br J Ophthalmol.* 90(6), 784–792.

Malinow, M.R., Feeney-Burns, L., Peterson, L.H., Klien, M.L., and Neuringer, M. 1980. Diet-related macular abnormalities in monkeys. *IOVS.* 19, 857–863.

Mares, J.A., Voland, R.P., Sondel, S.A., et al. 2011. Healthy lifestyles related to subsequent prevalence of age-related macular degeneration. *Arch Ophthalmol.* 129(4), 470–480.

Margrain, T.H., Boulton, M., Marshall, J., and Sliney, D.H. 2004. Do blue light filters confer protection against age-related macular degeneration. *Prog Ret Eye Res.* 23, 523–531.

McLellan, J.S., Marcos, S., Prieto, P.M., and Burns, S.A. 2002. Imperfect optics may be the eye's defence against chromatic blur. *Nature.* 417, 174–176.

Mortensen, A., and Skibsted, L.H. 1997. Relative stability of carotenoid radical cations and homologue tocopheroxyl radicals. A real time kinetic study of antioxidant hierarchy. *FEBS Lett.* 417(3), 261–266.

Neelam, K., Nolan, J., Loane, E., et al. 2006. Macular pigment and ocular biometry. *Vision Res.* 46, 2149–2156.

Neitz, J., Carroll, J., Yamauchi, Y., Neitz, M., and Williams, D.R. 2002. Color perception is mediated by a plastic neural mechanism that remains adjustable in adults. *Neuron.* 35, 783–792.

Nussbaum, J.J., Pruett, R.C., and Delori, F.C. 1981. Historic perspectives. Macular yellow pigment. The first 200 years. *Retina.* 1(4), 296–310.

O'Brien, J.S., and Sampson, E.L. 1965. Lipid composition of the normal human brain: gray matter, white matter, and myelin. *J Lipid Res.* 6, 537–544.

Olmedilla, B., Granado, F., Blanco, I., and Vaquero, M. 2003. Lutein, but not alpha-tocopherol, supplementation improves visual function in patients with age-related cataracts: a 2-y double-blind, placebo-controlled pilot study. *Nutrition.* 19(1), 21–24.

O'Sullivan, M., Jones, D.K., Summers, P.E., Morris, R.G., Williams, S.C.R., and Markus, H.S. 2001. Evidence for cortical "disconnection" as a mechanism of age-related cognitive decline. *Neurology.* 57, 632–638.

Park, D.C., and Reuter-Lorenz, P. 2009. The adaptive brain: aging and neurocognitive scaffolding. *Annu Rev Psychol.* 60, 173–196.

Pérez, M.J., Puell, M.C., Sánchez, C., and Langa, A. 2003. Effect of a yellow filter on mesopic contrast perception and differential light sensitivity in the visual field. *Ophthalmic Res.* 35, 54–59.

Phipps, J.A., Dang, T.M., Vingrys, A.J., and Guymer, R.H. 2004. Flicker perimetry losses in age-related macular degeneration. *IOVS.* 45, 3355–3360.

Powell, R.R. 1983. Flicker fusion as a typological index of nervous system 'reactivity'. *Percept Mot Skills.* 57, 701–702.

Ray, N.J., Fowler, S., and Stein, J.F. 2005. Yellow filters can improve magnocellular function: motion sensitivity, convergence, accommodation, and reading. *Ann N Y Acad Sci.* 1039, 283–293.

Reading, V.M., and Weale, R.A. 1974. Macular pigment and chromatic aberration. *J Am Optom Assoc.* 64(2), 231–234.

Renzi, L., and Hammond, B.R. 2010a. The effect of macular pigment on heterochromatic luminance contrast. *Exp Eye Res.* 91(6), 896–900.

Renzi, L.M., and Hammond, B.R. 2010b. The relation between the macular carotenoids, lutein and zeaxanthin, and temporal vision. *Ophthalmic Physiol Opt.* 30(4), 351–357.

Richer, S., Stiles, W., Statkute, L., et al. 2004. Double-masked, placebo-controlled, randomized trial of lutein and antioxidant supplementation in the intervention of atrophic age-related macular degeneration: the Veterans LAST Study (Lutein Antioxidant Supplementation Trial). *Optometry.* 75(4), 216–230.

Richer, S.P., Stiles, W., Graham-Hoffman, K., et al. 2011. Randomized, double-blind, placebo-controlled study of zeaxanthin and visual function in patients with atrophic age-related macular degeneration: the Zeaxanthin and Visual Function Study (ZVF) FDA IND# 78, 973. *Optomet J Am Optom Assoc.* 82(11), 667–680.

Richer, S., Park, D.W., Epstein, R., Wrobel, J.S., and Thomas, C. 2012. Macular re-pigmentation enhances driving vision in elderly adult males with macular degeneration. *J Clin Exp Ophthalmol.* 3, 217. doi:10.4172/2155-9570.1000217

Rosinski, R.R., Mulholland, T., Degelman, D., and Farber, J. 1980. Picture perception: an analysis of visual compensation. *Atten Percept Psychophys.* 28(6), 521–526.

Salthouse, T.A. 1996. The processing-speed theory of adult age differences in cognition. *Psychol Rev.* 103(3), 403.

Salthouse, T.A. 2000. Aging and measures of processing speed. *Biol Psychol.* 54, 35–54.

Sandberg, M.A., and Gaudio, A.R. 1995. Slow photostress recovery and disease severity in age-related macular degeneration. *Retina.* 15, 407–412.

SanGiovanni, J.P., and Neuringer, M. 2012. The putative role of lutein and zeaxanthin as protective agents against age-related macular degeneration: promise of molecular genetics for guiding mechanistic and translational research in the field. *Am J Clin Nutr.* 96(5), 1223S–1233S.

Sasamoto, Y., Gomi, F., Sawa, M., Tsujikawa, M., and Nishida, K. 2011. Effect of 1-year lutein supplementation on macular pigment optical density and visual function. *Graefe's Arch Clin Exp Ophthalmol.* 249(12), 1847–1854.

Schalch, W., Dayhaw–Barker, P., and Barker, F.M. 1999. The carotenoids of the human retina. In: *Nutritional and Environmental Influences on the Eye*, Taylor, A. (ed). Boca Raton, FL: CRC Press, 215–250.

Schultze, M. 1866. *Ueber den gelben Fleck der Retina, seinen Einfluss auf normales Sehen und auf Farbenblindheit* (About the yellow spot in the retina, its influence on normal vision and on color blindness). Bonn: Von Max Cohen & Sohn.

Schwartz, S.H. 2010. *Visual Perception: A Clinical Orientation*, 4th edition. New York: McGraw-Hill.

Seddon, J.M., Ajani, U.A., Sperduto, R.D., et al. 1994. Dietary carotenoids, vitamin A, C, and E, and advanced age-related macular degeneration. Eye Disease Case-Control Study Group. *JAMA.* 272, 1413–1420.

Seddon, J.M., Cote, J., Page, W.F., Aggen, S.H., and Neale, M.C. 2005. The US twin study of age-related macular degeneration: relative roles of genetic and environmental influences. *Arch Ophthalmol.* 123, 321–327.

Shen, J., Yang, X., Dong, A., et al. 2005. Oxidative damage is a potential cause of cone cell death in retinitis pigmentosa. *J Cell Physiol.* 203, 457–464.

Sparrow, J., and Boulton, M. 2005. RPE lipofuscin and its role in retinal pathobiology. *Exp Eye Res.* 80, 595–606.

Stahl, W., Nicolai, S., Briviba, K., et al. 1997. Biological activities of natural and synthetic carotenoids: induction of gap junctional communication and singlet oxygen quenching. *Carcinogenesis.* 18, 89–92.

Stringham, J.M., Fuld, K., and Wenzel, A.J. 2003. Action spectrum for photophobia. *J Opt Soc Am A.* 20, 1852–1858.

Stringham, J.M., Fuld, K., and Wenzel, A.J. 2004. Spatial properties of photophobia. *IOVS.* 45, 3838–3848.

Stringham, J., and Hammond, B.R. 2007. The glare hypothesis of macular pigment function. *Optom Vis Sci.* 84, 859–864.

Stringham, J., and Hammond, B.R. 2008. Macular pigment and visual performance under glare conditions. *Optom Vis Sci.* 85, 82–88.

Stringham J.M., Bovier E.R., Wong J.C., and Hammond B.R. 2010. The influence of dietary lutein and zeaxanthin on visual performance. *J Food Sci.* 75(1), R24–9.

Sujak, A., Gabrielska, J., Grudzinski, W., Borc, R., Mazurek, P., and Gruszecki, W.I. 1999. Lutein and zeaxanthin as protectors of lipid membranes against oxidative damage: the structural aspects. *Arch Biochem Biophys.* 371, 301–307.

Thomson, L.R., Toyoda, Y., Langner, A., et al. 2002. Elevated retinal zeaxanthin and prevention of light-induced photoreceptor cell death in quail. *Invest Ophthalmol Vis Sci.* 43(11), 3538–3549.

Thurnham, D.I., Tremel, A., and Howard, A. 2008. A supplementation study in human subjects with a combination of meso-zeaxanthin, (3R, 30R)-zeaxanthin and (3R, 30R, 60R)-lutein. *Br J Nutr.* 100(6), 1307–1314.

Uc, E.Y., Rizzo, M., Anderson, S.W., Qian, S., Rodnitzky, R.L., and Dawson, J.D. 2005. Visual dysfunction in Parkinson disease without dementia. *Neurology.* 65, 1907–1913.

VandenLangenberg, G.M., Mares-Perlman, J.A., Klein, R., Klein, B.E.K., Brady, W.E., and Palta, M. 1998. Associations between antioxidant and zinc intake and the 5-year incidence of early age-related maculopathy in the Beaver Dam Eye Study. *Am J Epidemiol.* 148, 204–214.

van Norren, D., and Schellekens, P. 1990. Blue light hazard in rat. *Vision Res.* 30(10), 1517–1520.

Verhaeghen, P., and De Meersman, L. 1998. Aging and the Stroop effect: a meta-analysis. *Psychol Aging.* 13, 120–126.

Vishwanathan, R., Kuchan, M.J., and Johnson, E.J. (2011). Lutein is the predominant carotenoid in the infant brain. In *16th International Symposium on Carotenoids*, Krakow, Poland. No. 1.23. 2011.

Walls, G.L., and Judd, H.D. 1933. Intra-ocular color filters of vertebrates. *Br J Ophthalmol.* 17, 641–725.

Wenzel, A.J., Stringham, J.M., Fuld, K., and Curran-Celentano, J. 2006. Macular pigment optical density and photophobia light threshold. *Vision Res.* 46, 4615–4622.

Wolffsohn, J.S., Cochrane, A.L., Khoo, H., Yochimitsu, Y., and Wu, S. 2000. Contrast is enhanced by yellow lenses because of selective reduction of short-wavelength light. *Optom Vis Sci.* 77(2), 73–81.

Wong, J.C., Engles, M., Wooten, B.R., and Hammond, B.R. 2009. Testing the visibility hypothesis of macular pigment using a filter that simulates changes in macular pigment density. ARVO Abstracts. *Invest Ophthalmol Vis Sci.* 1704.

Wooten, B.R., and Hammond, B.R. 2002. Macular pigment: influences on visual acuity and visibility. *Prog Retinal Eye Res.* 21, 225–240.

Wu, J., Seregard, S., and Algvere, P.V. 2006. Photochemical damage of the retina. *Surv Ophthalmol.* 51(5), 461–481.

Zheng, W., Zhang, Z., Jiang, K., Zhu, J., He, G., and Ke, B. 2012. Macular pigment optical density and its relationship with refractive status and foveal thickness in Chinese school-aged children. *Curr Eye Res.* 1–6, 38(1), 168–173.

Zimmer, P., and Hammond, B.R. 2007. Lutein and zeaxanthin and the developing retina. *Clin Ophthalmol.* 1, 181–189.

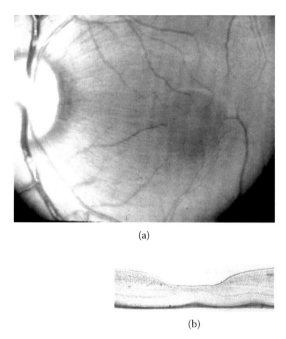

**FIGURE 1.1** Microscopic cross sections of macula aligned with a fundus photograph of the macula. (a) The macaque fundus and (b) microscopic cross section of macaque fovea in natural light. (Photos courtesy of M. Neuringer and D. M. Snodderly. With permission.)

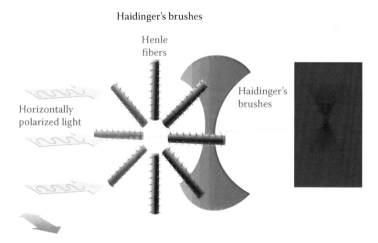

**FIGURE 1.3** The macular carotenoids are organized by their preference to align perpendicularly to the membranes of the axons that are themselves radially displayed within the fovea. Polarized light is absorbed only by those molecules that are aligned generally parallel with the plane of polarization resulting in attenuation of the beam in the zone depicted with the hourglass shape producing the Haidinger's brush entoptical effect.

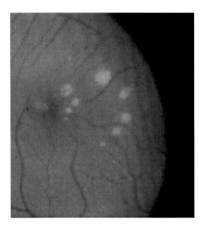

**FIGURE 1.6** Exposure of the macaque retina to blue light produces lesions that are dependent on the total energy of the exposure. Two separate sets of five exposures were made in arcs: (a) the outer arc produced larger lesions at each exposure than the identical energy exposures associated with the (b) inner arc where the macular pigment attenuates the intensity of the damaging light reaching the photoreceptors and retinal pigment epithelium. The lowest energy exposure within the central yellow pigmented region failed to produce a detectable lesion. (From Barker, F.M. et al., *Invest Ophthalmol Vis Sci.*, 52: 3934, 2011. With permission.)

**FIGURE 7.5** Top: Microscopic pictures of the two tested lutein products, A (left) and B (right). Bottom: Macroscopically evident differences in dispersion behavior: product A (left) disperses readily in artificial gastric juice (0.1 N HCl) while product B (right) does disperse very little if at all (pictures are taken within 1 minute after addition of 20 mg of the respective lutein product).

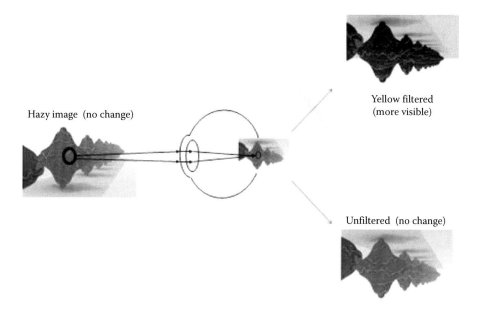

**FIGURE 8.9** The example shows an achromatic test target and surround of nearly the same luminance. The central target becomes visible when either the target is just slightly brighter or the surrounding is just slightly darker. This very small change in brightness is enough to create a luminance edge (a phenomenon called brightness induction). These experiments are typically done with achromatic stimuli like those shown above but a similar effect would hold for colored stimuli.

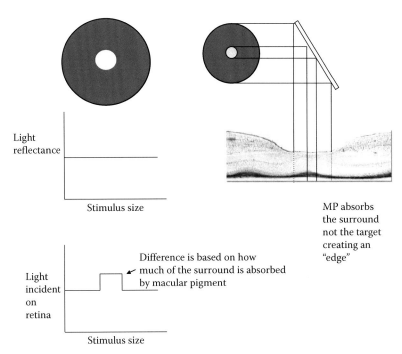

**FIGURE 8.10** This example shows a blue surround (black in illustration) with a yellow (white in illustration) stimulus of equal luminance. Note that what defines the edge is based only on the wavelength or color difference. The luminance difference itself, however, will be exaggerated by differential absorption by macular pigment.

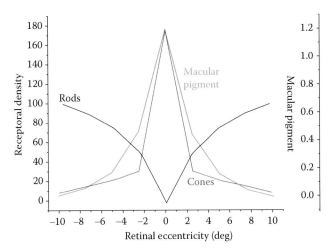

**FIGURE 8.12** Data for the rod and cone densities were obtained from the original data by Osterberg (1935). The MP distribution was obtained from Werner et al. (2000) who measured MP density with HFP using a 12-degree reference. Note that for this example, MP is screening a significant number of rods in the central macula (around 10 degrees in diameter).

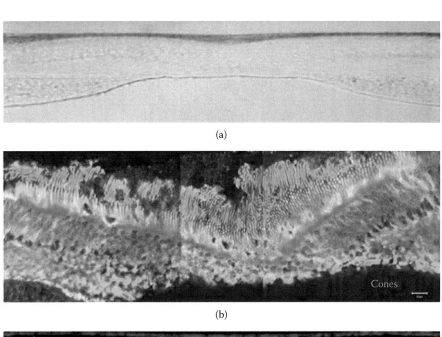

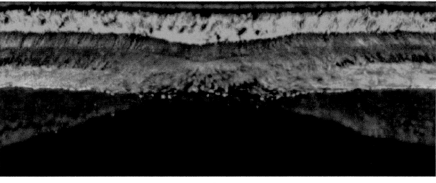

**FIGURE 9.1** (a) Vertical section (vitreous side down) through a monkey fovea showing the distribution of the yellow macular carotenoids. (Courtesy of Dr. Max Snodderly.) (b) GSTP1 labeling of foveal cones in the macula of a 3-year-old monkey. The orientation of the section is the same as in Figure 9.1a. This montage shows strongest labeling by antibody against GSTP1 (red) over the myoid and ellipsoid regions of cones identified by monoclonal antibody 7G6 (green). (c) A low-magnification view of a near-foveal retina section in which N-62 StAR (red) identifies StARD3, an anti-cone arrestin monoclonal antibody, 7G6 (green) identifies monkey cones. (Courtesy of Dr. Jeanne M. Frederick.)

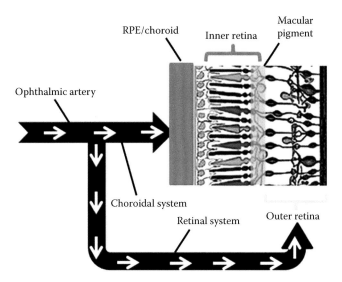

**FIGURE 9.2** Diagram of the human retina circulation. White arrow stands for blood flow.

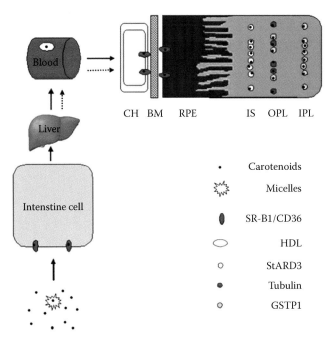

**FIGURE 9.3** Possible pathway for MP carotenoid uptake, transport, and accumulation in the human retina. BM, Bruch's membrane; CH, choroicapillaris; IPL, inner plexiform layer; IS, inner segments; OPL, outer plexiform layer; RPE, retinal pigment epithelium; ⇢, retinal transport pathway.

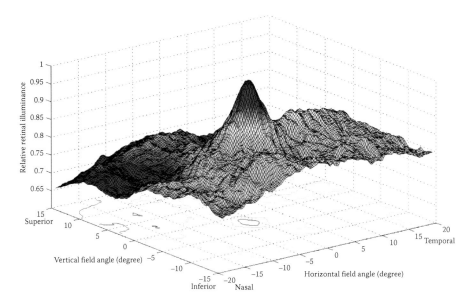

**FIGURE 12.2** Sample mesh plot showing the relative cumulative light distribution on the retina for one of the informed subjects who was instructed to look at bright objects in a sequence of photographic images.

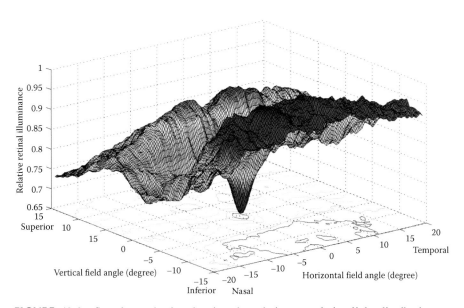

**FIGURE 12.3** Sample mesh plot showing the relative cumulative light distribution on the retina for one of the informed subjects who was instructed to look at dark objects in a sequence of photographic images.

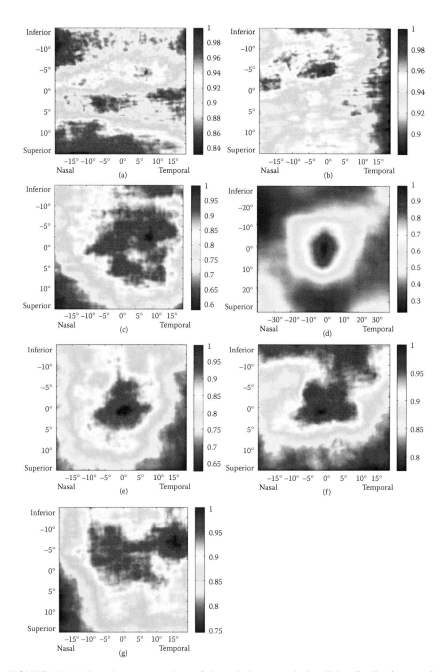

**FIGURE 12.4** Sample contour plots of the relative cumulative light distribution on the retina for one of the naive subjects under different viewing situations: (a) and (b) viewing a sequence of photographic images; (c) viewing a video presentation; (d) viewing a computer monitor; and (e), (f), and (g) free viewing while walking.

# 9 Transport and Retinal Capture of the Macular Carotenoids

*Binxing Li and Paul S. Bernstein*

## CONTENTS

9.1 Introduction ................................................................................................ 171
9.2 Macular Pigment Carotenoids Binding Proteins ...................................... 173
    9.2.1 Glutathione S-Transferase P1 ..................................................... 173
    9.2.2 Steroidogenic Acute Regulatory Domain Protein 3 .................. 174
9.3 Macular Pigment Uptake and Transport Proteins .................................... 175
    9.3.1 Plasma Lipoproteins .................................................................... 175
    9.3.2 Scavenger Receptor Proteins ...................................................... 176
    9.3.3 Metabolic Enzymes .................................................................... 177
9.4 Macular Pigment Uptake and Transport Pathways .................................. 177
    9.4.1 Retina Circulation ....................................................................... 178
    9.4.2 Pathways ...................................................................................... 179
9.5 Conclusion ................................................................................................. 180
References ....................................................................................................... 180

## 9.1 INTRODUCTION

In the past few decades, extensive investigations have been performed on the macular pigments' (MP) chemical structures (Bone et al. 1993), ocular distribution and locations (Snodderly et al. 1984), and possible photoprotective mechanisms (Beatty et al. 1999; Krinsky et al. 2003; Li et al. 2010), and the relationships of the MPs with several eye diseases such as age-related macular degeneration (AMD) have been studied, culminating in age-related eye disease study 2 (AREDS2), the first large, randomized, nationwide clinic trial to test the efficacy of lutein and zeaxanthin in the prevention of visual loss from AMD (Beatty et al. 1999; Landrum et al. 1997; Seddon et al. 1994; Whitehead et al. 2006). As discussed in Chapters 1, 2, and 4, the major MPs in human retina are two carotenoids, lutein and zeaxanthin, along with lutein's metabolite *meso*-zeaxanthin. The macular carotenoid pigments are concentrated in human fovea in the Henle fiber layer at by far the highest carotenoid concentration encountered anywhere in the human body (Bone and Landrum 1984; Bone et al. 1997; Landrum and Bone 2001; Rapp et al. 2000; Snodderly et al. 1984) (Figure 9.1a). Moreover, the uptake of the MP is chemically

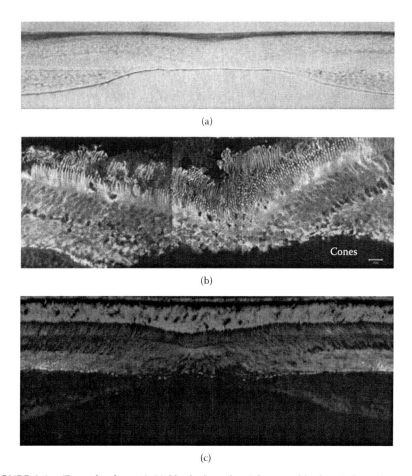

**FIGURE 9.1** **(See color insert.)** (a) Vertical section (vitreous side down) through a monkey fovea showing the distribution of the yellow macular carotenoids. (Courtesy of Dr. Max Snodderly.) (b) GSTP1 labeling of foveal cones in the macula of a 3-year-old monkey. The orientation of the section is the same as in Figure 9.1a. This montage shows strongest labeling by antibody against GSTP1 (red) over the myoid and ellipsoid regions of cones identified by monoclonal antibody 7G6 (green). (c) A low-magnification view of a near-foveal retina section in which N-62 StAR (red) identifies StARD3, an anti-cone arrestin monoclonal antibody, 7G6 (green) identifies monkey cones. (Courtesy of Dr. Jeanne M. Frederick.)

selective because of the 15–20 carotenoids detectable in human plasma, only lutein and zeaxanthin are delivered into human retina (Khachik et al. 1991, 1997, 1998). The specific distribution and selective uptake of lutein and zeaxanthin suggest that the MPs are bound to specific carotenoid-binding proteins that provide a stable destination for circulating lutein and zeaxanthin (Bhosale et al. 2004, 2009; Li et al. 2011). Here, we review the identification of the binding proteins of the MP carotenoids, summarize the current knowledge of the pathways of carotenoid transport from the diet to the eye, and examine the potential participation of metabolic enzymes in regulating ocular carotenoid concentrations.

## 9.2 MACULAR PIGMENT CAROTENOIDS BINDING PROTEINS

The *macula lutea* is the visible yellow spot at the fovea of the human retina. Two xanthophyll carotenoids, lutein and zeaxanthin, are responsible for this yellow color (Home 1798; Wald 1945). These two carotenoids are also called MP. The concentration of MP in the human macula is around 100 times higher than that of the peripheral retina (Khachik et al. 2002). Cross-sectionally, the retina is composed of 10 layers of neural cells, and the MP is located only in the outer plexiform (Henle fiber) layer in the fovea, while in the parafovea the MP is also located in another retina layer, the inner plexiform layer (Bone and Landrum 1984; Snodderly et al. 1984). The carotenoid concentration of lens is 1/200th of macula (Bernstein et al. 2001). The total carotenoid concentrations in the retinal pigment epithelium–choroid (RPE–choroid), ciliary body, and iris are 10 times lower than in the human macula (Bernstein et al. 2002; Khachik et al. 2002). Whenever a tissue exhibits highly selective uptake for a compound, it is likely that one or more specific binding proteins are involved in the process. Several carotenoid-binding proteins that are responsible for the capture of the MP have been identified in our laboratory (Bernstein et al. 1997; Bhosale et al. 2004; Li et al. 2011). From the water-soluble proteins of human retina, tubulin has been identified as a low specificity carotenoid-binding protein that shows affinity for lutein, zeaxanthin, and other carotenoids. Using human macular membranes, glutathione S-transferase P1 (GSTP1) was purified and identified to be a zeaxanthin-binding protein. More recently, steroidogenic acute regulatory domain protein 3 (StARD3) was identified as a lutein-binding protein. More details about the two specific carotenoid-binding proteins, GSTP1 and StARD3, are presented in Sections 9.2.1 and 9.2.2.

### 9.2.1 GLUTATHIONE S-TRANSFERASE P1

Glutathione S-transferases (GSTs) are members of a family of phase II detoxification enzymes that catalyze the conjugation of glutathione (GSH) to a wide variety of endogenous and exogenous electrophilic compounds. GSTs are classified into two superfamilies: the membrane-bound microsomal family and the cytosolic family. Human cytosolic GSTs can be divided into six classes: $\alpha$, $\mu$, $\omega$, $\pi$, $\theta$, and $\zeta$ (Townsend and Tew 2003). Only the pi isoform of GST, GSTP1 (formerly GST $\pi$), is known to be expressed abundantly in human tissue, and its gene is localized to chromosomal locus 11q13-qter (Di Ilio et al. 1990). GSTP1 cDNA contains an open reading frame of 630 nucleotides, encoding 210 amino acids (Ali-Osman et al. 1997). GSTP1 is widely expressed in human epithelial tissue and is composed of two identical 23-kDa subunits (Gulick et al. 1992).

Recombinant human GSTP1 exhibited high affinity for macular zeaxanthins with an equilibrium two-site average $K_D$ of 0.33 µM for (3R, 3′R)-zeaxanthin and 0.52 µM for (3R, 3′S-*meso*)-zeaxanthin and only low-affinity interactions with lutein. When closely related human GST proteins were tested, GSTM1 and GSTA1 exhibited no appreciable affinity for lutein or zeaxanthin, further confirming the specificity of interaction between GSTP1 and macular zeaxanthin. Immunolocalization of GSTP1 in the human and monkey retina revealed that GSTP1 was concentrated in the outer and inner plexiform layers of the fovea and in the photoreceptor inner segment ellipsoid region (Figure 9.1b) (Bhosale et al. 2004).

It has been reported that GSTP1 can act as a retinoic acid *cis-trans* isomerase in a GSH-independent manner (Chen and Juchau 1997). The *GSTP1* gene is upregulated during the early stages of oncogenesis, and it is the most significantly overexpressed *GST* gene in many human tumors (Lo and Ali-Osman 1997). Recently, GSTP1 was identified as a nitric oxide scavenger protein, and it also can serve as a regulator of protein S-glutathionylation caused by reactive oxygen and nitrogen species (Pedersen et al. 2007; Townsend et al. 2009). Our identification of GSTP1 as a zeaxanthin-binding protein in the macula of human eye and our subsequent finding that they can synergistically protect lipid membranes from oxidation assign additional important roles to this well-known protein (Bhosale and Bernstein 2005; Bhosale et al. 2004).

Three polymorphic *GSTP1* genes have been cloned from malignant glioma cells (AliOsman et al. 1997). More recently, it has been suggested that certain gene polymorphisms of *GSTs* including *GSTP1* may be associated with the subsequent developments of neovascular AMD and cortical cataracts (Juronen et al. 2000a; 2000b, Oz et al. 2006).

### 9.2.2 Steroidogenic Acute Regulatory Domain Protein 3

StARD3, also known as MLN64, belongs to a lipid transfer–related protein family composed of 15 identified protein members in humans (Alpy and Tomasetto 2005; Sierra 2002). These proteins can be divided into six subfamilies (Soccio et al. 2002). StARD3 was first found as a gene in breast and ovarian cancer cells (Strauss et al. 2003). The gene chromosomal location of StARD3 is 17q11-q12. StARD3 has a wide expression pattern in tissues (Watari et al. 1997). The subcellular localization of StARD3 is in late endosomes unlike its subfamily member, StARD1, which is a mitochondrial protein (Alpy et al. 2001; Neufeld et al. 1999; Zhang et al. 2002). The StARD3 protein consists of 445 amino acid residues, containing two distinct domains: an N-terminal MENTAL domain and a C-terminal StARD Domain (Bose et al. 2000). The 2.2-Å crystal structure of the StARD domain of StARD3 with cholesterol bound to it reveals an a/b fold built around a U-shaped incomplete B-barrel, and the ratio of cholesterol molecule to StARD domain is 1:1 (Tsujishita and Hurley 2000).

StARD3 manifests several properties expected of a lutein-binding protein (Li et al. 2011). Shown macula-enriched by immunoblot analysis, StARD3 binds lutein selectively with high affinity, it induces a spectral shift of lutein's absorption spectrum in a manner that corresponds well with the *in vivo* MP spectrum, and it reveals an immunolocalization overlapping with our previously measured resonance Raman distribution of MP carotenoids. A specific antibody to StARD3, N-62 StAR, localizes to all neurons of monkey macular retina and it is especially present in foveal cone inner segments and axons, but it does not colocalize with the Müller cell marker, glutamine synthetase (Figure 9.1c). Recombinant StARD3 selectively binds lutein with high affinity ($K_D$ = 0.45 mM) when assessed by surface plasmon resonance binding assays. Thus, StARD3 and GSTP1 proteins provide abundant lutein- and zeaxanthin-binding sites, respectively, that account for the unique distribution and stability of carotenoids found in the primate *macula lutea*.

The function of StARD3 is still not clear, but it is thought to participate in the transmembrane transport process of cholesterol based on the presence of StAR domain (Alpy and Tomasetto 2005; Strauss et al. 2002). Interestingly, full-length StARD3 shows no activity in steroidogenesis assays, while the StAR domain released from StARD3 protein can deliver cholesterol to mitochondria. Overexpression of a StARD3 mutant whose StAR domain has been deleted results in cholesterol accumulation in enlarged endosomes (Zhang et al. 2002); however, mice lacking the StARD3–STAR domain show no defect in steroidogenesis (Caron et al. 1998). In addition, StARD3 is likely through a N-terminally truncated form to participate in the cholesterol transport process (Miller 2007).

StARD3 exhibits a specific, high affinity for lutein molecules. Although the binding site of lutein in StARD3 is not yet known, it shows that lutein can bind to the StAR domain of StARD3, suggesting that lutein may reduce the risk of oxidation to cholesterol in the human retina as lutein itself is a natural antioxidant.

## 9.3 MACULAR PIGMENT UPTAKE AND TRANSPORT PROTEINS

The uptake and transport processes of MP are selective. There are about 600 carotenoids that exist in nature, 30–50 of which are found in the human diet and about 15 of which are detectable in the serum, but only lutein, zeaxanthin, and their metabolites are present in the human macula. The ratio of lutein:zeaxanthin:*meso*-zeaxanthin is 3:1:0 in blood and liver, 2:1:0.5 in peripheral retina, and 1:1:1 in macula (Khachik et al. 2002). Some plasma lipoproteins such as high-density lipoprotein (HDL), low-density lipoprotein (LDL), and their receptors such as CD36 and SR-BI are involved in the MP uptake process (Li et al. 2010; Loane et al. 2008). MP may travel to the retina from the serum or through the interphotoreceptor space by sharing some of the retinoid transporters, such as interphotoreceptor retinoid–binding protein (IRBP) and retinol binding protein 4 (RBP4). Carotenoid cleavage enzymes may regulate ocular carotenoid levels by binding lutein and/or zeaxanthin and cleaving them to their metabolites that no longer function as the MP.

### 9.3.1 PLASMA LIPOPROTEINS

Plasma lipids such as cholesterol, triglycerides, and phospholipids along with low levels of other hydrophobic compounds such as vitamin E and retinoids are carried in the blood by water-soluble lipoproteins, and the same holds true for the various carotenoids (Rigotti et al. 2003). Lipoproteins have a polar outer shell of protein and phospholipid and an inner core of neutral lipid. Lipoproteins can be divided into six groups: chylomicrons, chylomicron remnants, very low-density lipoproteins, LDLs, intermediate-density lipoproteins, and HDLs (Mahley et al. 1984). HDL is the smallest and densest of all plasma lipoproteins. It plays a critical function in cholesterol metabolism with an important role in removing cholesterol from peripheral tissues, a process known as "reverse cholesterol transport," and in directly delivering cholesteryl esters to other lipoproteins and to tissues (Trigatti et al. 2000). HDLs are a group of lipoprotein particles containing nearly equal amounts of lipid and protein. The major apolipoproteins of HDL are apoA-I and apoA-II. It has been observed in many large-scale prospective studies that there is an inverse relationship between HDL levels and premature cardiovascular disease (Rye et al. 2009). HDL can inhibit the oxidation of LDL, promote endothelial repair, and improve endothelial functions (Tso et al. 2006). HDL also plays a role in the promotion of lesion regression in animals (Tangirala et al. 1999). In the bloodstream, all carotenoids are detectable in all lipoprotein classes to varying degrees, but lutein and zeaxanthin are primarily associated with HDL (Olson 1994), consistent with their less hydrophobic nature relative to the carotenes. However, the specific components of HDL responsible for carotenoid binding remain to be identified. The Wisconsin hypoalpha mutant chicken has very low levels of HDL due to a mutation in the ABCA1 transporter gene (Connor et al. 2007). When these chickens are fed a high-lutein diet, lutein levels increase in plasma, heart, and liver, but not in retina, suggesting that HDL is critical for delivery of carotenoids to retinal tissue.

### 9.3.2 SCAVENGER RECEPTOR PROTEINS

Scavenger receptor class B member 1 (SR-BI), a cell surface glycoprotein that binds HDL, mediates selective cholesteryl ester uptake from lipoprotein into liver and steroidogenic tissues as well as cholesterol efflux from macrophages (Acton et al. 1996; Pagler et al. 2006). SR-BI is a member of CD36 superfamily (Oquendo et al. 1989). The SR-BI gene is located on chromosome 12q24, encoding a 509–amino acid protein (Rigotti et al. 2003). SR-BI contains two trans-membrane domains on both the N- and the C-terminals and a large extracellular domain (Rigotti et al. 1995). The predicted size of SR-BI is 57 kDa, but it usually behaves as an 82-kDa membrane protein on sodium dodecyl sulfate–polyacrylamide gel electrophoresis as it is heavily N-glycosylated (Babitt et al. 1997). SR-BI is highly expressed in liver, adrenals, and ovaries, with the highest amounts in the liver (Krieger and Kozarsky 1999). Female SR-BI knockout mice are infertile, suggesting SR-BI may play an important role in normal ovarian function (Trigatti et al. 1999). As SR-BI is involved in the initial step of reverse cholesterol transport, it could influence atherosclerosis and coronary artery disease (Krieger and Kozarsky 1999). It has also been shown that SR-BI participates in intestinal cholesterol absorption, embryogenesis, and vitamin E transport (During et al. 2005). Recently, there have been several reports that SR-BI is involved in the process of carotenoid uptake and transport to human and fly retina. During et al. (2008) demonstrated that macular carotenoids lutein and zeaxanthin can be better taken up by RPE cells than beta-carotene through an SR-BI-dependent mechanism. When macular carotenoids or beta-carotene were incubated with fully differentiated ARPE-19 cells, the quantity of the macular carotenoids taken up by the cells was two times higher than beta-carotene. Blocking SR-BI by its antibody or knocking down SR-BI expression by small interfering RNA reduced the absorption of carotenoids by RPE cells, especially for zeaxanthin. Similarly, Kiefer et al. (2002) showed that the molecular basis for the blindness of a *Drosophila* mutant, *NinaD*, is a defect in the cellular uptake of carotenoids caused by a mutation in the *NinaD* gene, which has high similarity to mammalian SR-BI. While it seems likely that SR-BI is important for carotenoid uptake into the RPE in living humans, its role in the retinal uptake of the macular carotenoids is less clear, as its immunolocalization in the primate retina does not match lutein's or zeaxanthin's distribution in the retina (Tserentsoodol et al. 2006).

Its scavenger receptor relative, CD36, is a better match. Interestingly, Cameo2, a CD36 homolog in silkworms, is required for uptake of lutein into the silk gland (Sakudoh et al. 2010). CD36 was isolated and identified as a platelet integral membrane glycoprotein in 1973 (Green et al. 1990; Greenwalt et al. 1992). The genetic location of CD36 is 7q11.2. The human cDNA encodes a 471–amino acid protein, and its molecular weight in tissues ranges from 70 to 88 kDa due to posttranslational modifications (Lynes and Widmaier 2011). CD36 has two putative transmembrane segments (one lies near the C-terminal and the other near the N-terminal), and between them is an extracellular loop containing ligand-binding and phosphorylation sites (Glatz et al. 2010). CD36 is distributed in a wide variety of cell types, and it can bind numerous ligands such as thrombospondin, collagen, anionic phospholipids, and oxidized LDL; CD36 has another name of FAT (fatty acid translocase) because it can bind long-chain free fatty acids and transport them into cells (Febbraio et al. 2001; Silverstein and Febbraio 2009; Silverstein et al. 2010). The catabolism of photoreceptor outer segments is mediated by CD36 (Ryeom et al. 1996). The components of rod outer segments, such as rhodopsin and phospholipids, including anionic phospholipids, are shown to be the ligands of CD36. More recently, it was reported that genetic variants of CD36 is associated with serum lutein levels and MP optical density (MPOD) in AMD patients (Borel et al. 2011), suggesting that CD36 is likely to be involved in the MP uptake process.

### 9.3.3 Metabolic Enzymes

Beta-carotene cleavage oxygenase 1 (BCO1, also known as BCMO1 or BCDO1) and beta-carotene cleavage oxygenase2 (BCO2, also known as BCMO2 or BCDO2) are the two carotenoid cleavage enzymes found in animals, and they are immunolocalized to human retina and RPE (Bhatti et al. 2003; Lindqvist and Andersson 2004; Lindqvist et al. 2005). The published literature uses a variety of names for these two enzymes. "BC" refers to beta-carotene, the carotenoid substrate employed during the enzymes' initial characterization. The "MO" term refers to monooxygenase cleavage mechanism and "DO" refers to a dioxygenase cleavage mechanism, but currently the simplified names of BCO1 and BCO2 are preferred in many of the gene databases.

BCO1 cleaves carotenes symmetrically at 15-15′ double carbon bond, an essential step for generation of vitamin A. BCO2 catalyzes eccentric cleavage of carotenes at 9, 10 double carbon bonds, generating 10′-apo-β-carotenal ($C_{27}$), β-ionone ($C_{13}$), and $C_9$ dialdehyde as three possible cleavage products (Kloer et al. 2006; Von Lintig et al. 2005; Wang 2009). It has been shown that ferret and mouse BCO2 can cleave xanthophylls such as lutein and zeaxanthin (Amengual et al. 2011; Mein et al. 2011), while published reports indicate that BCO1 cannot cleave xanthophylls (Amengual et al. 2011); these distinct cleavage specificities and differential tissue localizations could help to mediate differential tissue targeting of the various carotenoids.

## 9.4 MACULAR PIGMENT UPTAKE AND TRANSPORT PATHWAYS

Carotenoids cannot be synthesized by animals; they have to obtain carotenoids from their diets. Therefore, the diet is a "carotenoid source" while the carotenoid target tissues such as the human macula can be considered as a "carotenoid sink." The "carotenoid flow" from source to sink is the carotenoid uptake and transport process (Li et al. 2010). Although not completely understood, it showed that the uptake and transport process of carotenoids is similar to the lipid transport pathways: carotenoids first are released into micelles from food matrix. They are taken up into intestinal cells and then moved into the lymphatic system. Next, they enter liver and finally they are released into the bloodstream to be transported to various tissues (Goti et al. 1998; Rigotti and Krieger 1999; Toyoda et al. 2002). The transport pathway of MP is quite similar to the cholesterol transport pathway, especially the reverse transport process. From liver to retina, MP may share some of the retinoid transport pathways too. MP enters the human eye through retina circulation system. Just like the majority of retinal nutrition, they need to pass through choroid and RPE to get to the retina. When they reach the retina, MP can be held in the macula area by the relevant carotenoid-binding proteins such as GSTP1, StARD3, and tublin for the very long term.

### 9.4.1 Retina Circulation

Nutrition and oxygen are delivered into human retina through the ophthalmic artery (Anderson and Mcintosh 1967). Anatomically and functionally, the ophthalmic artery can be separated into two circulatory systems: the choroidal system and the retinal system. The choroid receives about 65%–85% of the blood flow from ophthalmic artery and is vital for the maintenance of the outer retina, particularly the photoreceptors, and the remaining 20%–30% of blood flow is mainly to the inner retina through the retinal system (Henkind et al. 1979). The majority of the MP is localized to the Henle fiber layer (the axons of photoreceptor cells), and MPs have also been detected in human RPE–choroid and in the outer segments of the photoreceptor layers (Figure 9.2). Therefore, it is likely that MPs enter the human retina mainly

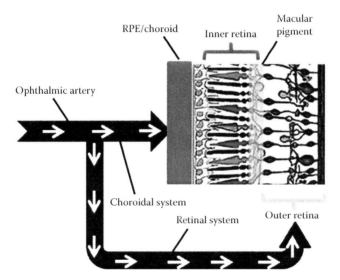

**FIGURE 9.2** **(See color insert.)** Diagram of the human retina circulation. White arrow stands for blood flow.

Transport and Retinal Capture of the Macular Carotenoids

initially through the choroidal circulation in a manner similar to retinoids. From the choriocapillaris, they pass through Bruch's membrane and then enter RPE cells. From the RPE they must traverse the interphotoreceptor space. Ultimately, they enter the photoreceptor cells and are concentrated in the axons of foveal photoreceptor cells. Since only two carotenoids, lutein and zeaxanthin, are selectively taken up into the retina from around 20 carotenoids of human serum, this transport process is thought to be an active transport process. Although not yet proven, we speculate that carotenoids are delivered from the RPE to the retina by a pathway analogous to the one used for retinoid transport that employs IRBP to facilitate transport of hydrophobic ligands across the interphotoreceptor space.

### 9.4.2 Pathways

In Figure 9.3, we provide a brief schematic to describe our current understanding of the whole process of transport and retinal capture of MP carotenoids by connecting the known MP carotenoid transport and binding proteins. Dietary carotenoids are released from ingested foods after ester saponification (if necessary) and incorporated into lipid micelles. SR-BI and CD36 located on the surface of intestinal cells facilitate uptake and transport to the portal circulation in the chylomicron fraction. Although it is still not known if carotenoids are modified in the liver before release into the bloodstream, it is clear that supplying carotenoids to animals can increase their content in

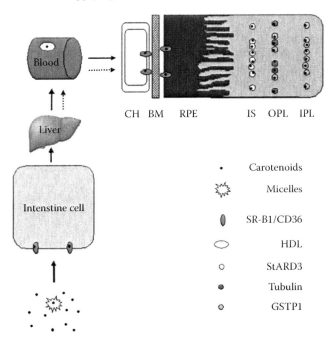

**FIGURE 9.3** (See color insert.) Possible pathway for MP carotenoid uptake, transport, and accumulation in the human retina. BM, Bruch's membrane; CH, choroicapillaris; IPL, inner plexiform layer; IS, inner segments; OPL, outer plexiform layer; RPE, retinal pigment epithelium; ⤑, retinal transport pathway.

the liver. Most hydrophobic carotenoids such as lycopene and beta-carotene are transported on LDL, whereas the more hydrophilic xanthophyll carotenoids, such as lutein and zeaxanthin, are primarily carried by HDL. A possible transport pathway of MP from liver to retina is through sharing the cargoes of retinal molecules, RBP4. RPE SR-BI facilitates uptake of lutein, zeaxanthin, and other carotenoids into the cell. IRBP may facilitate transport of lutein and zeaxanthin to the retinal cells through CD36, but specificity and uptake are ultimately driven by selective binding proteins such as GSTP1 and StARD3, possibly in conjunction with tubulin.

## 9.5 CONCLUSION

The MP carotenoids are concentrated at the human fovea area at very high concentrations through the combined efforts from the selective uptake and transport process of MP carotenoids and the selective retinal capture process of MP carotenoids. Although the number of identified MP carotenoid transport and capture proteins is small, almost all of them are associated with retinal diseases. Variants of GSTP1 combined with GSTPM1 (null) and GSTT1 (null) are reported to associate with exudative AMD (Henkind et al. 1979). SR-BI has been found to be a potential new AMD genetic factor (Zerbib et al. 2009), and the gene for CD36 is associated with the lutein content of serum, the MPOD of the macula, and may also be an AMD genetic marker (Borel et al. 2011). More recently, it is reported that acquired deficiencies and abnormal distributions of MPs are prominent features of macular telangiectasia, an eye disease that can result in loss of visual acuity, reading difficulties, and/or metamorphopsia, suggesting that further investigations on human ocular carotenoid-binding proteins will also help to better understand the pathophysiological processes of other eye diseases as well (Issa et al. 2009). Of course, more efforts from biochemists, geneticists, and immunochemists are required to completely understand this process. In addition, the protective role of MP carotenoid in diseases related to lipid uptake or transport should be investigated since MP carotenoids share some of the same lipid transport proteins.

## REFERENCES

Acton, S., Rigotti, A., Landschulz, K. T., Xu, S., Hobbs, H. H. and Krieger, M. 1996. Identification of scavenger receptor SR-BI as a high density lipoprotein receptor. *Science*, 271, 518–20.

Ali-Osman, F., Akande, O., Antoun, G., Mao, J. X. and Buolamwini, J. 1997. Molecular cloning, characterization, and expression in *Escherichia* coli of full-length cDNAs of three human glutathione S-transferase Pi gene variants: Evidence for differential catalytic activity of the encoded proteinsevidence for differential catalytic activity of the encoded proteins. *J Biol Chem*, 272, 10004–12.

Alpy, F., Stoeckel, M. E., Dierich, A., et al. 2001. The steroidogenic acute regulatory protein homolog MLN64, a late endosomal cholesterol-binding protein. *J Biol Chem*, 276, 4261–9.

Alpy, F. and Tomasetto, C. 2005. Give lipids a START: The StAR-related lipid transfer (START) domain in mammals. *J Cell Sci*, 118, 2791–801.

Amengual, J., Lobo, G. P., Golczak, M., et al. 2011. A mitochondrial enzyme degrades carotenoids and protects against oxidative stress. *FASEB J*, 25, 948–59.

Anderson, B., Jr. and Mcintosh, H. D. 1967. Retinal circulation. *Annu Rev Med*, 18, 15–26.
Babitt, J., Trigatti, B., Rigotti, A., et al. 1997. Murine SR-BI, a high density lipoprotein receptor that mediates selective lipid uptake, is N-glycosylated and fatty acylated and colocalizes with plasma membrane caveolae. *J Biol Chem*, 272, 13242–9.
Beatty, S., Boulton, M., Henson, D., Kon, H. -H. and Murray, I. J. 1999. Macular pigment and age relate macular degeneration. *Br J Ophthalmol*, 83, 867–77.
Bernstein, P. S., Balashov, N. A., Tsong, E. D. and Rando, R. R. 1997. Retinal tubulin binds macular carotenoids. *Invest Ophthalmol Vis Sci*, 38, 167–75.
Bernstein, P. S., Khachik, F., Carvalho, L. S., Muir, G. J., Zhao, D. Y. and Katz, N. B. 2001. Identification and quantitation of carotenoids and their metabolites in the tissues of the human eye. *Exp Eye Res*, 72, 215–23.
Bernstein, P. S., Zhao, D. Y., Wintch, S. W., Ermakov, I. V., Mcclane, R. W. and Gellermann, W. 2002. Resonance Raman measurement of macular carotenoids in normal subjects and in age-related macular degeneration patients. *Ophthalmology*, 109, 1780–7.
Bhatti, R. A., Yu, S., Boulanger, A., et al. 2003. Expression of beta-carotene 15,15' monooxygenase in retina and RPE-choroid. *Invest Ophthalmol Vis Sci*, 44, 44–9.
Bhosale, P. and Bernstein, P. S. 2005. Synergistic effects of zeaxanthin and its binding protein in the prevention of lipid membrane oxidation. *Biochim Biophys Acta*, 1740, 116–21.
Bhosale, P., Larson, A. J., Frederick, J. M., Southwick, K., Thulin, C. D. and Bernstein, P. S. 2004. Identification and characterization of a Pi isoform of glutathione S-transferase (GSTP1) as a zeaxanthin-binding protein in the macula of the human eye. *J Biol Chem*, 279, 49447–54.
Bhosale, P., Li, B., Sharifzadeh, M., et al. 2009. Purification and partial characterization of a lutein-binding protein from human retina. *Biochemistry*, 48, 4798–807.
Bone, R. A. and Landrum, J. T. 1984. Macular pigment in Henle fiber membranes: A model for Haidinger's brushes. *Vision Res*, 24, 103–8.
Bone, R. A., Landrum, J. T., Friedes, L. M., et al. 1997. Distribution of lutein and zeaxanthin stereoisomers in the human retina. *Exp Eye Res*, 64, 211–8.
Bone, R. A., Laudrum, J. T., Hime, G. W., Cains, A. and Zamor, J. 1993. Stereochemistry of the human macular carotenoids. *Inves Ophthalmol Vis Sci*, 34, 2033–40.
Borel, P., De Edelenyi, F. S., Vincent-Baudry, S., et al. 2011. Genetic variants in BCMO1 and CD36 are associated with plasma lutein concentrations and macular pigment optical density in humans. *Ann Med*, 43, 47–59.
Bose, H. S., Whittal, R. M., Huang, M. C., Baldwin, M. A. and Miller, W. L. 2000. N-218 MLN64, a protein with StAR-like steroidogenic activity, is folded and cleaved similarly to StAR. *Biochemistry*, 39, 11722–31.
Caron, K. M., Soo, S. C. and Parker, K. L. 1998. Targeted disruption of StAR provides novel insights into congenital adrenal hyperplasia. *Endocr Res*, 24, 827–34.
Chen, H. and Juchau, M. R. 1997. Glutathione S-transferases act as isomerases in isomerization of 13-cis-retinoic acid to all-trans-retinoic acid *in vitro*. *Biochem J*, 327, 721–6.
Connor, W. E., Duell, P. B., Kean, R. and Wang, Y. 2007. The prime role of HDL to transport lutein into the retina: Evidence from HDL-deficient WHAM chicks having a mutant ABCA1 transporter. *Invest Ophthalmol Vis Sci*, 48, 4226–31.
Di Ilio, C., Aceto, A., Bucciarelli, T., et al. 1990. Glutathione transferase isoenzymes from human prostate. *Biochem J*, 271, 481–5.
During, A., Dawson, H. D. and Harrison, E. H. 2005. Carotenoid transport is decreased and expression of the lipid transporters SR-BI, NPC1L1, and ABCA1 is downregulated in Caco-2 cells treated with ezetimibe. *J Nutr*, 135, 2305–12.
During, A., Doraiswamy, S. and Harrison, E. H. 2008. Xanthophylls are preferentially taken up compared with beta-carotene by retinal cells via a SRBI-dependent mechanism. *J Lipid Res*, 49, 1715–24.

Febbraio, M., Hajjar, D. P. and Silverstein, R. L. 2001. CD36: A class B scavenger receptor involved in angiogenesis, atherosclerosis, inflammation, and lipid metabolism. *J Clin Invest*, 108, 785–91.

Glatz, J. F., Luiken, J. J. and Bonen, A. 2010. Membrane fatty acid transporters as regulators of lipid metabolism: Implications for metabolic disease. *Physiol Rev*, 90, 367–417.

Goti, D., Reicher, H., Malle, E., Kostner, G. M., Panzenboeck, U. and Sattler, W. 1998. High-density lipoprotein (HDL3)-associated tocopherol is taken up by HepG2 cells via the selective uptake pathway and resecreted with endogenously synthesized apo-lipoprotein B-rich lipoprotein particles. *Biochem J*, 332, 57–65.

Green, D. W., Aykent, S., Gierse, J. K. and Zupec, M. E. 1990. Substrate specificity of recombinant human renal renin: Effect of histidine in the P2 subsite on pH dependence. *Biochemistry*, 29, 3126–33.

Greenwalt, D. E., Lipsky, R. H., Ockenhouse, C. F., Ikeda, H., Tandon, N. N. and Jamieson, G. A. 1992. Membrane glycoprotein CD36: A review of its roles in adherence, signal transduction, and transfusion medicine. *Blood*, 80, 1105–15.

Gulick, A. M., Goihl, A. L. and Fahl, W. E. 1992. Studies on human glutathione S-transferase pi. Family of native-specific monoclonal antibodies used to block catalysis. *J Biol Chem*, 267, 18946–52.

Henkind, P., Hansen, R. I. and Szalay, J. 1979. *Ocular circulation*, Harper & Row, New York.

Home, E. 1798. An account of the orifice in the retina of the human eye, discovered by Professor Soemmering: To which are added proofs of this appearance being extended to the eyes of other animals. *Philos Trans R Soc Lond*, 2, 332–45.

Issa, P. C., Van Der Veen, R. L., Stijfs, A., Holz, F. G., Scholl, H. P. and Berendschot, T. T. 2009. Quantification of reduced macular pigment optical density in the central retina in macular telangiectasia type 2. *Exp Eye Res*, 89, 25–31.

Juronen, E., Tasa, G., Veromann, S., et al. 2000a. Polymorphic glutathione S-transferase M1 is a risk factor of primary open-angle glaucoma among Estonians. *Exp Eye Res*, 71, 447–52.

Juronen, E., Tasa, T., Veromann, S., et al. 2000b. Polymorphic glutathione S-transferases as genetic risk factors for senile cortical cataract in Estonians. *Invest Ophthalmol Visual Sci*, 41, 2262–7.

Khachik, F., Askin, F. B. and Lai, K. 1998. Distribution, bioavailability, and metabolism of carotenoids in humans. In: Waye R. Bidlack, Stanley T. Omaye, Mark S. Meskin and D. Janher (eds.), *Phytochemical a new paradigm*. Lancaster, PA: CRC Press LLC, pp. 77–96.

Khachik, F., Beecher, G. R. and Goli, M. B. 1991. Separation, identification and quantification of carotenoids in fruits, vegetables and human plasma by high performance liquid chromatography. *Pure Appl Chem*, 63, 71–80.

Khachik, F., De Moura, F. F., Zhao, D. Y., Aebischer, C. P. and Bernstein, P. S. 2002. Transformations of selected carotenoids in plasma, liver, and ocular tissues of humans and in nonprimate animal models. *Invest Ophthalmol Vis Sci*, 43, 3383–92.

Khachik, F., Spangler, C. J., Smith, J. C. Jr., Canfield, L. M., Pfander, H. and Steck, A. 1997. Identification, quantification, and relative concentrations of carotenoids and their metabolites in human milk and serum. *Anal Chem*, 69, 1873–81.

Kiefer, C., Sumser, E., Wernet, M. F. and Von Lintig, J. 2002. A class B scavenger receptor mediates the cellular uptake of carotenoids in Drosophila. *Proc Natl Acad Sci USA*, 99, 10581–6.

Kloer, D. P., Welsch, R., Beyer, P. and Schulz, G. E. 2006. Structure and reaction geometry of geranylgeranyl diphosphate synthase from *Sinapis alba*. *Biochemistry*, 45, 15197–204.

Krieger, M. and Kozarsky, K. 1999. Influence of the HDL receptor SR-BI on atherosclerosis. *Curr Opin Lipidol*, 10, 491–7.

Krinsky, N. I., Landrum, J. T. and Bone, R. A. 2003. Biologic mechanisms of the protective role of lutein and zeaxanthin in the eye. *Annu Rev Nutr*, 23, 171–201.

Landrum, J. T. and Bone, R. A. 2001. Lutein, zeaxanthin, and the macular pigment. *Arch Biochem Biophys*, 385, 28–40.

Landrum, J. T., Bone, R. A., Joa, H., Kilburn, M. D., Moore, L. L. and Sprague, K. E. 1997. A one year study of the macular pigment: The effect of 140 days of a lutein supplement. *Exp Eye Res*, 65, 57–62.

Li, B., Vachali, P. and Bernstein, P. S. 2010. Human ocular carotenoid-binding proteins. *Photochem Photobiol Sci*, 9, 1418–25.

Li, B., Vachali, P., Frederick, J. M. and Bernstein, P. S. 2011. Identification of StARD3 as a lutein-binding protein in the macula of the primate retina. *Biochemistry*, 50, 2541–9.

Lindqvist, A. and Andersson, S. 2004. Cell type-specific expression of beta-carotene 15,15'-mono-oxygenase in human tissues. *J Histochem Cytochem*, 52, 491–9.

Lindqvist, A., He, Y. G. and Andersson, S. 2005. Cell type-specific expression of beta-carotene 9',10'-monooxygenase in human tissues. *J Histochem Cytochem*, 53, 1403–12.

Lo, H. W. and Ali-Osman, F. 1997. Genomic cloning of hGSTP1*C, an allelic human Pi class glutathione S-transferase gene variant and functional characterization of its retinoic acid response elements. *J Biol Chem*, 272, 32743–9.

Loane, E., Nolan, J. M., O'donovan, O., Bhosale, P., Bernstein, P. S. and Beatty, S. 2008. Transport and retinal capture of lutein and zeaxanthin with reference to age-related macular degeneration. *Surv Ophthalmol*, 53, 68–81.

Lynes, M. D. and Widmaier, E. P. 2011. Involvement of CD36 and intestinal alkaline phosphatases in fatty acid transport in enterocytes, and the response to a high-fat diet. *Life Sci*, 88, 384–91.

Mahley, R. W., Innerarity, T. L., Rall, S. C. Jr. and Weisgraber, K. H. 1984. Plasma lipoproteins: Apolipoprotein structure and function. *J Lipid Res*, 25, 1277–94.

Mein, J. R., Dolnikowski, G. G., Ernst, H., Russell, R. M. and Wang, X. D. 2011. Enzymatic formation of apo-carotenoids from the xanthophyll carotenoids lutein, zeaxanthin and beta-cryptoxanthin by ferret carotene-9',10'-monooxygenase. *Arch Biochem Biophys*, 506, 109–21.

Miller, W. L. 2007. Mechanism of StAR's regulation of mitochondrial cholesterol import. *Mol Cell Endocrinol*, 265–266, 46–50.

Neufeld, E. B., Wastney, M., Patel, S., et al. 1999. The Niemann-Pick C1 protein resides in a vesicular compartment linked to retrograde transport of multiple lysosomal cargo. *J Biol Chem*, 274, 9627–35.

Olson, J. A. 1994. Absorption, transport, and metabolism of carotenoids in humans. *Pure & Appl Chem*, 66, 1011–6.

Oquendo, P., Hundt, E., Lawler, J. and Seed, B. 1989. CD36 directly mediates cytoadherence of Plasmodium falciparum parasitized erythrocytes. *Cell*, 58, 95–101.

Oz, O., Aras Ates, N., Tamer, L., Yildirim, O. and Adiguzel, U. 2006. Glutathione S-transferase m1, t1, and p1 gene polymorphism in exudative age-related macular degeneration: A preliminary report. *Eur J Ophthalmol*, 16, 105–10.

Pagler, T. A., Rhode, S., Neuhofer, A., et al. 2006. SR-BI-mediated high density lipoprotein (HDL) endocytosis leads to HDL resecretion facilitating cholesterol efflux. *J Biol Chem*, 281, 11193–204.

Pedersen, J. Z., De Maria, F., Turella, P., et al. 2007. Glutathione transferases sequester toxic Dinitrosyl-Iron complexes in cells: A protection mechanism against excess nitric oxide. *J Biol Chem*, 282, 6364–71.

Rapp, L. M., Maple, S. S. and Choi, J. H. 2000. Lutein and zeaxanthin concentrations in rod outer segment membranes from perifoveal and peripheral human retina. *Invest Ophthalmol Visual Sci*, 41, 1200–9.

Rigotti, A., Acton, S. L., and Krieger, M. 1995. The class B scavenger receptors SR-BI and CD36 are receptors for anionic phospholipids. *J Biol Chem*, 270, 16221–4.

Rigotti, A. and Krieger, M. 1999. Getting a handle on "good" cholesterol with the high-density lipoprotein receptor. *N Engl J Med*, 341, 2011–13.

Rigotti, A., Miettinen, H. E., and Krieger, M. 2003. The role of the high-density lipoprotein receptor SR-BI in the lipid metabolism of endocrine and other tissues. *Endocr Rev*, 24, 357–87.

Rye, K. A., Bursill, C. A., Lambert, G., Tabet, F. and Barter, P. J. 2009. The metabolism and anti-atherogenic properties of HDL. *J Lipid Res*, 50, S195–200.

Ryeom, S. W., Sparrow, J. R. and Silverstein, R. L. 1996. CD36 participates in the phagocytosis of rod outer segments by retinal pigment epithelium. *J Cell Sci*, 109 (Pt 2), 387–95.

Sakudoh, T., Iizuka, T., Narukawa, J., et al. 2010. A CD36-related transmembrane protein is coordinated with an intracellular lipid-binding protein in selective carotenoid transport for cocoon coloration. *J Biol Chem*, 285, 7739–51.

Seddon, J. M., Ajani, U. A., Sperduto, R. D., et al. 1994. Dietary carotenoids, vitamins A, C, and E, and advanced age-related macular degeneration. Eye Disease Case-Control Study Group. *JAMA*, 272, 1413–20.

Sierra, A. 2002. Neurosteroids: The StAR protein in the brain. *J Neuroendocrinol*, 16, 787–93.

Silverstein, R. L. and Febbraio, M. 2009. CD36, a scavenger receptor involved in immunity, metabolism, angiogenesis, and behavior. *Sci Signal*, 2, re3.

Silverstein, R. L., Li, W., Park, Y. M. and Rahaman, S. O. 2010. Mechanisms of cell signaling by the scavenger receptor CD36: Implications in atherosclerosis and thrombosis. *Trans Am Clin Climatol Assoc*, 121, 206–20.

Snodderly, D. M., Auran, J. D. and Delori, F. C. 1984. The macular pigment. II. Spatial distribution in primate retinas. *Invest Ophthalmol Visual Sci*, 25, 674–85.

Soccio, R. E., Adams, R. M., Romanowski, M. J., Sehayek, E., Burley, S. K. and Breslow, J. L. 2002. The cholesterol-regulated StarD4 gene encodes a StAR-related lipid transfer protein with two closely related homologues, StarD5 and StarD6. *Proc Natl Acad Sci USA*, 99, 6943–8.

Strauss, J. F., 3rd, Kishida, T., Christenson, L. K., Fujimoto, T. and Hiroi, H. 2003. START domain proteins and the intracellular trafficking of cholesterol in steroidogenic cells. *Mol Cell Endocrinol*, 202, 59–65.

Strauss, J. F., 3rd, Liu, P., Christenson, L. K. and Watari, H. 2002. Sterols and intracellular vesicular trafficking: lessons from the study of NPC1. *Steroids*, 67, 947–51.

Tangirala, R. K., Tsukamoto, K., Chun, S. H., Usher, D., Pure, E. and Rader, D. J. 1999. Regression of atherosclerosis induced by liver-directed gene transfer of apolipoprotein A-I in mice. *Circulation*, 100, 1816–22.

Townsend, D. M., Manevich, Y., He, L., Hutchens, S., Pazoles, C. J. and Tew, K. D. 2009. Novel role for glutathione S-transferase π: Regulator of protein s-glutathionylation following oxidative and nitrosative stress. *J Biol Chem*, 284, 436–45.

Townsend, D. M. and Tew, K. D. 2003. The role of glutathione-S-transferase in anti-cancer drug resistance. *Oncogene*, 22, 7369–75.

Toyoda, Y., Thomson, L. R., Langner, A., et al. 2002. Effect of dietary zeaxanthin on tissue distribution of zeaxanthin and lutein in quail. *Invest Ophthalmol Vis Sci*, 43, 1210–21.

Trigatti, B., Rayburn, H., Vinals, M., et al. 1999. Influence of the high density lipoprotein receptor SR-BI on reproductive and cardiovascular pathophysiology. *Proc Natl Acad Sci USA*, 96, 9322–7.

Trigatti, B. L., Rigotti, A. and Braun, A. 2000. Cellular and physiological roles of SR-BI, a lipoprotein receptor which mediates selective lipid uptake. *Biochim Biophys Acta*, 1529, 276–86.

Tserentsoodol, N., Gordiyenko, N. V., Pascual, I., Lee, J. W., Fliesler, S. J. and Rodriguez, I. R. 2006. Intraretinal lipid transport is dependent on high density lipoprotein-like particles and class B scavenger receptors. *Mol Vis*, 12, 1319–33.

Tso, C., Martinic, G., Fan, W. H., Rogers, C., Rye, K. A. and Barter, P. J. 2006. High-density lipoproteins enhance progenitor-mediated endothelium repair in mice. *Arterioscler Thromb Vasc Biol*, 26, 1144–9.

Tsujishita, Y. and Hurley, J. H. 2000. Structure and lipid transport mechanism of a StAR-related domain. *Nat Struct Biol*, 7, 408–14.

Von Lintig, J., Hessel, S., Isken, A., et al. 2005. Towards a better understanding of carotenoid metabolism in animals. *Biochim Biophys Acta*, 1740, 122–31.

Wald, G. 1945. Human vision and the spectrum. *Science*, 101, 653–8.

Wang, X.-D. 2009. Biological activities of carotenoid metabolites. In: G. Britton, S. Liaaen-Jensen, and H. Pfander (eds.), *Carotenoids. Volume 5: Nutrition and health*. Birkhäuser Verlag, Basel, pp. 383–404.

Watari, H., Arakane, F., Moog-Lutz, C., et al. 1997. MLN64 contains a domain with homology to the steroidogenic acute regulatory protein (StAR) that stimulates steroidogenesis. *Proc Natl Acad Sci USA*, 94, 8462–7.

Whitehead, A. J., Mares, J. A. and Danis, R. P. 2006. Macular pigment. *Arch Ophthalmol*, 124, 1038–45.

Zerbib, J., Seddon, J. M., Richard, F., et al. 2009. rs5888 variant of SCARB1 gene is a possible susceptibility factor for age-related macular degeneration. *PLoS One*, 4, e7341.

Zhang, M., Liu, P., Dwyer, N. K., et al. 2002. MLN64 mediates mobilization of lysosomal cholesterol to steroidogenic mitochondria. *J Biol Chem*, 277, 33300–10.

# 10 Measurement and Interpretation of Macular Carotenoids in Human Serum

*David I. Thurnham, Katherine A. Meagher, Eithne Connolly, and John M. Nolan*

## CONTENTS

10.1 Introduction ..................................................................................................... 188
10.2 Methods .......................................................................................................... 191
    10.2.1 Serum or Plasma Extraction ............................................................... 191
    10.2.2 Separation of Lutein and Zeaxanthin Isomer
           Fraction (Assay 1) ................................................................................ 191
    10.2.3 Separation of the Z and MZ Stereoisomers (Assay 2) ...................... 192
    10.2.4 Calculation of Xanthophyll Concentrations ....................................... 193
10.3 Results and Discussion .................................................................................. 193
    10.3.1 Method Development .......................................................................... 193
    10.3.2 Analysis of Macular Carotenoids in Food .......................................... 194
    10.3.3 Interpretation of Serum Carotenoid Concentrations ........................ 194
           10.3.3.1 Plasma Carotenoid Concentrations and
                    Dietary Intake ...................................................................... 194
           10.3.3.2 Seasonal Factors and Sources of Lutein and
                    Zeaxanthin in the Diet ........................................................ 195
           10.3.3.3 Sex ....................................................................................... 195
           10.3.3.4 Smoking and Subclinical Inflammation ........................... 196
           10.3.3.5 Dietary Intake of Lutein and Zeaxanthin and
                    Risk of AMD ........................................................................ 197
10.4 Conclusions ..................................................................................................... 198
Acknowledgment ..................................................................................................... 198
Conflicts of Interest ................................................................................................. 198
References ................................................................................................................ 198

## 10.1 INTRODUCTION

The structure and spectral characteristics of nearly 750 carotenoids present in our natural surroundings have been identified.[1] About 50 of these carotenoids are present in the human diet but mostly in small amounts.[2] The major carotenoids in human blood are lutein, β-cryptoxanthin, lycopene, and α- and β-carotene.[3,4] Their structures are shown in Figure 10.1, and isocratic, reverse-phase chromatography has been shown to separate these carotenoids by many researchers.[3–6] The carotenoids have at least two major functions in the human body: to provide a source of vitamin A and to act as antioxidants.[7,8] Of the principal carotenoids in serum, only α-carotene, β-carotene, α-cryptoxanthin, and β-cryptoxanthin are provitamin A carotenoids, that is, they contain at least one β-ionone ring linked to a hydrocarbon chain with a conjugated double bond structure, which can be cleaved by the oxidation of the 15,15′ double bond to form β-apo-15-carotenal. β-apo-15-carotenal is subsequently reduced to retinol. Most carotenoids are not vitamin A precursors. Although many carotenoids possess a β-ionone ring, they contain a variety of substituents that commonly include oxygen. These oxygen-containing carotenoids are known as xanthophylls. A few of these, such as β-cryptoxanthin (Figure 10.1), are vitamin A precursors, but most are not. All carotenoids have antioxidant capacity associated with their conjugated carbon

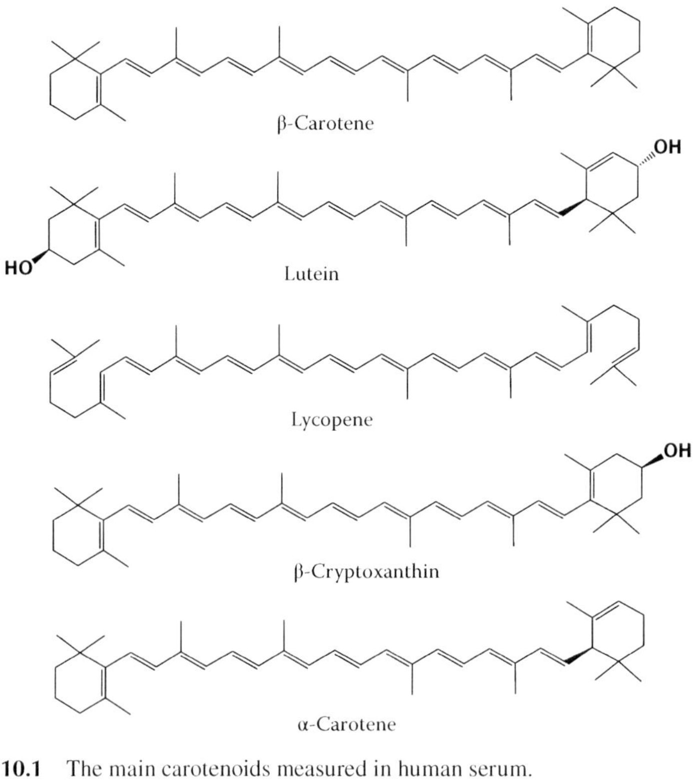

**FIGURE 10.1** The main carotenoids measured in human serum.

structure. Typically, the longer the polyene chain, the greater the ability of a carotenoid to deactivate reactive oxygen species, such as oxy radicals and singlet oxygen.

It is common for body tissues to contain most or all of the carotenoids that are found in blood. However, the retina is unique in containing principally only three xanthophylls, lutein and the two isomers of zeaxanthin (Figure 10.2).[9] Zeaxanthin has 11 conjugated double bonds and has particularly good antioxidant capacity.[10] We are interested in determining the effectiveness of xanthophyll supplements, particularly those containing *meso*-zeaxanthin (MZ), to increase macular pigment optical density (MPOD). Maintaining or increasing MPOD may prevent age-related macular disease (AMD) in the aging population, improve visual acuity in sportsmen and sportswomen, and reduce night-time glare in car drivers.[11,12] A number of high-pressure liquid chromatography (HPLC) methods have been reported to separate the carotenoids in human blood.[13–15] However, under conditions that rapidly separate the nonpolar carotenes (i.e., β-carotene, α-carotene, and lycopene), it is common for lutein and zeaxanthin to coelute.[3–5] Even when chromatography conditions are optimized to produce efficient separation of lutein from zeaxanthin (Figure 10.3), the stereoisomers Z and MZ do not separate. To quantitatively assess all three isomers (lutein, Z, and MZ) in the serum of subjects who are taking supplements to increase macular pigment concentrations, it is important to develop rapid and effective methods for their chromatographic separation.

At the time lutein, Z, and MZ were first reported in the macula,[9] most laboratories were incapable of measuring the total zeaxanthin concentration, principally reporting the combined L&Z concentration.[3,4,6,16,17] All three stereoisomers had been reported to be present in postmortem retinas.[9] However, at that time the analysis and determination of zeaxanthin stereoisomers was possible in only a few laboratories.[9,18]

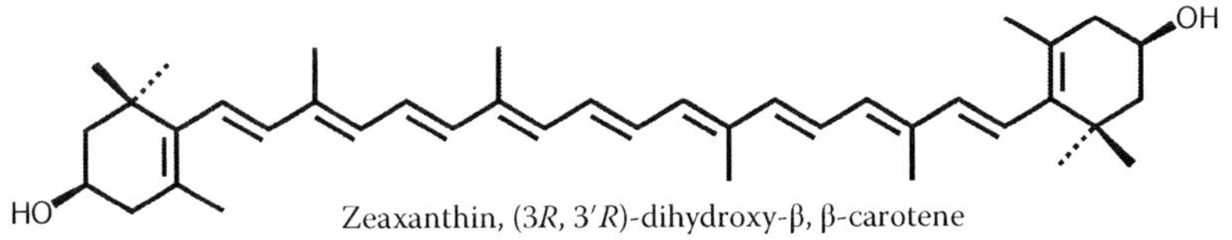

Zeaxanthin, (3*R*, 3′*R*)-dihydroxy-β, β-carotene

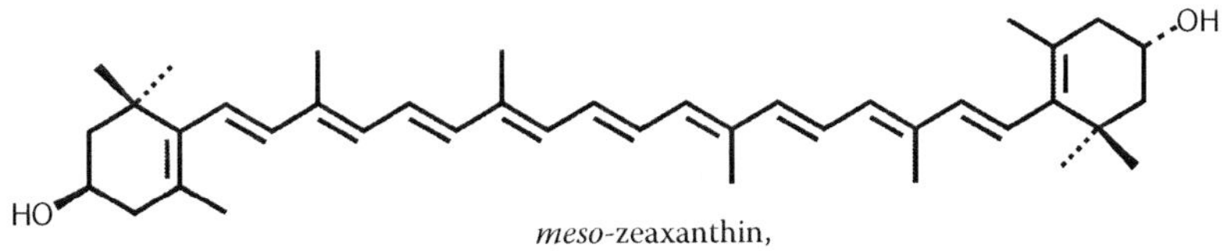

*meso*-zeaxanthin,
3*R*, 3′*S*-dihydroxy-β, β-carotene

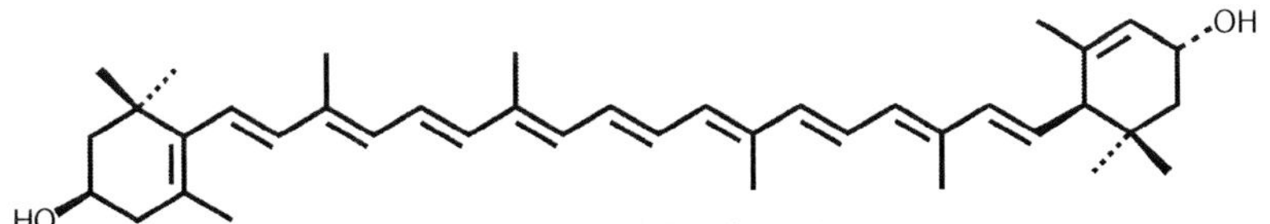

Lutein, (3*R*, 3′*R*, 6′*R*)-dihydroxy-β, ε-carotene

**FIGURE 10.2** Structures of the main macular carotenoids.

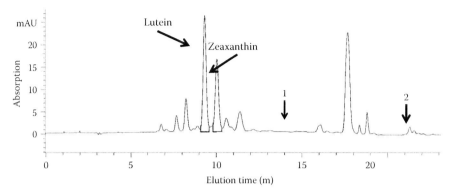

**FIGURE 10.3** Separation of lutein and zeaxanthin fractions using high-performance liquid chromatography at 450 nm: carotenoids were separated on a 3 μm Ultracarb ODS column (Phenomenex; 250 and 4.6 cm). The initial solvent was acetonitrile–methanol (85:15 containing 1% triethylamine) pumped at 1 mL/min. At 15 minutes (arrow 1), pump speed increased to 2 mL/min and dichloromethane (DCM) was introduced rapidly to raise the concentration to 10% DCM. At 25 minutes (arrow 2), DCM was raised to 50% to rapidly elute any remaining carotenoids. The internal standard eluted at approximately 18 minutes and was measured at 292 nm.

Strikingly, the principal interest in lutein until the mid-1990s was as a food colorant,[19] and it was not until lutein and zeaxanthin were recognized as the principal carotenoids in the macula[20] that interest in their blood concentrations began to increase. The two zeaxanthin stereoisomers Z [(3R,3′R)-dihydroxy-β,β-carotene] and MZ [(3R,3′S)-dihydroxy-β,β-carotene)] have been identified in fish and other marine organisms that are known to be present in the human diet,[21] but the methods described for analysis were not in regular use by nutritionists.[18] The two principal forms of zeaxanthin (Z and MZ) identified in the retina are, in fact, diastereomers. Although in many instances diastereomers may be separable by conventional chromatographic methods, this is not so for Z and MZ. The chiral stationary phases needed for their effective separation have only become commonly available during the past two decades.

Early efforts to separate the zeaxanthin stereoisomers required that they be derivatized to enhance the differences between them and, in an early paper on their natural occurrence in food sources, the dibenzoate derivatives of Z and MZ were prepared from tissue extracts prior to separation using HPLC.[21] It is now possible to separate Z and MZ without derivatization by HPLC using a chiral column (Figure 10.4). Usually, the fraction containing the zeaxanthin isomers must first be separated from the other serum carotenoids to prevent the overloading of the chiral column and to eliminate interference. We and others[22,23] have developed these methods to enable quantification of the concentrations of zeaxanthin stereoisomers in human serum during supplementation studies.

In Section 10.2, we describe the three stages of the method used at the Waterford Institute of Technology to quantify lutein, Z, and MZ in human serum.

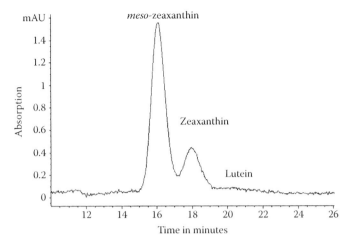

**FIGURE 10.4** Separation of *meso*-zeaxanthin and zeaxanthin using a chiral column: figure shows the separation of zeaxanthin stereoisomers from the zeaxanthin fraction obtained from assay 1 followed by the appearance of any lutein at approximately 20 to 21 minutes if present.

## 10.2 METHODS

### 10.2.1 Serum or Plasma Extraction

All reagents or solvents were AnalaR® or HPLC grade and obtained from Sigma-Aldrich Ireland, Ltd., Arklow, Ireland, unless otherwise indicated.

Many methods have been described for the extraction of carotenoids from serum.[3,24–26] All of these, and probably many others, provide a sample that can be used to prepare the fraction containing the Z stereoisomers. In our laboratory, serum (0.4 mL) was pipetted into clear 1.5 mL polyethylene (Eppendorf) tubes together with 0.2 mL of ethyl alcohol containing α-tocopherol acetate as internal standard (25 mg/100 mL) and 0.3 mL of ethyl alcohol containing butylated hydroxytoluene (BHT; 25 mg/100 mL). Heptane (0.5 mL, containing 500 mg BHT/L) was then added as the extraction solvent. The resulting mixture was vortexed for 2 minutes (Scientific Industries Genie-2, Biosciences, Dublin, Ireland) at the highest setting, followed by centrifugation at 400 × gravity for 5 minutes at room temperature (Fisher Scientific AccuSpin Micro 17, Dublin, Ireland).

A 0.4 mL aliquot of the upper heptane layer was removed to a 1.5 mL opaque or amber Eppendorf tube. The heptane extraction was repeated once more after the addition of a further 0.5 mL of heptane to the original residue. The combined extracts were dried under nitrogen at room temperature and then stored at −70°C until analysis.

### 10.2.2 Separation of Lutein and Zeaxanthin Isomer Fraction (Assay 1)

The HPLC system used was an Agilent 1200 Series (Agilent Technologies, Ltd., Dublin, Ireland) system consisting of a quaternary pump, an autosampler, a thermostatic-column compartment, and a photodiode array detector monitoring at a wavelength of 450 nm for carotenoids and 292 nm for the internal standard, α-tocopherol acetate. Lutein and the Z stereoisomers were separated using isocratic chromatography followed by the rapid introduction of dichloromethane into the system to accelerate the elution of the internal standard and other plasma carotenoids.

The stored, dried samples were reconstituted in 0.2 mL of the initial mobile phase and vortexed at the lowest setting for 1 minute. The reconstituted sample (~0.2 mL) was pipetted into 2.5 mL vials containing 0.35 mL glass inserts, and 0.1 mL was injected via an autosampler. The column used was a Phenomenex Ultracarb ODS 3 μm C18 column, 250 × 4.6 mm (part number: 00G-025-E0, Macclesfield, Cheshire, United Kingdom), with an in-line guard column and a stainless steel filter (pore size of 0.5 μm, Upchurch precolumn filter, Sigma-Aldrich, United Kingdom). The initial eluting solvent was a mobile phase of acetonitrile–methanol (85:15 by volume containing 0.1% triethylamine) at 1 mL/min. At 15 minutes, the flow rate was increased to 2 mL/min and dichloromethane (100%) was introduced in a gradient increasing from 0% to 10% of the mobile phase over a 30-second interval. At 25 minutes, after the internal standard had eluted, the dichloromethane concentration in the mobile phase was increased to 50% to elute all remaining carotenoids. At 30 minutes, the dichloromethane was removed to return to starting conditions and starting conditions were maintained until 34 minutes when the next injection took place. The column temperature was maintained at 15°C throughout.

Lutein and the Z fraction (zeaxanthin peak contains coeluting Z and MZ fractions) eluted at approximately 9.9 and 10.5 minutes, respectively (Figure 10.3). The Z fraction was manually collected from the waste tube into a small glass collection tube a couple of seconds after the starting point of the zeaxanthin peak was first observed on the monitor. The eluent was dried under nitrogen at room temperature and the tube sealed with several layers of Parafilm® (Alpha Laboratories, Eastleigh, Hampshire, United Kingdom) and stored at −70°C for a few days until analysis by assay 2.

### 10.2.3 Separation of the Z and MZ Stereoisomers (Assay 2)

The Z and MZ stereoisomers present in the zeaxanthin fraction were separated using normal phase chromatography on a 10 μm Chiralpak™ AD column (250 × 4.6 mm; Chiral Technologies Europe, Illkirch, France) with an in-line Chiralpak guard column (Chiral Technologies Europe) and a 2 μm filter. The dry Z fraction from assay 1 was reconstituted by vortexing with 0.1 mL of the starting mobile phase ($n$-hexane–isopropanol, 90:10 by volume), and 0.05 mL was injected into a linear gradient that increased to $n$-hexane–isopropanol, 95:5 by volume, over 30 minutes. The solvents were pumped at 0.8 mL/min; column temperature was maintained at 15°C throughout; and MZ, Z, and lutein eluted at approximately 15.8, 18, and 20 minutes, respectively (Figure 10.4). The starting mobile phase was pumped for 5 minutes to equilibrate the column before the next injection.

## 10.2.4 CALCULATION OF XANTHOPHYLL CONCENTRATIONS

Absorption of individual lutein and Z standards was determined by ultraviolet–visible light spectroscopy analysis at 445 and 450 nm in absolute ethanol, and concentrations were calculated using 145.1 and 156.2 L/mmol/cm extinction coefficients, respectively. Pure standards of lutein and Z were injected into assay 1, and their peak areas were used to calculate response factors. The response factors were then used to obtain xanthophyll concentrations from the peak areas of test samples obtained in assay 1. Sample recovery was calculated from the recovery of the internal standard in assay 1 by comparison with the peak area of a known volume of a nonextracted internal standard.

Individual concentrations of Z and MZ were then quantified from assay 2, by calculating the percentage proportions of Z and MZ from their peak areas and applying the resulting ratio to the corresponding zeaxanthin fraction concentration obtained in assay 1. All chromatographic peaks of interest were manually integrated using Agilent ChemStation software.

## 10.3 RESULTS AND DISCUSSION

### 10.3.1 METHOD DEVELOPMENT

Although some of the earlier methods used to fractionate serum carotenoids on Vydak® 201TP[27] or Spherisorb ODS[3,28] columns can display a specific peak for Z, the separation from the preceding lutein fraction is sometimes poorly resolved and difficult to quantify accurately. Bone and colleagues found that 3 μm Ultracarb ODS columns (Phenomenex, Torrance, California) provided a good separation of lutein and Z in extracts of macular pigment[29] and blood.[30] In a study on 21 subjects who received a supplement of lutein, Z, and MZ (10.8:1.2:8.0 mg Lutein Plus®, Holland & Barrett, Nuneaton, Warwickshire, United Kingdom), we confirmed that baseline separation of lutein and the Z fraction is achieved using the 3 μm 250 × 4.6 mm Ultracarb ODS column.[24] The disadvantage of the method, however, was that it took approximately 75 minutes for the last serum carotenoid (β-carotene) to emerge from the column and the next fractionation to start.

Manually collecting the Z fraction in each run demanded the constant attention of the operator and became very tedious; hence, we sought ways to improve the procedure. Rapid methods to separate carotenoids usually include a dipolar solvent in the eluting solution such as chloroform, methylene chloride, or dichloromethane.[3,14,28] Our objective was to retain the advantages of using the methanol–acetonitrile mixture (assay 1 conditions) until lutein and the zeaxanthin fractions had eluted and then add dichloromethane to rapidly elute the remaining carotenoids. We found that continuous linear gradients were not successful; even the smallest amount of dichloromethane in the eluting solvent at the start interfered with the separation of lutein and the zeaxanthin fractions. We therefore adopted a step-gradient approach in which a small amount of dichloromethane was rapidly introduced after the zeaxanthin fraction had eluted to accelerate the elution of the internal standard, tocopherol acetate, followed by a large flush of dichloromethane to remove all remaining carotenoids from the column. The procedure sacrifices all the other carotenoids in plasma but is useful where the main interest is on lutein and the Z stereoisomers (Figure 10.3).

We attempted to shorten the entire elution sequence to 20 minutes and experimented with different proportions of dichloromethane and different temperatures. Adding dichloromethane can cause baseline noise and spurious peaks. Below-ambient temperatures improved peak resolution and reduced noise but slowed elution. The final run time of 30 minutes was a compromise to gain the advantage of a lower temperature, 15°C, and the two-step gradient described for assay 1. This method gave a stable baseline at 292 nm around the time of elution of the internal standard (17.8 minutes for tocopherol acetate; not shown in Figure 10.3) or in the case of food extracts (19.2 minutes for echinenone; see Section 10.3.2).

### 10.3.2 Analysis of Macular Carotenoids in Food

The chromatography for assays 1 and 2 mentioned in Sections 10.2.2 and 10.2.3 is also suitable for extracts of food samples. The only difference in the procedures described relates to the internal standard. For food, we add echinenone (DSM Nutritional Products, Basel, Switzerland) at the start and saponify before the extraction procedure. In assay 1, echinenone is monitored at 450 nm and elutes at 19.2 minutes. Procedures for food extraction will be described in future publications.

### 10.3.3 Interpretation of Serum Carotenoid Concentrations

#### 10.3.3.1 Plasma Carotenoid Concentrations and Dietary Intake

There is a direct relationship between the dietary intake of carotenoids and plasma concentrations,[31,32] but often the relationship is weak[33,34] because of difficulties in measuring the dietary intake.[35] Although the amounts of carotenoids in different foods can vary greatly, the bioavailability of carotenoids can also vary enormously depending on how a food is prepared. This is especially important with vegetable foods where the cellulose cytoskeleton has to be disrupted for carotenoids to be released. For example, it has been shown that the amount of β-carotene released from carrots can vary from 2% to 95%.[36] For maximum bioavailability, foods must be finely divided, for example, pureed, cooked to soften the cell wall, and consumed with oil to stimulate fat absorption in the gut.

Supplementation studies provide information on the plasma carotenoid response to a supplementary carotenoid. Data obtained from two 42 day oral feeding studies with lutein[37] and Z[38] where plasma concentrations were determined at frequent intervals provide useful information with which to interpret carotenoid responses in other studies. In the two studies, 20 male and female volunteers were fed either lutein (4.1 or 20.5 mg) or zeaxanthin (1 or 10 mg). Regular analysis of the xanthophylls showed that a >90% fraction of the steady state plasma concentrations was achieved by days 18 and 17 for lutein[37] and Z,[38] respectively. Dose response to the two levels of the xanthophyll supplements was almost linear, and when the response line was forced through 0 the mean increases in serum carotenoids were 0.10 and 0.075 μmol per milligram daily for Z and lutein, respectively, calculated from the total increase over baseline observed between day 0 and days 17 or 18.[39]

However, these data should really only be used to compare the relative bioavailabilities of xanthophyll supplements in supplementation studies. If they are used to calculate the carotenoid intakes corresponding to plasma concentrations of people in the community, the results will probably be inaccurate since steady state concentrations have been achieved over long periods in the presence of many foods that might accelerate or depress absorption. For example, using the response figure for lutein (0.075 µmol/mg) to interpret lutein intake in the U.S. Women's Health Initiative where the plasma concentration was approximately 0.29 µmol/L[40] suggests that the lutein intake is approximately 3.9 mg/day under steady state conditions. This is higher than the lutein intake reported in white Americans, approximately 1 mg/day,[41] or Spaniards, 0.5 mg/day,[42] and indicates the difficulty in interpreting plasma carotenoid concentrations in a free-living population.

### 10.3.3.2 Seasonal Factors and Sources of Lutein and Zeaxanthin in the Diet

Carotenoids are found in the blood as a consequence of the food we eat. The total concentration of carotenoids present in the blood of people in the United Kingdom is approximately 2 µmol/L.[43] The median concentrations of the main carotenoids in men and women, respectively, were 0.24 and 0.32 µmol/L of β-carotene, 0.25 and 0.25 µmol/L of lycopene, and 0.29 and 0.29 µmol/L of lutein.[44] Most dietary carotenoid intakes derive from fruits and vegetables. Carotenoids in fruits tend to be more bioavailable than those in vegetables because carotenoids in fruit are present in lipid droplets.[45] The seasonal glut of mangoes in many tropical countries is an important source of vitamin A as mangoes contain large amounts of α- and β-carotene.[46] Several authors have also noted seasonal variations in plasma carotenoids including lutein as consumption patterns change during the year.[34,46–48] Green leafy vegetables are the principal source of lutein and maize and maize products are the main source of Z in the diet.[49] Z is generally present with lutein in the diet, although the amount in dark green vegetables is very low (0%–3%).[50] Sweet orange and red peppers[51] are also important sources of Z.[49] Egg yolks are an especially important source of both lutein and Z for they are rich in both pigments and absorption is three to four times higher from egg yolk than from green vegetables.[39] On average, it has been reported that the amount of Z in the diet varies between 10% and 20% of the lutein content.[38,52] To date, there has been no satisfactory investigation of MZ content in foods. One study performed by Maoka et al.[21] did report MZ to be present in some fish and seafood.

### 10.3.3.3 Sex

In the United Kingdom, we found that concentrations of the provitamin A carotenoids β-carotene, β-cryptoxanthin, and α-carotene were higher in women than men,[43,44] but there were no differences between lutein and lycopene. Similar observations have been made by others.[53,54] Currently, there is no good explanation for the differences in plasma concentrations, which may be due to differences in intake, variations during the menstrual cycle[55] or hormonal differences between men and women, or different distributions of body fat. We investigated the response to a mixed xanthophyll supplement in a recent study with Lutein Plus, which contains lutein, Z, and MZ. We found no differences between young men and women in serum lutein and Z concentrations

## TABLE 10.1
### Relative Increases in Plasma Concentrations of Lutein, Zeaxanthin, and MZ to Doses of the Respective Xanthophylls in Lutein Plus

| | Change in Plasma Concentration of Xanthophyll | | | |
|---|---|---|---|---|
| | Increase over 21 Days (µmol/L) | | | Increase (µmol/L/mg[a]) |
| | Women (n = 9) | Men (n = 10) | | Combined Sexes |
| Xanthophyll | Mean (SD) | Mean (SD) | P[b] | Mean (SD) |
| Lutein | 0.78 (0.35) | 0.45 (0.15) | .013 | 0.06 (0.028) |
| Zeaxanthin | 0.11 (0.08) | 0.11 (0.06) | NS | 0.09 (0.06) |
| MZ | 0.30 (0.12) | 0.12 (0.05) | .001 | 0.03 (0.02) |

*Source:* Thurnham et al., *Brit. J. Nutr.*, 100, 1307–1314, 2008. With permission.

[a] Consumptions of respective xanthophylls were 10.8, 1.2, and 8 mg for lutein, zeaxanthin, and MZ from one capsule of Lutein Plus daily.
[b] Difference between the sexes (*t* test).

at baseline, but there were very clear differences in the uptake of lutein and MZ from the supplement, although not in Z; women obtained 30%–50% higher plasma concentrations at 21 days than men (Table 10.1).[24] The changes in plasma lutein (0.06 µmol/L/mg) and Z over 21 days for the sexes combined (0.09 µmol/L/mg) were slightly lower than those previously reported for lutein and Z (0.075 and 0.10 µmol/L/mg, respectively)[37,38] but of a similar magnitude. In contrast, the increase in plasma MZ was only half that of lutein (0.03 µmol/L/mg) for the combined sexes. The reason for the lower uptake of MZ into the plasma by both sexes is not known, but other workers have reported similar observations.[56] The reason why the concentrations of lutein and MZ were lower in men than women may be associated with the differences in body fat between the sexes. The proportion of body fat in women is greater than in men[57] and may assist women to retain more dietary carotenoids more rapidly than men. Thurmann et al.[37] did not report their plasma lutein data by sex, but the standard deviation (SD) for the distribution of results was large even at 42 days, approximately 66% of the mean lutein concentration 1.5 µmol/L.[37]

### 10.3.3.4 Smoking and Subclinical Inflammation

The presence of low plasma concentrations of carotenoids in people who smoked tobacco was an early observation,[58,59] and smokers are often reported to consume less carotenoid-containing foods than nonsmokers.[31] However, even when carotenoid intakes are accounted for, circulating concentrations in smokers are lower than those in nonsmokers.[31] We previously reported that smokers tended to absorb less carotenoids than nonsmokers following meals containing 42 g of fat and either 31.2 mg of lutein, 40 mg of β-carotene, or 38 mg of lycopene. The differences between smokers and non-smokers for β-carotene and lycopene were not significant but the amount of lutein in the triglyceride-rich fraction of plasma was significantly lower in smokers (median, range: 32.2, 9–41.8 nmol/h/L) compared with

## TABLE 10.2
## Influence of Subclinical Inflammation on Plasma Carotenoid Concentrations in Apparently Healthy HIV-Seropositive Kenyan Adults

| | Control | Inflammation Positive | | | Analysis of |
|---|---|---|---|---|---|
| Carotenoid | No Raised APP | Increased CRP | Increased CRP and AGP | Increased AGP | Variance $P$ |
| Lutein | $0.39^a$ (0.25, 0.52) | $0.34^{a,b}$ (0.17, 0.47) | $0.22^b$ (0.12, 0.37) | $0.28^{a,b}$ (0.21, 0.49) | .005 |
| β-Carotene | $0.32^a$ (0.18, 0.50) | $0.32^{a,c}$ (0.18, 0.46) | $0.14^b$ (0.05, 0.29) | $0.16^{b,c}$ (0.04, 0.29) | <.001 |
| α-Carotene | $0.10^a$ (0.04, 0.15) | $0.09^{a,b}$ (0.04, 0.15) | $0.04^b$ (0.01, 0.08) | $0.04^b$ (0.01, 0.08) | <.001 |
| Lycopene | $0.12^a$ (0.08, 0.22) | $0.09^{a,b}$ (0.05, 0.18) | $0.06^b$ (0.02, 0.11) | $0.09^{a,b}$ (0.04, 0.14) | .001 |

*Source:* Data from Thurnham DI et al., *Brit J Nutr* 2008; 100:174–182.

*Note:* Inflammation was assessed using CRP (>5 mg/L) and/or AGP (>1 g/L). Statistics were done on logarithm to the base 10 data using analysis of variance followed by Scheffé test where medians with unlike superscripts indicate significant differences ($P < .05$).

nonsmokers (96.0, 68.5–177.3; $P < .05$).[60] However, low plasma carotenoid concentrations in smokers may also be a result of inflammation.[61] Disease depresses the concentrations of serum carotenoids,[62–64] and low concentrations may be found in people with subclinical inflammation.[65] In a study of Kenyan men and women who had tested positive on two occasions for the human immunodeficiency virus (HIV) but who were still apparently healthy, there was evidence of low plasma carotenoids. Subclinical inflammation was detected by the presence of raised acute phase proteins (APPs), namely, C-reactive protein (CRP) (>5 mg/L) and α1-acid glycoprotein (AGP) (>1g/L). Table 10.2 shows lutein and other plasma carotenoid concentrations depressed by 30%–40% in the groups with inflammation.[65]

### 10.3.3.5 Dietary Intake of Lutein and Zeaxanthin and Risk of AMD

In spite of all the factors influencing plasma carotenoid concentrations, workers have reported significant relationships between dietary intakes of lutein and Z and AMD in the Third National Health and Nutrition Examination Survey (NHANES III)[41] and the Carotenoids in Age-Related Eye Disease Study (CAREDS).[66] In NHANES III, for subjects over 40 years of age higher dietary intakes of lutein and Z were related to lower levels of pigment abnormalities based on photographic evidence of early and late AMD after adjustment for age, gender, alcohol use, hypertension, smoking, and body mass index.[41] In CAREDS, MPOD at 0.5° from the foveal center was 30% higher in data from 1698 women aged 53–86 years in the highest quintile for lutein and Z intake [(±SD): 0.40 ± 0.21] than in women in the lowest quintile [(±SD): 0.31 ± 0.21]. It was concluded that MPOD was directly related to the dietary intake of lutein and Z, and the relationship with MPOD was further strengthened by including data on plasma concentrations of the same carotenoids obtained 4–7 years earlier. The authors suggested that the contribution from the serum carotenoids may reflect unmeasured physical and medical factors that influence the uptake, distribution, and utilization of lutein and Z.[40]

## 10.4 CONCLUSIONS

The separation of the three macular carotenoids lutein, Z, and MZ using the two HPLC systems is relatively quick and suitable for batch processing of many samples. It is currently not fully automated, but we believe this may change with the addition of a fully automatic fraction collector to identify and process the Z fraction obtained from the first column (assay 1). The method has been successfully used to identify and quantify the changes in concentration of the three macular carotenoids following supplementation of subjects with these carotenoids. Our studies suggest that the uptake of the three carotenoids into the blood from supplements is different, with Z > lutein > MZ, and that women may respond better than men to lutein and MZ. Some of the factors influencing carotenoid concentrations in blood are described in this chapter, including dietary sources, seasonality, gender, and inflammation.

## ACKNOWLEDGMENT

We thank the Howard Foundation, Cambridge, United Kingdom, for financial assistance in support of this work.

## CONFLICTS OF INTEREST

Thurnham receives payment from the Howard Foundation for consultancy. Nolan acknowledges financial assistance in the form of research grants from the Howard Foundation. Meagher and Connolly have no conflicts to declare.

## REFERENCES

1. Maoka T. Carotenoids in marine animals. *Marine Drugs* 2011; 9:278–293.
2. Khachik F. Distribution and metabolism of dietary carotenoids in humans as a criterion for development of nutritional supplements. *Pure Appl Chem* 2006; 78:1551–1557.
3. Thurnham DI, Smith E, Flora PS. Concurrent liquid-chromatographic assay of retinol, α-tocopherol, β-carotene, α-carotene, lycopene and β-cryptoxanthin in plasma with tocopherol acetate as internal standard. *Clin Chem* 1988; 34:377–381.
4. Bieri JG, Brown ED, Smith JC Jr. Determination of individual carotenoids in human plasma by high performance chromatography. *J Liq Chromatogr* 1985; 8:473–484.
5. Stahl W, Schwarz W, Sies H. Human serum concentrations of all-*trans* beta- and alpha-carotene but not 9-*cis* beta-carotene increase upon ingestion of a natural isomer mixture obtained from *Dunaliella salina* (Betatene). *J Nutr* 1993; 123:847–851.
6. Granado F, Olmedilla B, Gil-Martinez E, et al. Lutein ester in serum after lutein supplementation in human subjects. *Brit J Nutr* 1998; 80:445–449.
7. Bendich A, Olson JA. Biological action of carotenoids. *FASEB J* 1989; 3:1927–1932.
8. Edge R, McGarvey DJ, Truscott TG. The carotenoids as anti-oxidants. *J Photochem Photobiol B:Biol* 1997; 41:189–200.
9. Bone RA, Landrum JT, Hime GW, et al. Stereochemistry of the human macular carotenoids. *Invest Ophthalmol Vis Sci* 1993; 34:2033–2040.
10. Li B, Ahmed F, Bernstein PS. Studies on the singlet oxygen scavenging mechanism of human macular pigment. *Arch Biochem Biophys* 2010; 504:56–60.

11. Sabour-Pickett S, Nolan JM, Loughman J, et al. A review of the evidence germane to the putative protective role of the macular carotenoids for age-related macular degeneration. *Mol Nutr Food Res* 2011; 55 in press.
12. Stringhan JM, Hammond BR Jr. Macular pigment and visual performance under glare conditions. *Optom Vis Sci* 2008; 85:82–88.
13. Erdman JW Jr, Bierer TL, Gugger ET. Absorption and transport of carotenoids. *Ann N Y Acad Sci* 1993; 691:76–85.
14. Craft NE, Brown ED, Smith JC. Effects of storage and handling conditions on concentrations of individual carotenoids, retinol and tocopherol in plasma. *Clin Chem* 1988; 34:44–48.
15. Cantilena LR, Nierenberg DW. Simultaneous analysis of five carotenoids in human plasma by isocratic high performance liquid chromatography. *J Micronut Anal* 1989; 6:181–209.
16. Krinsky NI, Russett MD, Handelman GJ, et al. Structural and geometric isomers of carotenoids in human plasma. *J Nutr* 1990; 120:1654–1662.
17. Olmedilla B, Grando F, Rojas-Hidalgo E, et al. A rapid separation of ten carotenoids, three retinoids, alpha-tocopherol, and *d*-alpha-tocopherol acetate by high performance liquid chromatography and its application to serum and vegetable samples. *J Liq Chromatogr* 1990; 13:1455–1485.
18. Rüttimann A, Schiedt K, Vecchi M. Separation of (3R,3'R)-, (3R,3'S; *meso*)-, (3S,3'S)-zeaxanthin, (3R,3'R,6'R)-, (3R,3'S,6'S)- and (3S,3'S,6'S)-lutein via the dicarbamates of (*S*)-(+)-a-(1-naphthyl) ethyl isocyanate. *J High Resolut Chromatogr* 1983; 6:612–616.
19. Bunnell RH, Bauernfeind JC. Chemistry, uses and properties of carotenoids in foods. *Food Technol* 1962; 16:36–42.
20. Bone RA, Landrum JT, Tarsis SL. Preliminary identification of the human macular pigment. *Vision Res* 1985; 25:1531–1535.
21. Maoka T, Arai A, Sinuzu M, et al. The first isolation of enantiomeric and *meso*-zeaxanthin in nature. *Comp Biochem Physiol B* 1986; 83:121–124.
22. Landrum JT, Bone RA, Moore LL, et al. Analysis of zeaxanthin within individual human retinas. *Methods Enzymol* 1999; 299:457–467.
23. Johnson EJ, Neuringer M, Russell RM, et al. Nutritional manipulation of primate retinas, III: effects of lutein or zeaxanthin supplementation on adipose tissue and retina of xanthophyll-free monkeys. *Invest Ophthalmol Vis Sci* 2005; 46:692–702.
24. Thurnham DI, Tremel A, Howard AN. A supplementation study in humans with a combination of *meso*-zeaxanthin, (3R,3'R)-zeaxanthin and (3R,3'R,6')-lutein. *Brit J Nutr* 2008; 100:1307–1314.
25. Britton G. General carotenoid methods. In: Law JH, Rilling HS, editors. *Steroids and Isoprenoids. Part B*. In: *Methods in Enzymology Volume 111*. New York: Academic Press; 1985. 113–149.
26. Wang G, Root M, Ye X, et al. Routine assay of plasma carotenoids by high performance liquid chromatography with an internal standard. *J Micronut Anal* 1989; 5:3–14.
27. Connolly EE, Beatty S, Thurnham DI, et al. Augmentation of macular pigment following supplementation with all three macular carotenoids: an exploratory study. *Curr Eye Res* 2010; 35:335–351.
28. Olmedilla-Alonso B, Granado-Lorencio F, Blanco-Navarro I. Carotenoids, retinol and tocopherols in blood: comparability between serum and plasma (Li-heparin) values. *Clin Biochem* 2005; 38:444–449.
29. Bone RA, Landrum JT, Friedes LM, et al. Distribution of lutein and zeaxanthin stereoisomers in the human retina. *Exp Eye Res* 1997; 64:211–218.
30. Bone RA, Landrum JT, Cao Y, et al. Macular pigment response to a xanthophyll supplement of lutein, zeaxanthin and *meso*-zeaxanthin. *Proc Nutr Soc* 2006; 65:105A.
31. Margetts BM, Jackson AA. The determinants of plasma β-carotene: interaction between smoking and other lifestyle factors. *Eur J Clin Nutr* 1996; 50:236–238.

32. Scott KJ, Thurnham DI, Hart DJ, et al. The correlation between intake of lutein, lycopene and β-carotene from vegetables and fruits and concentrations in blood plasma. *Proc Nutr Soc* 1994; 53:138A.
33. Brady WE, Mares-Perlman JA, Bowen P, et al. Human serum carotenoid concentrations are related to physiologic and lifestyle factors. *J Nutr* 1996; 126:129–137.
34. Scott KJ, Thurnham DI, Hart DJ, et al. The correlation between the intake of lutein, lycopene and β-carotene from vegetables and fruits and blood plasma concentrations in a group of women ages 50–65 years in the UK. *Brit J Nutr* 1996; 75:409–418.
35. Carroll Y, Corridan B, Morrissey PA. Carotenoids in young and elderly healthy humans: dietary intakes biochemical status and diet-plasma relationships. *Eur J Clin Nutr* 1999; 53:644–653.
36. FAO/WHO. *Requirements of Vitamin A, Thiamine, Riboflavin and Niacin.* Geneva: FAO/WHO. Expert Groups Technical Report Series No. 362; 1967.
37. Thurmann PA, Schalch W, Aebischer C-P, et al. Plasma kinetics of lutein, zeaxanthin, and 3'-dehydro-lutein after multiple doses of a lutein supplement. *Am J Clin Nutr* 2005; 82:88–97.
38. Hartmann D, Thurmann PA, Spitzer V, et al. Plasma kinetics of zeaxanthin and 3'-dehydro-lutein after multiple oral doses of synthetic zeaxanthin. *Am J Clin Nutr* 2004; 79:410–417.
39. Thurnham DI. Macular zeaxanthins and lutein—a review of dietary sources and bioavailability and some relationships with macular pigment optical density and age-related macular disease. *Nutr Res Rev* 2007; 20:163–179.
40. Mares JA, LaRowe TL, Snodderly DM, et al. Predictors of optical density of lutein and zeaxanthin in retinas of older women in the Carotenoids in Age-Related Eye Disease Study, an ancillary study of the Women's Health Initiative. *Am J Clin Nutr* 2006; 84:1107–1122.
41. Mares-Perlman JA, Fisher AI, Klein R, et al. Lutein and zeaxanthin in the diet and serum and their relation to age-related maculopathy in the Third National Health and Nutrition Examination Survey. *Am J Epidemiol* 2001; 153:424–432.
42. Granado F, Olmedilla B, Blanco I, et al. Major fruit and vegetable contributors to the main serum carotenoids in the Spanish diet. *Eur J Clin Nutr* 1996; 50:246–250.
43. Thurnham DI, Flora PS. Do higher vitamin A requirements in men explain the difference between the sexes in plasma provitamin A carotenoids and retinol. *Proc Nutr Soc* 1988; 47:181A.
44. Gregory JR, Foster K, Tyler H, Wiseman M. *The Dietary and Nutritional Survey of British Adults.* London: HMSO; 1990.
45. de Pee S, West CE, Permaesih D, et al. Orange fruit is more effective than are dark-green, leafy vegetables in increasing serum concentrations of retinol and β-carotene in schoolchildren in Indonesia. *Am J Clin Nutr* 1998; 68:1058–1067.
46. Cooney RV, Franke AA, Hankin JH, et al. Seasonal variations in plasma micronutrients and antioxidants. *Cancer Epidemiol Biomark Prev* 1995; 4:207–215.
47. Olmedilla B, Granado F, Blanco I, et al. Seasonal and sex-related variations in six serum carotenoids, retinol, and α-tocopherol. *Am J Clin Nutr* 1994; 60:106–110.
48. Takagi S, Kishi F, Nakajima K, et al. A seasonal variation of carotenoid composition in green leaves and effect of environmental factors on it. *Sci Rep Fac Agr Okayama Uni* 1990; 75:1–7.
49. Perry A, Rasmussen H, Johnson EJ. Xanthophyll (lutein, zeaxanthin) content of fruits, vegetables and corn and egg products. *J Fd Comp Anal* 2009; 22:91–15.
50. Sommerburg O, Keunen JEE, Bird AC, et al. Fruits and vegetables that are sources of lutein and zeaxanthin: the macular pigment in the human eye. *Br J Ophthalmol* 1998; 82:907–910.

51. Granado F, Olmedilla B, Blanco I, et al. Carotenoid composition of raw and cooked Spanish vegetables. *J Agric Food Chem* 1992; 40:2135–2140.
52. Yeum K-J, Booth SL, Sadowski JA, et al. Human plasma carotenoid response to the ingestion of controlled diets high in fruits and vegetables. *Am J Clin Nutr* 1996; 64:594–602.
53. Curran-Celentano J, Hammond BR Jr, Ciulla TA, et al. Relation between dietary intake, serum concentrations, and retinal concentrations of lutein and zeaxanthin in a Midwest population. *Am J Clin Nutr* 2001; 74:796–802.
54. Ito Y, Ochiai J, Sasaki R, et al. Serum concentrations of carotenoids, retinol, and a-tocopherol in healthy persons determined by high-performance liquid chromatography. *Clin Chim Acta* 1990; 194:131–144.
55. Rock CL, Demitrack MA, Rosenwald EN, et al. Carotenoids and menstrual cycle phase in young women. *Cancer Epidemiol Biomark Prev* 1995; 4:283–288.
56. Schiedt K, Leuenberger FJ, Vecchi M, et al. Absorption, retention and metabolic transformation of carotenoids in rainbow trout, salmon and chicken. *Pure Appl Chem* 1985; 57:685–692.
57. Forbes GB. *Human Body Composition*. New York: Springer Verlag; 1987.
58. Connett JE, Kuller LH, Kjelsberg MO, et al. Relationship between carotenoids and cancer. The multiple risk factor intervention trial (MRFIT) study. *Cancer* 1989; 64:126–134.
59. Northrop-Clewes CA, Thurnham DI. Monitoring micronutrients in cigarette smokers. *Clin Chim Acta* 2007; 377:14–38.
60. O'Neill ME, Thurnham DI. Difference between smokers and non-smokers in the intestinal absorption of carotenoids. *Proc Nutr Soc* 1998; 57:26A.
61. Das I. Raised C-reactive protein levels in serum from smokers. *Clin Chim Acta* 1985; 153:9–13.
62. Baranowitz SA, Starrett B, Brookner AR. Carotene deficiency in HIV patients. *AIDS* 1996; 10:115 (letter).
63. Ullrich R, Schneider T, Heise W, et al. Serum carotene deficiency in HIV-infected patients. *AIDS* 1994; 8:661–665.
64. Cser MA, Majchrzak D, Rust P, et al. Serum carotenoid and retinol levels during childhood infections. *Ann Nutr Metab* 2004; 48:156–162.
65. Thurnham DI, Mburu ASW, Mwaniki DL, et al. Using plasma acute-phase protein concentrations to interpret nutritional biomarkers in apparently healthy HIV-1-seropositive Kenyan adults. *Brit J Nutr* 2008; 100:174–182.
66. Mares JA, Voland RP, Sondel SA, et al. Healthy lifestyles related to subsequent prevalence of age-related macular degeneration. *Arch Ophthalmol* 2011; 129:470–480.

# 11 Xanthophyll–Membrane Interactions
## Implications for Age-Related Macular Degeneration

*Witold K. Subczynski, Anna Wisniewska-Becker, and Justyna Widomska*

**CONTENTS**

11.1 Introduction ..................................................................................................204
11.2 Why Are Macular Xanthophylls Selected from Other Carotenoids?...........205
11.3 What Is the Mechanism of the Selective Accumulation of Macular Xanthophylls in the Retina?..........................................................................206
11.4 Why Are Macular Xanthophylls So Stably Stored in the Retina? ...............207
11.5 Can Xanthophyll–Membrane Interactions Enhance a Protective Activity of Macular Xanthophylls Against AMD? .....................................208
    11.5.1 High Incorporation Yield................................................................208
    11.5.2 Transmembrane Orientation Enhancing Antioxidant Properties of Carotenoids..................................................................208
    11.5.3 Two Orientations of Lutein in Retina Membranes .........................209
    11.5.4 Indirect Antioxidant Action of Macular Xanthophylls by Changing Membrane Properties................................................. 210
    11.5.5 *Cis* Isomers in Retina Membranes................................................. 211
    11.5.6 Macular Xanthophylls Located in the Most Vulnerable Regions of Photoreceptor Outer-Segment Membranes ................................. 212
    11.5.7 Colocalization of Lutein in Membrane Domains Together with Unsaturated and Polyunsaturated Phospholipids Significantly Enhancing Antioxidant Activity ....................................................... 214
11.6 Concluding Remarks ................................................................................... 215
Acknowledgment .................................................................................................. 216
References............................................................................................................. 216

## 11.1 INTRODUCTION

Only two carotenoids (namely, lutein and zeaxanthin) are selectively accumulated in the retina of the human eye from blood plasma, where more than 20 other carotenoids are available (Khachik et al. 1997). A third carotenoid, *meso*-zeaxanthin, is formed directly in the human retina (where it is found) from lutein (Landrum and Bone 2001). All of these dipolar, terminally dihydroxylated carotenoids (named "macular xanthophylls") are accumulated in nerve fibers and photoreceptor outer segments (POSs) and impede the onset of age-related macular degeneration (AMD) (Mares-Perlman et al. 2002; Gale et al. 2003; Cho et al. 2004; Mares 2004). The distribution of macular xanthophylls in the retina is not uniform. Their concentration in the fovea can reach values as high as 0.1 to 1 mM, which is about 100 times greater than the values in peripheral retina and about 1000 times greater than in other organs and tissues (Bone et al. 1985; Landrum et al. 1999). Additionally, the ratio of total lutein to total zeaxanthin differs when comparing the central fovea to the more peripheral regions of the retina; zeaxanthin is predominant in the former and lutein in the latter (Bone et al. 1988; Handelman et al. 1988).

The retina of the macaque monkey possesses a macula that closely resembles the macula in the human retina and, therefore, provides an optimal model for nutritional studies of macular biology. It has been shown that after xanthophyll supplementation in the macaque monkey, the serum level of these carotenoids increases rapidly (in less than 2 weeks), whereas the optical density of macular pigment increases slowly (in 24–32 weeks) (Neuringer et al. 2004). Similar fast and slow accumulations were observed in humans (Hammond et al. 1997; Landrum et al. 1997). The unusual behavior of macular xanthophylls raises a few basic questions: why are macular xanthophylls selected from other carotenoids? What is the mechanism of their selective accumulation in the retina? Why are macular xanthophylls so firmly stored in the retina? Lastly, can xanthophyll–membrane interaction enhance the protective capability of xanthophylls against AMD?

The precise location of macular xanthophylls in the Henle's fiber layer of photoreceptor axons and POSs is not known. There are two major hypotheses about their location. According to the first argument, macular xanthophylls are transversely incorporated in the lipid-bilayer portion of membranes of the human retina (Bone and Landrum 1984; Bone et al. 1992). Macular xanthophylls are thought to combat light-induced damage, mediated by reactive oxygen species, by absorbing the most damaging incoming wavelength of light prior to the formation of reactive oxygen species (a function expected of carotenoids in Henle's fibers) and by chemically and physically quenching reactive oxygen species once they are formed (a function expected of carotenoids in POSs). The location of macular xanthophylls in the membrane is ideal for these actions.

According to the second hypothesis, macular xanthophylls are protein bound by membrane-associated, xanthophyll-binding proteins (Bhosale and Bernstein 2007; Loane et al. 2008). Bernstein's group has identified and characterized a zeaxanthin-binding protein (Bhosale et al. 2004; Bhosale and Bernstein 2005) and a lutein-binding protein (Bhosale et al. 2009) in human macula. The interactions between macular xanthophylls and both lipid-bilayer membranes and xanthophyll-binding proteins are significant. However, in this chapter we focus on xanthophyll–membrane interactions, which help us to answer the aforementioned basic questions and to better understand the protective actions of xanthophylls against AMD.

## 11.2 WHY ARE MACULAR XANTHOPHYLLS SELECTED FROM OTHER CAROTENOIDS?

After determining that the abilities of macular xanthophylls are similar to those of other carotenoids (the abilities to filter out blue light and quench singlet oxygen [Cantrell et al. 2003] and scavenge free radicals [Stahl et al. 1998] in organic solvents), Krinsky and coauthors (Krinsky 2002; Krinsky et al. 2003) concluded that there must be some specific property of xanthophylls that helps to explain their selective presence in the retina of primates. The authors determined that this property was the behavior of macular xanthophylls in biological membranes. Thus, unique lutein– and zeaxanthin–membrane interactions distinguish these macular xanthophylls from the other carotenoids available from blood plasma.

Macular xanthophylls have high membrane solubility, which is a major characteristic that distinguishes them from other dietary carotenoids. Reported xanthophyll solubility thresholds (the concentration of xanthophylls up to which they exist mainly as monomers) in fluid-phase model membranes lie in the area of 10 mol% (Milon et al. 1986). However, lower and higher values of xanthophyll solubility were also indicated (see the appendix in the work by Wisniewska and Subczynski [1998]). Nonpolar β-carotene starts to aggregate at concentrations as low as 0.5–1 mol% (Kennedy and Liebler 1992; Woodall et al. 1995).

Macular xanthophylls are oriented perpendicular to the membrane surface, which ensures their high solubility, stability, and significant effects on membrane properties (Wisniewska et al. 2006) (see also Sections 11.5.3 and 11.5.5). It has been recently shown that they are selectively accumulated in membrane domains that contain unsaturated phospholipids and, thus, are located in the most vulnerable regions of the membrane (Wisniewska and Subczynski 2006a,b; Subczynski et al. 2010). Nonpolar carotenoids are oriented in the lipid bilayer rather randomly (van de Ven et al. 1984) and are distributed between membrane domains more evenly (Subczynski et al. 2010). The specific orientation and location of macular xanthophylls distinguishes them from nonpolar carotenoids and maximizes their protective action in the membranes of the eye retina.

Figure 11.1 shows the distribution of macular xanthophylls between coexisting unsaturated (bulk) and saturated (rafts) membrane domains and their preferential transmembrane orientation. The direct and indirect antioxidant actions of macular xanthophylls are indicated, as well as specific xanthophyll–membrane interactions, which can modulate or enhance antioxidant actions in retina membranes. This diagram is used as a guideline in this review.

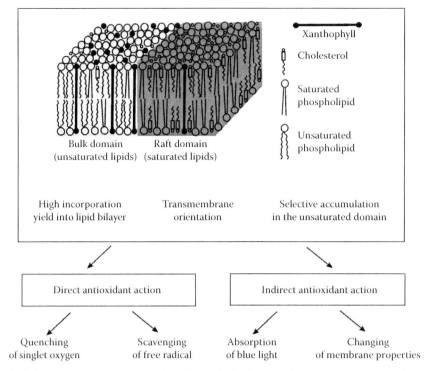

**FIGURE 11.1** Diagram showing location and distribution of macular xanthophylls in the lipid-bilayer membrane: direct and indirect antioxidant actions of xanthophylls are indicated. Changing of membrane properties includes hydrophobicity, oxygen diffusion–concentration product, permeability of metal ions and metal ion complexes, alkyl chain bending (vertical fluctuations), and membrane order (fluidity).

## 11.3 WHAT IS THE MECHANISM OF THE SELECTIVE ACCUMULATION OF MACULAR XANTHOPHYLLS IN THE RETINA?

The presence of xanthophylls in biological membranes is ideal if they are to act as lipid antioxidants (although lipid association alone cannot explain the extraordinarily specific uptake of xanthophylls into the retina). Whenever a tissue or an organ exhibits highly selective uptake and deposition of biological molecules, it is likely that specific binding proteins are involved. This explanation was given by Bhosale and Bernstein (2005), who identified and characterized xanthophyll-binding proteins in the human retina that enhance antioxidant activity. These membrane-associated proteins bind xanthophylls with high specificity and affinity. However, there is the question of whether the amount of these proteins is sufficient to bind and store all xanthophyll molecules, which accumulate in the retina in extremely high concentrations (up to 1 mM) (Landrum et al. 1999). This suggests that the function of xanthophyll-binding proteins can be traced to the selective collection of macular xanthophylls from blood plasma and their deposition into the membranes of the eye retina, where they are then stably stored.

Some data from the literature suggest that the segregation of polar and nonpolar carotenoids already occurs at the level of carotenoid transport. Carotenoids are transported in human blood plasma exclusively by lipoproteins. Nonpolar carotenoids are transported primarily in low-density lipoproteins (LDLs), whereas polar carotenoids are more evenly distributed between LDLs and high-density lipoproteins (HDLs) (Loane et al. 2008; Parker 1996). It is thought that most tissues obtain carotenoids via the LDL receptor route (Parker 1996). However, in the case of lutein and zeaxanthin transport, we hypothesize that receptors for HDL should be involved instead. It has been suggested that this role can be played by receptors that are similar to those found in the central nervous system for HDL particles containing apolipoprotein E (ApoE) (Thomson et al. 2002; Loane et al. 2008). Nevertheless, there is a lack of convincing evidence that lipoproteins containing ApoE are responsible for the delivery and accumulation of macular xanthophylls in the retina.

Adipose tissue is considered to be the main storage site for carotenoids (Moussa et al. 2011). Both polar and nonpolar carotenoids appear to be transported similarly to cells of this tissue, and it has been shown that lutein and lycopene can be transported by facilitated diffusion involving the cell-surface glycoprotein CD36. The same receptor proved to be involved in $\beta$-carotene uptake by the intestine (van Bennekum et al. 2005; Moussa et al. 2011). It is also present in different cell types of the monkey retina (Tserentsoodol et al. 2006). Tserentsoodol et al. (2006) also suggested the existence of an HDL-based intraretinal lipid transport mechanism. However, they presented no data on carotenoid transport within the retina. Even if the mechanism of the transport facilitated by special receptors and transporters is true for carotenoids, this does not explain the selective transport of only xanthophylls into the retina. On the other hand, canthaxanthin, which is also a dipolar carotenoid with two keto groups not normally present in the retina, can be accumulated in the monkey retina if the animal receives an excess of canthaxanthin-containing food (Goralczyk et al. 2000). Additionally, experiments performed on humans have shown that canthaxanthin can accumulate in the retina where it forms crystals (Daicker et al. 1987).

Thus, canthaxanthin did not replace physiological macular xanthophylls. Also, treatment with canthaxanthin showed no effect on the content of macular xanthophylls in the retina (Goralczyk et al. 2000). Astaxanthin, a dipolar carotenoid with two hydroxyl and two keto groups that is present in serum, was not found in the eye retina. Based on experiments on model membranes, it was shown that canthaxanthin (like zeaxanthin and lutein) is located perpendicular to the membrane surface and has a strong effect on membrane properties (Sujak et al. 2005; Sujak 2009). These results suggest that the mechanism of xanthophyll transport and accumulation in the retina may be less specific and depend on the presence of two (not four) polar groups at the ends of the xanthophyll molecule.

## 11.4 WHY ARE MACULAR XANTHOPHYLLS SO STABLY STORED IN THE RETINA?

The transmembrane location of a significant portion of macular xanthophylls in the retinal cells seems obvious (Bone and Landrum 1984; Gruszecki and Sielewiesiuk 1990; Subczynski et al. 1992, 1993; Gruszecki 1999; Gruszecki and Strzalka 2005).

Such a location explains their very slow removal from the retina, which was observed in a study by Landrum et al. (1997) that followed the discontinuation of a lutein supplement given to healthy volunteers. After stopping a 140-day lutein supplement, Landrum et al. (1997) observed a relatively fast decrease in lutein concentration in the serum, whereas the level of lutein in the retina remained unchanged and at a high concentration for up to 6 months. Similar effects were observed by Hammond et al. (1997). In comparison, it has been shown that rat retina can completely replace cholesterol, which is anchored at the membrane surface by hydroxyl groups, every 6 to 7 days. In humans, this replacement may be even more rapid (Tserentsoodol et al. 2006). The incorporation yield of macular xanthophylls into liposomes and retinal pigment epithelial cells is 5–10 times greater than that of canthaxanthin and 20–40 times greater than that of β-carotene (Shafaa et al. 2007). These observations suggest that the strong anchoring of xanthophyll molecules at opposite membrane surfaces by polar hydroxyl groups is significant not only in enhancing their effects on membrane properties (Subczynski et al. 1992, 1993; Gabrielska and Gruszecki 1996; Wisniewska et al. 2006) but also in the stabilization of these molecules in membranes of the human retina. Thus, transmembrane orientation can also be included as a characteristic that distinguishes macular xanthophylls from other dietary carotenoids.

## 11.5 CAN XANTHOPHYLL–MEMBRANE INTERACTIONS ENHANCE A PROTECTIVE ACTIVITY OF MACULAR XANTHOPHYLLS AGAINST AMD?

### 11.5.1 HIGH INCORPORATION YIELD

The extremely high level of macular xanthophylls in the retina, compared with other tissues, does not reflect their content in POS membranes. Macular xanthophylls in POSs constitute about 10% of the amount in the entire retina (Rapp et al. 2000), although values as high as 25% have also been reported in the outer segments (Sommerburg et al. 1999). Despite the lower percentage, the local concentration of macular xanthophylls in membranes of rod outer segments is approximately 70% higher than in residual retina membranes (Sommerburg et al. 1999). But even there, the carotenoid concentration in the lipid-bilayer portion of the membrane is much lower than 1 mol% (Bone and Landrum 1984). However, this concentration is high enough for effective blue-light filtration, quenching of singlet oxygen and molecular triplet states, and effective antioxidant action. For example, due to the presence of xanthophylls in the membranes of Henle's fiber layer (which is composed of photoreceptor axons), the reduction of blue-light intensity that reaches the back of the eye can be as great as 90% (Wooten and Hammond 2002).

### 11.5.2 TRANSMEMBRANE ORIENTATION ENHANCING ANTIOXIDANT PROPERTIES OF CAROTENOIDS

Orientation of polar and nonpolar carotenoids (including their *cis* forms) in the lipid-bilayer membrane has been investigated by many laboratories (Bone and Landrum 1984; van de Ven et al. 1984; Gruszecki and Sielewiesiuk 1990; Sujak et al. 1999;

Widomska and Subczynski 2008). It has been suggested that the presence of polar hydroxyl groups at the ends of carotenoid molecules and their transmembrane orientation (as in the case of zeaxanthin and lutein) enhance their antioxidant properties, compared with the antioxidant properties of monopolar (β-cryptoxanthin) and nonpolar (β-carotene) carotenoids (Woodall et al. 1995, 1997a). Although dipolar zeaxanthin and nonpolar β-carotene show similar antioxidant properties in organic solutions, they differ when incorporated into membranes. Zeaxanthin was shown to react with free radicals slightly more effectively than β-cryptoxanthin and much more effectively than β-carotene (Woodall et al. 1995, 1997b). β-carotene and lycopene can react efficiently only with radicals generated in the inner part of the membrane, whereas zeaxanthin and lutein (with their end groups exposed to an aqueous environment) can also scavenge free radicals generated in the aqueous phase (Britton 1995).

### 11.5.3 Two Orientations of Lutein in Retina Membranes

Relatively more lutein than zeaxanthin was found in the outer segments of the peripheral rods (Sommerburg et al. 1999; Rapp et al. 2000). It is unknown what causes this difference in the distribution of lutein and zeaxanthin throughout the retina. It could be the result of their different uptake and/or storage by specific xanthophyll-binding proteins (Bhosale et al. 2004, 2009) or the result of enzymatic activity that transforms a significant fraction of lutein into *meso*-zeaxanthin (Johnson et al. 2005). These possibilities suggest that these two xanthophylls play significant but somewhat different functions in vision. Lutein and zeaxanthin differ only in the position of one double bond in one of their ionone rings and in the stereochemistry at C(3′).

Gruszecki's group report (Sujak et al. 1999) states that in model phosphatidylcholine (PC) membranes, zeaxanthin and lutein adopt essentially different orientations (see Figure 11.2). Zeaxanthin was found to adopt a roughly vertical orientation with respect to the plane of the membrane. The relatively large orientation angle found in the case of lutein was interpreted as a representation of the existence of two orthogonally oriented pools of lutein, one following the orientation of zeaxanthin and the second being parallel with respect to the plane of the membrane. These unusual results were critically evaluated by Krinsky (2002), who cited as possibilities contamination of the sample by *cis* isomers of lutein or dehydration of some amount of

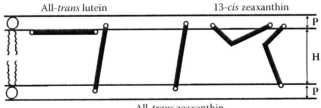

**FIGURE 11.2** Schematic drawing of the location of lutein and different isomers of zeaxanthin in the lipid-bilayer membrane: horizontal orientation of the *cis* isomers of zeaxanthin is also proposed (Gruszecki 2004) but has not been confirmed experimentally. P: polar head group region; H: hydrophobic, hydrocarbon region.

lutein, resulting in the formation of anhydrolutein with only a single –OH group. Interestingly, recent molecular dynamics simulations of the behavior of lutein in the PC bilayer showed that these two orientations of lutein molecules are very probable (Pasenkiewicz-Gierula et al. 2012). Pasenkiewicz-Gierula et al. (2012) observed that lutein can change orientation from perpendicular to parallel and from parallel to perpendicular in the lipid bilayer.

The fact that lutein can be oriented horizontally and along hydrophobic lipid chains was examined in terms of its biological function in protecting different membrane regions from oxidative damage (Sujak and Gruszecki 2000). Sujak and Gruszecki (2000) suggested that a different orientation of lutein in the membrane may enhance its physiological role as a blue-light filter since light absorption depends on the orientation of the dipole transition moment (close to the molecular axis of polyenes) with respect to the electric vector of the electromagnetic wave. The existence of two orthogonally oriented pools increases the chances that light will be captured by a carotenoid molecule (Sujak and Gruszecki 2000).

### 11.5.4 Indirect Antioxidant Action of Macular Xanthophylls by Changing Membrane Properties

For the effective physical interaction of carotenoid molecules and lipids in the lipid-bilayer membrane, the presence of polar groups on both ends of the long, stick-like carotenoid molecule and the ratio of carotenoid-molecule length to membrane thickness are very important. Monopolar and nonpolar carotenoids (like β-cryptoxanthin and β-carotene) affect membrane properties much less than dipolar carotenoids (Wisniewska et al. 2006). Macular xanthophylls, with polar hydroxyl groups on both ends of the molecule, affect membrane properties very strongly. Membrane properties that should be considered to have indirect antioxidant action are hydrophobicity, the oxygen diffusion–concentration product, permeability of metal ions and metal ion complexes, alkyl chain bending (vertical fluctuations), and membrane order (fluidity) (see Figure 11.1).

Incorporation of macular xanthophylls causes a considerable increase in hydrophobicity in the central region of the lipid bilayer. However, the hydrophobicity in the polar head group region decreases (Wisniewska and Subczynski 1998). Xanthophylls reduce the oxygen diffusion-concentration product at all locations in the membrane (Subczynski et al. 1991) and also ion penetration into the membrane (Wisniewska and Subczynski 1998). Additionally, the frequency of vertical fluctuations of the terminal methyl groups of alkyl chains toward the membrane surface is strongly reduced by the presence of xanthophylls (Yin and Subczynski 1996). Finally, xanthophylls increase the order of alkyl chains and decrease membrane fluidity (Subczynski et al. 1992, 1993). These changes in the membrane microenvironment (all induced by xanthophylls) should affect the chemical reactions that occur within the lipid bilayer (see also the study by Subczynski et al. [2009] for more explanation), making the membrane less sensitive to oxidative damage.

McNulty et al. (2007) related the physical interaction of carotenoids with the membrane to their antioxidant effectiveness. According to this group, an ordering effect of carotenoids is accompanied by strong antioxidant action, as seen for the dipolar xanthophyll astaxanthin. On the contrary, nonpolar carotenoids like β-carotene and lycopene disorder the membrane and act as prooxidants. β-carotene—because of its low membrane solubility as a monomer and low incorporation yield (and, therefore, weak effects on membranes)—does not protect membranes against lipid peroxidation. Moreover, at a high oxygen concentration it may act as a prooxidant (Palozza et al. 2006). It has to be emphasized, however, that possible indirect antioxidant action of carotenoids via their alteration of membrane properties would require a high concentration of carotenoids in the membrane. Moreover, unsaturated alkyl chains, which are abundant in the membranes of POSs, decrease the effects of xanthophylls on membrane properties.

### 11.5.5 CIS ISOMERS IN RETINA MEMBRANES

The polyene-chain double bonds present in carotenoids can exist in all-*trans*, mono-*cis*, or poly-*cis* configurations. The *cis* isomers of macular xanthophylls were isolated from the retina of the human eye (Bernstein et al. 2001). The 9-*cis* and 13-*cis* isomers of zeaxanthin were present in the greatest amounts. It has been established that *cis* isomers are mainly produced directly in the eye retina under intensive light exposure from all-*trans* isomers (Krinsky 2002). It is unknown whether *cis* isomers serve any function in the eye retina or if they are only the products of light-induced isomerization.

Recent measurements (Widomska and Subczynski 2008) indicate that in the lipid bilayer the transmembrane orientation of the *cis* isomers of zeaxanthin is prevalent (see Figure 11.2); however, the existence of a horizontal orientation cannot be excluded. An orientation that places the *cis* isomers of zeaxanthin horizontally with respect to the plane of the membrane and with polar hydroxyl groups anchored in the same polar head group region (the same leaflet) of the bilayer was proposed based on the molecular structure of the *cis* isomers, the location of the polar and hydrophobic parts of the molecule, and the "fit" to the membrane hydrophobic thickness (see the review by Gruszecki [2004]). However, this has never been confirmed experimentally.

The greater effect of *cis* isomers in the membrane center, compared with *trans* isomers, is most likely a result of their transmembrane orientation and the introduction of a larger number of conjugated double bonds into the region (due to their bent structure) (Figure 11.2). The effects of isomers of zeaxanthin on membrane properties increase as all-*trans* < 9-*cis* ≤ 13-*cis* isomers (Widomska and Subczynski 2008). Similar observations were reported by Widomska et al. (2009), who also detected greater effects of 9-*cis* and 13-*cis* isomers of zeaxanthin on thermodynamic characteristics of the main phase transition of the dipalmitoylphosphatidylcholine (DPPC) membrane (measured with differential scanning calorimetry) than effects caused by the all-*trans* isomer. The transmembrane orientation of the *cis* isomer of zeaxanthin is most likely enhanced in thin lipid-bilayer membranes in which the thickness of the hydrophobic core is smaller than the distance between the hydroxyl groups at the 3 and 3′ positions of zeaxanthin. However, in thick membranes the horizontal orientation of *cis* isomers may be prevalent (see the discussions by Widomska and Subczynski [2008] and Widomska et al. [2009]).

The organization of *cis*-zeaxanthin in the membrane has been investigated less. However, some conclusions can be made based on measurements with monomolecular layers on the air–water interface formed from the mixture of DPPC and all-*trans*, 9-*cis*, or 13-*cis* zeaxanthin (Sujak and Gruszecki 2000; Milanowska et al. 2003). Results showed that the concentration of zeaxanthin in the monomolecular layer at which aggregation starts is approximately 5 mol% for *trans*-zeaxanthin, approximately 20 mol% for 9-*cis* zeaxanthin, and higher than 20 mol% for 13-*cis* zeaxanthin. Additionally, *cis* isomers of zeaxanthin do not show a tendency to organize into aggregates, even in a very polar solvent like an ethanol/water mixture (5:95 v/v) where *trans*-isomers aggregate easily (Milanowska et al. 2003). It can be concluded that the formation of *cis* isomers protects xanthophylls from aggregation and enhances their effect on membrane properties.

### 11.5.6 Macular Xanthophylls Located in the Most Vulnerable Regions of Photoreceptor Outer-Segment Membranes

For systems with a high carotenoid concentration (e.g., bacteria and plants in which the local carotenoid concentration in membranes can reach a few mole percentages), it is most important to understand how carotenoids affect the physical properties, structure, and dynamics of the membrane (see also Section 11.5.4). In animals, the highest concentration of carotenoids is found in the retina of primates. But even there, the xanthophyll concentration in the lipid-bilayer portion of the membrane is much lower than 1 mol% (Bone and Landrum 1984). For systems with a low carotenoid concentration, it is especially important to understand how the membrane itself—its composition, structure, and lateral organization—affects the organization of carotenoids in the lipid bilayer, including orientation (transmembrane vs. parallel) and location (distribution between membrane domains). In spite of the need for these studies, only the orientation of polar and nonpolar carotenoids in the lipid-bilayer membrane has been investigated extensively (Bone and Landrum 1984; van de Ven et al. 1984; Gruszecki and Sielewiesiuk 1990; Sujak et al. 1999; Widomska and Subczynski 2008) (see also Sections 11.5.2; 11.5.3; and 11.5.5). Only two recent papers describe the distribution of carotenoids between domains in membranes made of raft-forming mixtures (Wisniewska and Subczynski 2006a) and in models of the POS membrane (Wisniewska and Subczynski 2006b).

The lipid-bilayer portion of biological membranes is currently depicted not as a passive matrix in which membrane proteins are immersed (like in a Singer–Nicolson model [Singer and Nicolson 1972]) but as an active membrane component that controls a variety of biological functions and processes through selective accumulation or exclusion of certain proteins and lipids from specific membrane domains (Simons and Ikonen 1997; Brown and London 2000). The crucial role in these processes is played by raft domains, which are formed within bulk lipids at certain lipid compositions (Wang et al. 2000; Dietrich et al. 2001; De Almeida et al. 2003) and/or under certain conditions (Kusumi et al. 2004). Using Triton X-100 extraction at 4°C, rafts were isolated as a detergent-resistant fraction (DRM) from POS membranes (Boesze-Battaglia et al. 2002; Martin et al. 2005). The DRM fraction was enriched in saturated lipids and cholesterol, whereas the detergent-soluble fraction (DSM) was rich in docosahexaenoic acid (DHA) (Martin et al. 2005).

The basic lipid composition of POS membranes is very similar to that of the raft-forming mixture, which usually consists of three types of lipids—unsaturated glycerophospholipids, saturated sphingolipids, and cholesterol—mixed at equimolar concentrations (De Almeida et al. 2003). POS disk membranes contain nearly equimolar concentrations of unsaturated phospholipids (mainly, containing a polyunsaturated DHA), saturated phospholipids (myristoyl, palmitoyl, and stearoyl), and cholesterol (Boesze-Battaglia and Schimmel 1997). In a model of POS membranes, consisting of palmitoyl-docosahexaenoylphospatidylcholine/distearoylphosphatidylcholine cholesterol, macular xanthophylls are about 14 times more concentrated in the unsaturated bulk domain (enriched in polyunsaturated DHA and isolated as DSM) (see Figure 11.3b) and excluded from the raft domain (Wisniewska and Subczynski 2006b). A similar distribution was found in membranes made of raft-forming mixtures consisting of dioleoylphosphatidylcholine/sphingomyelin/cholesterol, where xanthophylls were about eight times more concentrated in the bulk domain than in the saturated raft domain (Figure 11.3a) (Wisniewska and Subczynski 2006a). It is worth mentioning that in contrast to xanthophylls, nonpolar β-carotene was distributed more uniformly between domains (Figure 11.3a). Such selective accumulation of macular xanthophylls in domains rich in vulnerable unsaturated lipids seems to be ideal for their antioxidant action (Subczynski et al. 2010). These results are significant in understanding the role of macular xanthophylls in protecting against lipid peroxidation (Wrona et al. 2004; Pintea et al. 2011) in membranes of the eye retina and in preventing AMD (Landrum et al. 1997; Bone et al. 2001).

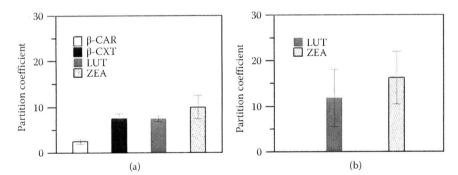

**FIGURE 11.3** Partition coefficient of dipolar xanthophylls, lutein (LUT) and zeaxanthin (ZEA); monopolar xanthophyll, β-cryptoxanthin (β-CXT); and nonpolar carotenoid, β-carotene (β-CAR), between the bulk (unsaturated) and the raft (saturated) domains in (a) the membrane made of raft-forming mixture and (b) the model of photoreceptor outer segment (POS) membranes: the unsaturated domain in the model of POS membranes is abundant in docosahexaenoic acid, which has six double bonds. (From Wisniewska, A., and W. K. Subczynski, *Free Radic. Biol. Med.*, 40, 1820–1826, 2006; Wisniewska, A., and W. K. Subczynski, *Free Radic. Biol. Med.*, 41, 1257–1265, 2006; Subczynski et al., *Arch. Biochem. Biophys.*, 504, 61–66, 2010.)

## 11.5.7 COLOCALIZATION OF LUTEIN IN MEMBRANE DOMAINS TOGETHER WITH UNSATURATED AND POLYUNSATURATED PHOSPHOLIPIDS SIGNIFICANTLY ENHANCING ANTIOXIDANT ACTIVITY

Based on the results presented in Section 11.5.6, it was hypothesized that the accumulation of macular xanthophylls in domains rich in vulnerable unsaturated lipids may enhance the antioxidant properties of these xanthophylls and be an additional mechanism to prevent oxidative damage to the retina. To prove this hypothesis, experiments were performed in which the protective role of lutein against lipid peroxidation in membranes made of raft-forming mixtures and in models of photoreceptor membranes was compared to lutein antioxidant activity in homogeneous membranes composed of unsaturated lipids (Wisniewska-Becker et al. 2012). Lipid peroxidation was induced by photosensitized reactions using rose bengal and monitored by a malondialdehyde–thiobarbituric acid (MDA-TBA) adduct formation test, an iodometric assay, and oxygen consumption. The rates of lipid hydroperoxide accumulation, oxygen consumption, and MDA-TBA adduct accumulation were inhibited in the presence of lutein. However, inhibition was significantly greater in membranes containing raft domains than in homogeneous membranes (Figure 11.4). Thus, the selective accumulation of macular xanthophylls in domains rich in unsaturated lipids significantly enhances their effectiveness as antioxidants in lipid membranes. This suggests that the selective accumulation of macular xanthophylls in the most vulnerable regions of photoreceptor membranes may play an important role in enhancing

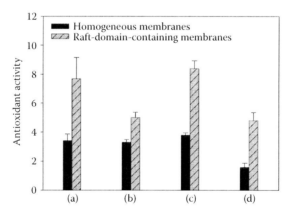

**FIGURE 11.4** Comparison of the antioxidant activity of the macular xanthophyll lutein in raft-domain-containing and homogeneous membranes: antioxidant activity is expressed as (a) a ratio of the rate of lipid hydroperoxide accumulation in membranes in the absence and presence of 0.1 mol% lutein, as a ratio of the oxygen consumption rate in membrane suspension in the absence and presence of (b) 0.3 mol% and (c) 0.5 mol% lutein, and as (d) a ratio of the malondialdehyde–thiobarbituric acid adduct accumulation rate in the absence and presence of 0.5 mol% lutein. Homogeneous membranes were made of dioleoylphosphatidylcholine (DOPC) (a, b, and c) and didocosahexaenoylphosphatidylcholine (DHAPC) (d). Raft-domain-containing membranes were made of DOPC/sphingomyelin/cholesterol equimolar mixture (a, b, and c) and DHAPC/distearoylphosphatidylcholine/cholesterol equimolar mixture (d). (From Wisniewska-Becker et al., *Acta Biochim Pol.* ;59:119–24, 2012.)

their antioxidant properties and ability to prevent age-related macular diseases (such as AMD).

## 11.6 CONCLUDING REMARKS

The retina is abundant in long-chain polyunsaturated phospholipids, which are necessary to maintain normal development and function (SanGiovanni and Chew 2005; Bazan 2007). The entire phototransduction cascade—from rhodopsin activation to the opening and closing of nucleotide-gated channels—has been shown to be modulated by long-chain polyunsaturated phospholipids in the membrane (Mitchell et al. 2007). These phospholipids can be isolated as a DSM fraction, which is formed by the bulk lipid domain surrounding the raft domain in POS disk membranes. Interestingly, rhodopsin (the main protein of POS membranes that is responsible for the first stages of visual signal transduction) is also located in the bulk domain of the POS membranes (Polozova and Litman 2000) and can be isolated mostly with the DSM fraction (Seno et al. 2001; Boesze-Battaglia et al. 2002; Martin et al. 2005). The rhodopsin molecule not only is colocalized in the same membrane domain with polyunsaturated phospholipids (Figure 11.5) but also possesses specific sites that bind the acyl chain of DHA (Soubias and Gawrisch 2005). The retina contains long-chain polyunsaturated phospholipids (C18–C24), as well as very-long-chain polyunsaturated phospholipids (>C24) with 3–9 double bonds (Rezanka 1989; Liu et al. 2010). POS membranes, particularly, are highly enriched in PC-containing very-long-chain polyunsaturated fatty acids (VLCPUFAs). It has been suggested that VLCPUFAs likely play a unique, important role in the retina because they are necessary for cell survival and their loss leads to cell death (Agbaga et al. 2008, 2010). Also, it has been suggested that a tight binding exists between VLCPUFAs and rhodopsin and that the unusually long chains of VLCPUFAs may partially surround the α-helical segments of rhodopsin (Aveldano 1988).

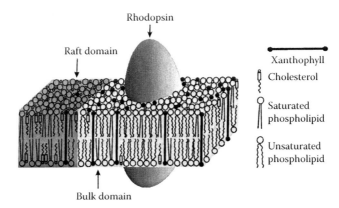

**FIGURE 11.5** Schematic drawing showing the distribution of macular xanthophylls between domains formed in membranes of photoreceptor outer segments: location of macular xanthophylls with vulnerable phospholipids containing polyunsaturated and very long polyunsaturated fatty acids and rhodopsin should significantly enhance protective activity.

Epidemiological studies of long-chain polyunsaturated phospholipid intake suggest a protective role against the incidence of advanced AMD (SanGiovanni and Chew 2005; Parekh et al. 2009). It was recently shown that the amount of lipids containing VLCPUFAs is reduced in the retina of aged eyes and severely reduced in the retina of eyes showing AMD (Liu et al. 2010). The causes of these changes remain to be elucidated. However, the results presented here support the potential value of interventions to increase retinal VLCPUFAs in the prevention and treatment of AMD.

Oxidative damage is postulated to be involved in AMD (Hollyfield et al. 2008). Additionally, phospholipids containing long-chain polyunsaturated fatty acids and VLCPUFAs, which are involved in preventing AMD, are also prone to oxidative damage. Also, colocalization of rhodopsin with polyunsaturated phospholipids (which has its functional purpose) creates a dangerous situation for both, especially during illumination when reactive oxygen species can be produced by photosensitizers. To protect the retina against oxidative damage, nature has used xanthophylls as an effective protector that can absorb damaging blue light, neutralize photosensitizers and reactive oxygen species, and scavenge free radicals (see Figure 11.1). Colocalization of these molecules, together with polyunsaturated phospholipids and rhodopsin (see Figure 11.5), should significantly enhance their effectiveness as protectors, especially when the local concentration of xanthophylls in the membrane is not very high. This is possible due to the domain structure of the POS membrane and the ability of these domains to select and exclude specific classes of lipids and proteins.

As indicated in the previous section, the effective protection of vulnerable molecules (polyunsaturated phospholipids) and processes (phototransduction cascade) that occur in POS membranes is possible by colocalizing them with protective molecules (macular xanthophylls) and processes (antioxidant action). Thus, the membrane domain structure plays a significant role in the protection of retina membranes against oxidative damage. We consider that this function of membrane domains—namely, the colocalization of vulnerable molecules and processes with protective molecules—should be added to currently accepted domain functions such as signal transduction, lipid sorting, and protein trafficking/recycling. Nevertheless, domain structure allows location of macular xanthophylls in the most vulnerable regions of POS membranes, which should significantly enhance their ability to prevent AMD.

## ACKNOWLEDGMENT

This chapter was supported by grants EY015526, EB002052, and EB001980 of the National Institutes of Health.

## REFERENCES

Agbaga, M. P., R. S. Brush, M. N. Mandal, K. Henry, M. H. Elliott, and R. E. Anderson. 2008. Role of Stargardt-3 macular dystrophy protein (ELOVL4) in the biosynthesis of very long chain fatty acids. *Proc. Natl. Acad. Sci. USA*. 105:12843–12848.

Agbaga, M. P., M. N. Mandal, and R. E. Anderson. 2010. Retinal very long-chain PUFAs: new insights from studies on ELOVL4 protein. *J. Lipid Res*. 51:1624–1642.

Aveldano, M. I. 1988. Phospholipid species containing long and very long polyenoic fatty acids remain with rhodopsin after hexane extraction of photoreceptor membranes. *Biochem.* 27:1229–1239.

Bazan, N. G. 2007. Homeostatic regulation of photoreceptor cell integrity: significance of the potent mediator neuroprotectin D1 biosynthesized from docosahexaenoic acid: the Proctor Lecture. *Invest. Ophthalmol. Vis. Sci.* 48:4866–4881; *Biography* 4864–4865.

Bernstein, P. S., F. Khachik, L. S. Carvalho, G. J. Muir, D. Y. Zhao, and N. B. Katz. 2001. Identification and quantitation of carotenoids and their metabolites in the tissues of the human eye. *Exp. Eye Res.* 72:215–223.

Bhosale, P., and P. S. Bernstein. 2005. Synergistic effects of zeaxanthin and its binding protein in the prevention of lipid membrane oxidation. *Biochim. Biophys. Acta.* 1740:116–121.

Bhosale, P., and P. S. Bernstein. 2007. Vertebrate and invertebrate carotenoid-binding proteins. *Arch. Biochem. Biophys.* 458:121–127.

Bhosale, P., A. J. Larson, J. M. Frederick, K. Southwick, C. D. Thulin, and P. S. Bernstein. 2004. Identification and characterization of a Pi isoform of glutathione S-transferase (GSTP1) as a zeaxanthin-binding protein in the macula of the human eye. *J. Biol. Chem.* 47:49447–49454.

Bhosale, P., B. Li, M. Sharifzadeh, W. Gellermann, J. M. Frederick, K. Tsuchida, and P. S. Bernstein. 2009. Purification and partial characterization of a lutein-binding protein from human retina. *Biochem.* 48:4798–4807.

Boesze-Battaglia, K., J. Dispoto, and M. A. Kahoe. 2002. Association of a photoreceptor-specific tetraspanin protein, ROM-1, with triton X-100-resistant membrane rafts from rod outer segment disk membranes. *J. Biol. Chem.* 277:41843–41849.

Boesze-Battaglia, K., and R. Schimmel. 1997. Cell membrane lipid composition and distribution: implications for cell function and lessons learned from photoreceptors and platelets. *J. Exp. Biol.* 200:2927–2936.

Bone, R. A., and J. T. Landrum. 1984. Macular pigment in Henle fiber membranes a model for Haidinger's brushes. *Vision Res.* 24:103–108.

Bone, R. A., J. T. Landrum, and A. Cains. 1992. Optical density spectra of the macular pigment *in vivo* and *in vitro*. *Vision Res.* 32:105–110.

Bone, R. A., J. T. Landrum, L. Fernandez, and S. L. Tarsis. 1988. Analysis of the macular pigment by HPLC: retinal distribution and age study. *Invest. Ophthalmol. Vis. Sci.* 29:843–849.

Bone, R. A., J. T. Landrum, S. T. Mayne, C. M. Gomez, S. E. Tibor, and E. E. Twaroska. 2001. Macular pigment in donor eyes with and without AMD: a case-control study. *Invest. Ophthalmol. Vis. Sci.* 42:235–40. Erratum in: *Invest. Ophthalmol. Vis. Sci.* 42:548.

Bone, R. A., J. T. Landrum, and S. L. Tarsis. 1985. Preliminary identification of the human macular pigment. *Vision Res.* 25:1531–1535.

Britton, G. 1995. Structure and properties of carotenoids in relation to function. *FASEB J.* 9:1551–1558.

Brown, D. A., and E. London. 2000. Structure and function of sphingolipid- and cholesterol-rich membrane rafts. *J. Biol. Chem.* 275:17221–17224.

Cantrell, A., D. J. MacGarvey, T. G. Truscott, F. Rancan, and F. Böhm. 2003. Singlet oxygen quenching in model membrane environment. *Arch. Biochem. Biophys.* 412:47–54.

Cho, E., J. M. Seddon, B. Rosner, W. C. Willett, and S. E. Hankinson. 2004. Prospective study of intake of fruits, vegetables, vitamins, and carotenoids and risk of age-related maculopathy. *Arch. Ophthalmol.* 122:883–892.

Daicker, B., K. Schiedt, J. J. Adnet, and P. Bermond. 1987. Canthaxanthin retinopathy. An investigation by light and electron microscopy and physicochemical analysis. *Graefes Arch. Clin. Exp. Ophthalmol.* 225:189–197.

De Almeida, R. F. M., A. Fedorov, and M. Prieto. 2003. Sphingomyelin/phosphatidylcholine/cholesterol phase diagram: boundaries and composition of lipid rafts. *Biophys. J.* 85:2406–2416.

Dietrich, C., L. A. Bagatolli, Z. N. Volovyk, N. L. Thompson, M. Levi, K. Jacobson, and E. Gratton. 2001. Lipid rafts reconstituted in model membranes. *Biophys. J.* 80:1417–1428.

Gabrielska, J., and W. I. Gruszecki. 1996. Zeaxanthin (dihydroxy-beta-carotene) but not beta-carotene rigidifies lipid membranes: a 1H-NMR study of carotenoid-egg phosphatidylcholine liposomes. *Biochim. Biophys. Acta.* 1285:167–174.

Gale, C. R., N. F. Hall, D. I. W. Phillips, and C. N. Martyn. 2003. Lutein and zeaxanthin status and risk of age-related macular degeneration. *Invest. Ophthalmol. Vis. Sci.* 44:2461–2465.

Goralczyk, R., F. M. Barker, S. Buser, H. Liechti, and J. Bausch. 2000. Dose dependency of canthaxanthin crystals in monkey retina and spatial distribution of its metabolites. *Invest. Ophthalmol. Vis. Sci.* 41:1513–1522.

Gruszecki, W. I. 1999. Carotenoids in membranes. In *The Photochemistry of Carotenoids*, eds. H. A. Frank, A. J. Young, and G. Britton, 363–379. Dordrecht: Kluwer Acad. Publ.

Gruszecki, W. I. 2004. Carotenoid orientation: role in membrane stabilization. In *Carotenoids in Health and Disease*, eds. N. I. Krinsky, S. T. Mayne, and H. Sies, 151–163. New York: Kluwer Marcel Dekker.

Gruszecki, W. I., and J. Sielewiesiuk. 1990. Orientation of xanthophylls in phosphatidylcholine multibilayer. *Biochim. Biophys. Acta.* 1023:405–412.

Gruszecki, W. I., and K. Strzalka. 2005. Carotenoids as modulators of lipid membrane physical properties. *Biochim. Biophys. Acta.* 1740:108–115.

Hammond, Jr., B. R., E. J. Johnson, R. M. Russell, N. I. Krinsky, K. J. Yeum, R. B. Edwards, and D. M. Snodderly. 1997. Dietary modification of human macular pigment density. *Invest. Ophthalmol. Vis. Sci.* 38:1795–1801.

Handelman, G. J., E. A. Dratz, C. C. Reay, and F. J. G. M. van Kuijk. 1988. Carotenoids in the human macula and whole retina. *Invest. Ophthalmol. Vis. Sci.* 29:397–406.

Hollyfield, J. G., V. L. Bonilha, M. E. Rayborn, X. Yang, K. G. Shadrach, L. Lu, R. L. Ufret, R. G. Salomon, and V. L. Perez. 2008. Oxidative damage-induced inflammation initiates age-related macular degeneration. *Nat. Med.* 14:194–198.

Johnson, E. J., M. Neuringer, R. M. Russell, W. Schalch, and D. M. Snodderly. 2005. Nutritional manipulation of primate retinas, III: effects of lutein or zeaxanthin supplementation on adipose tissue and retina of xanthophyll-free monkeys. *Invest. Ophthalmol. Vis. Sci.* 46:692–702.

Kennedy, T. A., and D. C. Liebler. 1992. Peroxyl radical scavenging by beta-carotene in lipid bilayers. Effect of oxygen partial pressure. *J. Biol. Chem.* 267:4658–4663.

Khachik, F., C. J. Spangler, J. C. Smith Jr, L. M. Canfield, A. Steck, and H. Pfander. 1997. Identification, quantification, and relative concentrations of carotenoids and their metabolites in human milk and serum. *Anal. Chem.* 69:1873–1881.

Krinsky, N. I. 2002. Possible biological mechanisms for protective role of xanthophylls. *J. Nutr.* 132:540S–542S.

Krinsky, N. I., J. T. Landrum, and R. A. Bone. 2003. Biological mechanisms of the protective role of lutein and zeaxanthin in the eye. *Annu. Rev. Nutr.* 23:171–201.

Kusumi, A., I. Koyama-Honda, and K. Suzuki. 2004. Molecular dynamics and interactions for creation of stimulation-induced stabilized rafts from small unstable steady-state rafts. *Traffic* 5:213–230.

Landrum, J. T., and R. A. Bone. 2001. Lutein, zeaxanthin, and the macular pigment. *Arch. Biochem. Biophys.* 385:28–40.

Landrum, J. T., R. A. Bone, H. Joa, M. D. Kilburn, L. L. Moore, and K. E. Sprague. 1997. A one year study of the macular pigment: the effect of 140 days of a lutein supplement. *Exp. Eye Res.* 65:57–62.

Landrum, J. T., R. A. Bone, L. L. Moore, and C. M. Gomea. 1999. Analysis of zeaxanthin distribution within individual human retinas. *Methods Enzymol.* 229:457–467.

Liu, A., J. Chang, Y. Lin, Z. Shen, and P. S. Bernstein. 2010. Long-chain and very long-chain polyunsaturated fatty acids in ocular aging and age-related macular degeneration. *J. Lipid Res.* 51: 3217–3229.

Loane, E., J. M. Nolan, O. O'Donovan, P. Bhosale, P. S. Bernstein, and S. Beatty. 2008. Transport and retinal capture of lutein and zeaxanthin with reference to age-related macular degeneration. *Surv. Ophthalmol.* 53:68–81.

Mares, J. A. 2004. Carotenoids and eye disease: epidemiological evidences. In *Carotenoids in Health and Disease*, eds. N. I. Krinsky, S. T. Mayne, and H. Sies, 427–444. New York: Marcel Dekker.

Mares-Perlman, J. A., A. E. Millen, T. L. Ficek, and S. Hankinson. 2002. The body of evidence to support a protective role for lutein and zeaxanthin in delaying chronic disease, overview. *J. Nutr.* 132:518–524.

Martin, R. E., M. H. Elliott, R. S. Brush, and R. E. Anderson. 2005. Detailed characterization of the lipid composition of detergent-resistant membranes from photoreceptor rod outer segment membranes. *Invest. Ophthalmol. Vis. Sci.* 46:1147–1154.

McNulty, H. P., J. Byun, S. F. Lockwood, R. F. Jacob, and R. P. Mason. 2007. Differential effects of carotenoids on lipid peroxidation due to membrane interactions: X-ray diffraction analysis. *Biochim. Biophys. Acta.* 1768:167–174.

Milanowska, J., A. Polit, Z. Wasylewski, and W. I. Gruszecki. 2003. Interaction of isomeric forms of xanthophyll pigment zeaxanthin with dipalmitoylphosphatidylcholine studied in monomolecular layers. *J. Photochem. Photobiol. B.* 72:1–9.

Milon, A., T. Lazrak, A. M. Albrecht, G. Wolff, G. Weill, G. Ourisson, and Y. Nakatani. 1986. Osmotic swelling of unilamellar vesicles by the stopped-flow light scattering method. Influence of vesicle size, solute, temperature, cholesterol and three $\alpha,\omega$-dihydroxycarotenoids. *Biochim. Biophys. Acta.* 859:1–9.

Mitchell, D. C., S. -L. Niu, S. Bennett, L. A. Greeley, K. G. Hines, and B. M. Andersen. 2007. Effects of ROS disk membrane phospholipids with extremely long polyunsaturated acyl chains on visual signalling. *Invest. Ophthalmol. Vis. Sci.* 48:2928.

Moussa, M., E. Gouranton, B. Gleize, C. El Yazidi, I. Niot, P. Besnard, P. Borel, and J. F. Landrier. 2011. CD36 is involved in lycopene and lutein uptake by adipocytes and adipose tissue cultures. *Mol. Nutr. Food Res.* 55:578–584.

Neuringer, M., M. M. Sandstrom, E. J. Johnson, and D. M. Snodderly. 2004. Nutritional manipulation of primate retinas. *Invest. Ophthalmol. Vis. Sci.* 45:3234–3243.

Palozza, P., S. Serini, S. Trombino, L. Lauriola, F. O. Ranelletti, and G. Calviello. 2006. Dual role of beta-carotene in combination with cigarette smoke aqueous extract on the formation of mutagenic lipid peroxidation products in lung membranes: dependence on pO2. *Carcinogenesis* 27:2383–2391.

Parekh, N., R. P. Voland, S. M. Moeller, B. A. Blodi, C. Ritenbaugh, R. J. Chappell, R. B. Wallace, and J. A. Mares. 2009. Association between dietary fat intake and age-related macular degeneration in the Carotenoids in Age-Related Eye Disease Study (CAREDS): an ancillary study of the Women's Health Initiative. *Arch. Ophthalmol.* 127:1483–1493.

Parker, R. S. 1996. Absorption, metabolism and transport of carotenoids. *FASEB J.* 10:542–551.

Pasenkiewicz-Gierula M., K. Baczyński, K. Murzyn, and M. Markiewicz. 2012. Orientation of lutein in a lipid bilayer—revisited. *Acta Biochim. Pol.* 59:115–8.

Pintea, A., D. Rugină, R. Pop, A. Bunea, C. Socaciu, and H. A. Diehl. 2011. Antioxidant effect of *trans*-resveratrol in cultured human retinal pigment epithelial cells. *J. Ocul. Pharmacol. Ther.* 27:315–21.

Polozova, A., and B. J. Litman. 2000. Cholesterol dependent recruitment of di22: 6-PC by a G protein-coupled receptor into lateral domains. *Biophys. J.* 79:2632–2643.

Rapp, L. M., S. S. Maple, and J. H. Choi. 2000. Lutein and zeaxanthin concentrations in rod outer segment membranes from perifoveal and peripheral human retina. *Invest. Ophthalmol. Vis. Sci.* 41:1200–1209.

Rezanka, T. 1989. Very-long-chain fatty acids from the animal and plant kingdoms. *Prog. Lipid Res.* 28:147–187.
SanGiovanni, J. P., and E. Y. Chew. 2005. The role of omega-3 long-chain polyunsaturated fatty acids in health and disease of the retina. *Prog. Retin. Eye Res.* 24:87–138.
Seno, K., M. Kishimoto, M. Abe, Y. Higuchi, M. Mieda, Y. Owada, W. Yoshiyama, H. Liu, and F. Hayashi. 2001. Light-and guanosine 5′-3-O-(thio) triphosphate-sensitive localization of a G protein and its effect on detergent-resistant membrane rafts in rod photoreceptor outer segments. *J. Biol. Chem.* 276:20813–20816.
Shafaa, M. W. I., H. A. Diehl, and C. Socaciu. 2007. The solubilisation pattern of lutein, zeaxanthin, canthaxanthin, and β-carotene differ characteristically in liposomes, liver microsomes and retinal epithelial cells. *Biophys. Chem.* 129:111–119.
Simons, K., and E. Ikonen. 1997. Functional rafts in cell membranes. *Nature* 387:569–72.
Singer, S. J., and G. L. Nicolson. 1972. The fluid mosaic model of the structure of cell membranes. *Science* 175:720–731.
Sommerburg, O. G., W. G. Siems, J. S. Hurst, J. W. Lewis, D. S. Kliger, and F. J. van Kuijk. 1999. Lutein and zeaxanthin are associated with photoreceptors in the human retina. *Curr. Eye Res.* 19:491–495.
Soubias, O., and K. Gawrisch. 2005. Probing specific lipid-protein interaction by saturation transfer difference NMR spectroscopy. *J. Am. Chem. Soc.* 127:13110–13111.
Stahl, W., A. Junghans, B. de Boer, E. S. Driomina, K. Briviba, and H. Sies. 1998. Carotenoid mixtures protect multilamellar liposomes against oxidative damage: synergistic effects of lycopene and lutein. *FEBS Lett.* 427:305–308.
Subczynski, W. K., E. Markowska, W. I. Gruszecki, and J. Sielewiesiuk. 1992. Effects of polar carotenoids on dimyristoylphosphatidylcholine membranes: spin-label studies. *Biochim. Biophys. Acta.* 1105:97–108.
Subczynski, W. K., E. Markowska, and J. Sielewiesiuk. 1991. Effect of polar carotenoids on the oxygen diffusion-concentration product in lipid bilayers. An EPR spin label study. *Biochim. Biophys. Acta.* 1068:68–72.
Subczynski, W. K., E. Markowska, and J. Sielewiesiuk. 1993. Spin-label studies on phosphatidylcholine-polar carotenoid membranes: effects of alkyl chain length and unsaturation. *Biochim. Biophys. Acta.* 1150:173–181.
Subczynski, W. K., J. Widomska, and J. B. Feix. 2009. Physical properties of lipid bilayers from EPR spin labeling and their influence on chemical reactions in a membrane environment. *Free Radic. Biol. Med.* 46:707–718.
Subczynski, W. K., A. Wisniewska, and J. Widomska. 2010. Macular xanthophylls are located in the most vulnerable regions of photoreceptor outer segment membranes. *Arch. Biochem. Biophys.* 504:61–66.
Sujak, A. 2009. Interactions between canthaxanthin and lipid membranes – possible mechanisms of canthaxanthin toxicity. *Cell. Mol. Biol. Lett.* 14:395–410.
Sujak, A., J. Gabrielska, W. Grudzinski, R. Borc, P. Mazurek, and W. I. Gruszecki. 1999. Lutein and zeaxanthin as protectors of lipid membranes against oxidative damage: the structural aspects. *Arch. Biochem. Biophys.* 371:301–307.
Sujak, A., J. Gabrielska, J. Milanowska, P. Mazurek, K. Strzalka, and W. I. Gruszecki. 2005. Studies on canthaxanthin in lipid membranes. *Biochim. Biophys. Acta.* 1712:17–28.
Sujak, A., and W. I. Gruszecki. 2000. Organization of mixed monomolecular layers formed with the xanthophyll pigments lutein or zeaxanthin and dipalmitoylphosphatidylcholine at the argon-water interface. *J. Photochem. Photobiol. B* 59:42–47.
Thomson, L. R., Y. Toyoda, A. Langner, F. C. Delori, K. M. Garnett, N. Craft, C. R. Nichols, K. M. Cheng, and C. K. Dorey. 2002. Elevated retinal zeaxanthin and prevention of light-induced photoreceptor cell death in quail. *Invest. Ophthalmol. Vis. Sci.* 43:3538–3549.

Tserentsoodol, N., N. V. Gordiyenko, I. Pascual, J. W. Lee, S. J. Fliesler, and I. R. Rodriguez. 2006. Intraretinal lipid transport is dependent on high density lipoprotein-like particles and class B scavenger receptors. *Mol. Vision* 12:1319–1333.

van Bennekum, A., M. Werder, S. T. Thuahnai, C. H. Han, et al. 2005. Class B scavenger receptor-mediated intestinal absorption of dietary beta-carotene and cholesterol. *Biochem.* 44:4517–4525.

van de Ven, M., M. Kattenberg, G. van Ginkel, and Y. K. Levine. 1984. Study of the orientational ordering of carotenoids in lipid bilayers by resonance-Raman spectroscopy. *Biophys. J.* 45:1203–1210.

Wang, T. Y., R. Leventis, and J. R. Silvius. 2000. Fluorescence-based evaluation of the partitioning of lipids and lipidated peptides into liquid-ordered lipid microdomains: a model for molecular partitioning into "lipid rafts." *Biophys. J.* 79:919–933.

Widomska, J., A. Kostecka-Gugała, D. Latowski, W. I. Gruszecki, and K. Strzałka. 2009. Calorimetric studies of the effect of *cis*-carotenoids on the thermotropic phase behavior of phosphatidylcholine bilayers. *Biophys. Chem.* 140:108–114.

Widomska, J., and W. K. Subczynski. 2008. Transmembrane localization of *cis*-isomers of zeaxanthin in the host dimyristoylphosphatidylcholine bilayer membrane. *Biochim. Biophys. Acta.* 1778:10–19.

Wisniewska, A., and W. K. Subczynski. 1998. Effects of polar carotenoids on the shape of the hydrophobic barrier of phospholipid bilayers. *Biochim. Biophys. Acta.* 1368:235–246.

Wisniewska, A., and W. K. Subczynski. 2006a. Accumulation of macular xanthophylls in unsaturated membrane domains. *Free Radic. Biol. Med.* 40:1820–1826.

Wisniewska, A., and W. K. Subczynski. 2006b. Distribution of macular xanthophylls between domains in model of photoreceptor outer segment membranes. *Free Radic. Biol. Med.* 41:1257–1265.

Wisniewska, A., J. Widomska, and W. K. Subczynski. 2006. Carotenoid-membrane interactions in liposomes: effect of dipolar, monopolar, and nonpolar carotenoids. *Acta Biochim. Polonica.* 53:475–484.

Wisniewska-Becker, A., G. Nawrocki, M. Duda, and W. K. Subczynski. 2012. Structural aspects of antioxidant activity of lutein in models of photoreceptors membranes. *Acta Biochim. Pol.* 59:119–24.

Woodall, A. A., G. Britton, and M. J. Jackson. 1995. Antioxidant activity of carotenoids in phosphatidylcholine vesicles: chemical and structural considerations. *Biochem. Soc. Trans.* 23:133S.

Woodall, A. A., G. Britton, and M. J. Jackson. 1997a. Carotenoids and protection of phospholipids in solution or in liposomes against oxidation by peroxyl radicals: relationship between carotenoid structure and protective ability. *Biochim. Biophys. Acta.* 1336:575–586.

Woodall, A. A., S. W.-M. Lee, R. J. Weesie, M. J. Jackson, and G. Britton. 1997b. Oxidation of carotenoids by free radicals: relationship between structure and reactivity. *Biochim. Biophys. Acta.* 1336:33–42.

Wooten, B. R., and B. R. Hammond. 2002. Macular pigment: influence on visual acuity and visibility. *Prog. Retin. Eye Res.* 21:225–240.

Wrona, M., M. Różanowska, and T. Sarna. 2004. Zeaxanthin in combination with ascorbic acid or alpha-tocopherol protects ARPE-19 cells against photosensitized peroxidation of lipids. *Free Radic. Biol. Med.* 36:1094–101.

Yin, J.-J., and W. K. Subczynski. 1996. Effect of lutein and cholesterol on alkyl chain bending in lipid bilayers: a pulse electron paramagnetic resonance spin labeling study. *Biophys. J.* 71:832–839.

# 12 Light Distribution on the Retina
## *Implications for Age-Related Macular Degeneration*

*Richard A. Bone, Jorge C. Gibert, and Anirbaan Mukherjee*

### CONTENTS

12.1 Introduction ........................................................................................................ 223
12.2 Materials and Methods ....................................................................................... 226
    12.2.1 Eye Tracker ............................................................................................. 226
    12.2.2 Subjects .................................................................................................. 226
    12.2.3 Experimental Procedures ........................................................................ 227
    12.2.4 Data Analysis .......................................................................................... 227
12.3 Results .................................................................................................................. 229
12.4 Discussion ............................................................................................................ 232
Acknowledgments ........................................................................................................ 233
References .................................................................................................................... 233

### 12.1 INTRODUCTION

Age-related macular degeneration (AMD) derives its name from the central area of the retina, the macula, where the destructive effects of this disease, ultimately its effects on photoreceptors, are the most pronounced. This is despite the presence of several defense mechanisms that are concentrated in the center of the retina, such as the carotenoids that form the *macula lutea* (yellow spot). Although both genetic and environmental risk factors appear to play a role in the etiology of the disease, their immediate effects may be the promotion of, or inability to counteract, oxidative damage and inflammation. The presence of photosensitizers in the retina, such as A2E, together with the relatively intense, short-wavelength, focused light to which the retina is subjected, means that much of the oxidative damage may be photooxidative in nature. Certainly, ambient light exposure in the long term has been associated with an increased risk of AMD in some studies. However, since AMD damage tends to be largely concentrated in the central retina, a legitimate question is whether the long-term light flux on the retina is uniformly distributed on the retina or, rather, tends to be disproportionately higher in the central macula.

To address this question, we considered the possibility of employing eye-tracking technology, essentially to capture a time sequence of retinal images under natural viewing conditions. By analyzing the spatial distribution of light in each image, our goal was to determine the cumulative light distribution on the retina over extended viewing periods. Our working hypothesis was that such cumulative retinal light distributions would exhibit a maximum in the center of the retina. The implication of this finding is that subject's gaze position tends to be attracted toward bright features in the visual field in preference to dark features, just as the sudden movement of an object in our peripheral visual field tends to grab our visual attention. Confirmation of our hypothesis would strengthen the proposal that light represents one of the risk factors for AMD. In addition, it would help us to understand the evolutionary advantage of the presence of macular carotenoids in the central retina, with their blue-light-absorbing and antioxidative properties. The importance of a diet providing an adequate intake of these nutrients is supported by a growing number of epidemiological studies and clinical trials involving AMD.

By 2020, the number of cases of AMD in the United States is expected to be around three million, up by 50% from the current two million sufferers (Friedman, O'Colmain et al. 2004). The worldwide number of cases has been estimated to be on the order of 25 million. Our understanding of AMD is complicated by the fact that there are both genetic and environmental risk factors at play, the latter fortunately largely modifiable by changes in lifestyle. Among the genetic factors, we find race (blacks are significantly less at risk than whites [Chang, Bressler et al. 2008]); a family history of AMD; and the presence of AMD in one eye, which raises the risk of AMD in the other eye. An important modifiable risk factor is smoking (Eye Disease Case-Control Study Group 1992; Chang, Bressler et al. 2008). Light is a probable risk factor according to Fletcher, Bentham et al. (2008) and possibly the most important one according to Chucair, Rotstein et al. (2007). Many studies on animals, reviewed by Young (1988), indicated that radiation damage from diffuse light sources is greatest at the posterior pole of the retina, that is, the location where, in humans, deterioration due to AMD is most pronounced. Another indication of the role of light in AMD comes from the observation that the risk of advanced AMD increases after cataract surgery (Klein, Kerri et al. 2012), and we may conjecture that this is the result of reduced light blocking by the cataractous lens.

Much attention has been given to oxidative damage in retinal tissue as a key contributing factor in the etiology of AMD (Young 1988; Mann 1993; Beatty, Koh et al. 2000). It is possible that certain people may be more susceptible to oxidative damage as a result of a genetic predisposition, because they are smokers, or because of poor diet. In addition, light can initiate photooxidation through the action of photosensitizers (Schalch 1992). Another degenerative process attributable to light is photoreceptor apoptosis. This has been shown to be mediated by the proto-oncogene *c-fos*, the absence of which, at least in mice, results in virtually zero photoreceptor apoptosis (Hafezi, Steinbach et al. 1997). Before photoreceptors can be regenerated, the damaged parts—the outer segments—are engulfed by the retinal pigment epithelium (RPE) cells in a process called phagocytosis and then removed via the choroidal circulation in another process called exocytosis. These processes, however, are not perfect, resulting over time in an accumulation of cellular debris within the RPE cells that manifests itself as lipofuscin and lipid-rich particles called drusen. The appearance of drusen is a hallmark of early maculopathy. If the RPE cell dies, its photoreceptor support function dies with it and photoreceptor degeneration is the inevitable consequence.

A number of epidemiological studies support the hypothesis that light is a contributing factor in the etiology of AMD. In a study of watermen in Chesapeake Bay, Maryland, an attempt was made to analyze 20 years of ocular exposure to sunlight, in particular the blue component. The conclusion was that light represented a probable risk factor (Taylor, West et al. 1992). In the Beaver Dam Eye Study, visible light exposure, based on the amount of time spent outdoors and the use of sunglasses and hats with brims, was again found to be associated with the incidence of AMD (Cruickshanks, Klein et al. 1993). Another study that produces a positive association was carried out on subjects drawn from the population of a small island in the Adriatic (Vojniković, Njirić et al. 2007). The incidence of AMD among 1300 fishermen and farmers was 18%, whereas for town dwellers it was only 2.5%. One would presume that the former would be more exposed to sunlight than their town-dwelling counterparts. The results, however, are not completely equivocal and sometimes mixed, as evidenced by the results of three Australian studies. In one of these, an increased risk of late AMD was found in subjects having blue irises. Light irises, generally, result in higher retinal light exposures due to their lower opacity (Mitchell, Smith et al. 1998). The researchers also considered skin sensitivity to sunlight as an indirect measure of how much time a subject was likely to spend exposed to strong sunlight. Here, the results were mixed, with an increased risk of late AMD appearing to be associated with both high and low skin sensitivity. No such association was found for early maculopathy. Another study muddying the waters indicated that it was the control subjects who had higher annual sunlight exposure (~20% greater) compared with the AMD cases (Darzins, Mitchell et al. 1997). A third study revealed a less than significant increase in the risk of age-related maculopathy in subjects having a higher annual ocular sunlight exposure, either averaged over a lifetime or during the previous 20 years (McCarty, Mukesh et al. 2001). Finally, in a study carried out in France, annual sunlight exposure was estimated from residential history and found not to be significantly associated with an increased risk of AMD (Delcourt, Carrière et al. 2001). Sunglass usage, on the other hand, showed a significant association with reduced risk of soft drusen.

In attempting to draw conclusions from these disparate results, we need to consider the difficulties in estimating a person's retinal exposure to light. If the cumulative retinal light flux, that is, total energy, is the relevant factor regarding the risk of AMD, then estimating that quantity from the amount of time that a person spends outdoors, lifestyle, and so on is fraught with difficulties. For example, glancing for just 1 second at the glint of sunlight reflecting from a shiny surface can expose the macula to roughly the same amount of light energy that it would receive in 1 hour inside a room illuminated with regular interior lighting. The purpose of the present study is to attempt to quantify the cumulative light distribution on the human retina through the use of eye-tracking technology.

## 12.2 MATERIALS AND METHODS

### 12.2.1 EYE TRACKER

An Arrington Research, Inc. (Scottsdale, Arizona), head-mounted eye tracker was used in this study. The hardware comprised two miniature cameras mounted on a lightweight spectacle frame. One of the cameras was a forward-pointing color video "scene camera" capable of capturing the scene in front of the subject at 30 frames per second. The other eye-tracking camera was infrared sensitive and directed toward the pupil of the right eye. A small infrared source was mounted alongside the camera to provide illumination of the exterior of the eye. This camera captured images of the eye at a rate of 60 frames per second. Through the use of image analysis software, and following a calibration routine, the subject's gaze position was automatically detected based on the location of the pupil (darkest feature in the image) and the corneal reflex (brightest feature in the image). The data that were generated consisted of a bitmap for each frame recorded by the scene camera and the $x$–$y$ coordinates, relative to the frame, of the subject's gaze position (on average, two gaze positions per frame). In addition, corresponding to each gaze position, the pupil dimensions were recorded, these being necessary to determine the pupil area, which controls the amount of light entering the eye. Automatic blink and saccade detection and suppression were incorporated in the software. The resolution of the scene camera was set at $320 \times 240$, and its purpose was to act in the capacity of an imaging photometer. The individual pixel values (0–255) provided a semiquantitative measurement of the relative light intensity in the visual field.

Calibration of the eye tracker was necessary to correctly associate the gaze position coordinates with each video frame captured by the scene camera. The procedure was to seat the subject facing a large projection screen. The subject's head movement was minimized through the use of forehead and chin rests and a head-restraining strap. The scene camera video was displayed on a computer monitor together with a two-dimensional array of 16 small, circular calibration targets. The first one to be calibrated was highlighted in color. Also appearing as a small circle was the uncalibrated gaze position. The operator directed a laser pointer toward the projection screen and positioned the laser dot so that on the monitor image it was centered on the calibration target. While the subject maintained visual fixation on the laser spot on the projection screen, the operator recorded the gaze position via a key on the computer keyboard. This resulted in the next circular target being highlighted in color, and the procedure was repeated. After repeating for all 16 targets, the operator checked the accuracy of the calibration by asking the subject to gaze at the laser spot as it was moved randomly around the projection screen. If the gaze position failed to track the laser spot satisfactorily, the software provided the operator with an opportunity to carry out a "slip correction."

### 12.2.2 SUBJECTS

We divided 20 volunteer subjects into a naive group of 15 subjects and an informed group of 5 subjects. The 5 informed subjects were familiar with the study goals and hypotheses. Their purpose was to aid in the development of the research protocols

# Light Distribution on the Retina 227

as well as to demonstrate the feasibility of using eye-tracking technology to quantify the distribution of light on the retina. On the other hand, the 15 naive subjects were uninformed about the study's hypotheses; they knew only that its purpose was to measure the distribution of light on the retina. What had to be avoided was any suggestion that they seek out bright features in the visual field as their preferred gaze positions during a test.

Subjects were provided with an informed consent form approved by Florida International University's institutional review board. The research procedures conformed to the tenets of the Declaration of Helsinki.

### 12.2.3 Experimental Procedures

For the feasibility experiments, the informed subjects, wearing the eye tracker, were seated facing the projection screen and a 10 minute Microsoft PowerPoint slideshow of photographic images was presented to them. Each photographic image was visible for 3 seconds and a 1 second blank screen was interspersed between the images. As soon as an image appeared, the subjects were required to quickly direct their gaze at the brightest feature. During a second experiment with the same slideshow, the subjects were required to direct their gaze at the darkest feature in each image.

The slideshow was also presented to the naive subjects, but for this experiment the only instruction given to them was to simply watch the slideshow. The naive subjects participated in three additional experiments designed to measure cumulative retinal light distribution under different viewing conditions. These were (1) to again sit in front of the projection screen but this time to view a short video presentation; (2) to sit in front of a computer monitor and carry out ordinary computer tasks such as reading and responding to e-mail, visiting web pages, and so on; and (3) free viewing while strolling around the university campus, the laboratory, and surrounding hallways. For these free-viewing experiments, the eye tracker was connected to the computer by a long cable. The computer, powered by a battery backup (uninterruptible power supply), was placed on a small cart and pushed along behind the subject. For outdoor use, it was necessary to shield the eye tracker from direct sunlight, which otherwise might interfere with the eye-tracking capability of the infrared-sensitive eye camera.

### 12.2.4 Data Analysis

The output data from the eye tracker that were required to determine the light distribution on the retina included the scene camera video data in the form of two-dimensional arrays of pixel values, one array for each video frame; the subject's gaze position coordinates, on average two per frame; and the pupil dimensions, on average two sets per frame. We made the assumption that each video frame, acting as an object for the eye's imaging system, resulted in an image on the retina positioned so that the gaze position coincided with the center of the fovea. The relative illuminance distribution across the image was assumed to be the same as the relative luminance distribution for the object, that is, the scene in front of the subject. This is an appropriate assumption provided that the object (and image) subtends reasonably small angles. For large angles, the off-axis incoming rays are presented with an effectively reduced pupil area, resulting in a reduced illuminance of the image. In theoretical analyses, Kooijman (1983) and Pflibsen, Pomerantzeff et al. (1988) calculated the retinal illuminance for Ganzfeld illumination (uniformly illuminated visual field) and found it to be quite uniform from the fovea out to about 25°. This is larger than the maximum visual angle used in all but one of our experiments (see Section 12.3). The pixel values for each frame were multiplied by a factor representing the area of the pupil since from simple optics the brightness of an image is proportional to the area of the entrance pupil. For this exploratory study, we did not include the effects of preretinal, intraocular, short-wavelength filters such as the lens (Van Norren and Vos 1974) or macular pigment (Bone, Landrum et al. 1992).

The procedure for determining the cumulative light distribution on the retina from the set of video frames can be understood if we picture the frames stacked one on top of the other like a deck of cards (see Figure 12.1). The edges of the frames were kept parallel to one another, but the frames were displaced sideways so that the gaze positions were lined up along a vertical line through the stack. The pixel at the gaze position was assigned the coordinates (0,0), and the coordinates of every other pixel were adjusted accordingly. Each frame was converted from color to gray scale and the scene camera's "gamma correction" was removed so that the individual pixel values now represented luminance (relative) in the gaze space. By multiplying the pixel values by a factor representing the area of the pupil for that frame, we obtained the illuminance distribution (relative) on the retina. To obtain the cumulative illuminance at a given point on the retina, we simply added the pixel values in all the frames for the pixels sharing the coordinates of that point.

It is apparent from Figure 12.1 that the process is equivalent to summing pixel values along vertical columns through the stack (assuming that the aforementioned frame adjustments have already been made). It is also apparent that not all columns will contain the same number of pixels. To eliminate this problem, we rejected those video frames

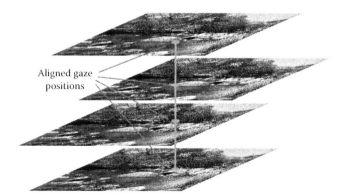

**FIGURE 12.1** (**See color insert.**) Illustration of the technique for the alignment of successive video frames captured by the scene camera and imaged on the retina: the images are stacked with the gaze positions (pink spots) lying one above the other. To obtain the relative cumulative light distribution on the retina, pixel values lying on the same vertical lines, for example, the pink line, are added and the entire light distribution is subsequently normalized to the maximum value.

for which the gaze position fell outside a central rectangular area with dimensions half those of the frame itself. We then added only those columns of pixel values within the rectangular area. These columns now contained the same number of pixels; however, the procedure reduced the area of the retina for which we could determine the cumulative light distribution. The area of the retina thus limited was represented by a 160 × 120 pixel map that was converted to visual angle using the optical characteristics of the scene camera, specifically its angular aperture. Since our technique provides only relative retinal light distributions, the pixel maps were normalized to a peak value of unity.

## 12.3 RESULTS

The experimental procedures and method of data analysis were validated by the initial experiments in which informed subjects, viewing a sequence of photographic images, attempted to hold their gaze on either the brightest or darkest feature in each image. Corresponding results for a typical observer are shown in Figures 12.2 and 12.3, where the vertical axis represents relative retinal illuminance (cumulative) and the horizontal axes specify position on the retina. The peak in Figure 12.2 at the horizontal coordinates (0,0) indicates that the maximum retinal illuminance occurred at the center of the fovea. Likewise, the dip in Figure 12.3 at (0,0) indicates that the minimum retinal illuminance occurred at this location. The uneven surfaces away from the centers represent approximately the averaging of the retinal illuminances produced by each photographic image.

When the naive subjects were presented with the same sequence of photographic images but without any instruction to direct their gaze at bright or dark features, the

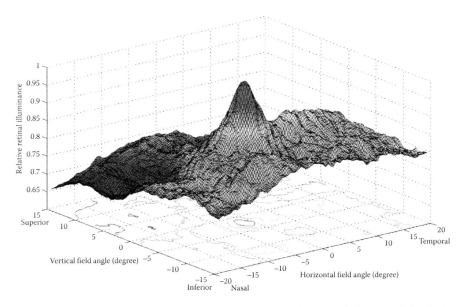

**FIGURE 12.2** (See color insert.) Sample mesh plot showing the relative cumulative light distribution on the retina for one of the informed subjects who was instructed to look at bright objects in a sequence of photographic images.

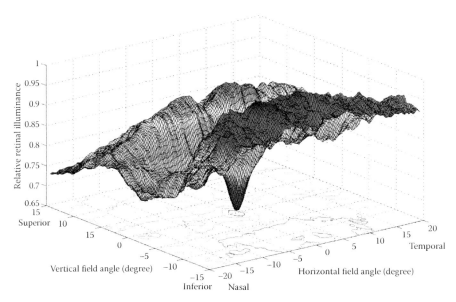

**FIGURE 12.3** (See color insert.) Sample mesh plot showing the relative cumulative light distribution on the retina for one of the informed subjects who was instructed to look at dark objects in a sequence of photographic images.

results were mixed. In Figure 12.4a, for example, a contour plot of cumulative retinal illuminance, there is no central maximum or minimum that would indicate that the subject was devoting more time to viewing bright or dark features, respectively. For some subjects, such as in Figure 12.4b, there was a region of increased retinal illuminance outside the fovea, in this case approximately 5° inferior to the fovea. (We cannot rule out the possibility that camera "slip" occurred after calibration and that this maximum really did represent the retinal illuminance at the fovea.)

In the second series of experiments with the naive subjects, they were provided the opportunity to freely view a short video presentation. The video had been recorded while panning the video camera over various scenes around the university campus. Compared with the results obtained with still images, the results of this experiment revealed a general tendency for increased retinal illuminance in and around the fovea, although the actual maximum was not necessarily at the foveal center. Figure 12.4c for one of the subjects exemplifies this situation.

This was not the situation when the naive subjects' task was to view a computer monitor while performing routine computer tasks (browsing, reading/writing e-mail, etc.). For every subject who participated, we found a very pronounced maximum in cumulative retinal illuminance centered on the fovea (see Figure 12.4d).

The final series of experiments provided what we consider to be the most natural viewing situation. Here, the naive subjects were allowed to walk around (laboratory, hallways, and campus) while wearing the eye tracker. Of the 15 subjects who took part, 2 exhibited retinal illuminance maxima at the fovea (see Figure 12.4e and f). For the remaining 13, we often found increased retinal illuminance in and around the fovea, but the maximum was elsewhere (see Figure 12.4g).

# Light Distribution on the Retina 231

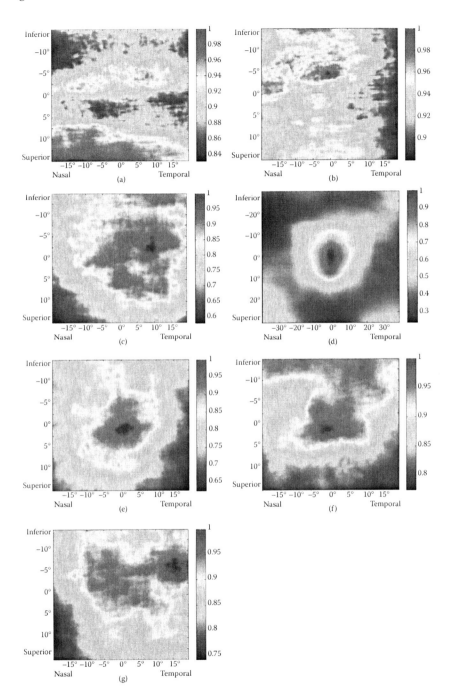

**FIGURE 12.4** (See color insert.) Sample contour plots of the relative cumulative light distribution on the retina for one of the naive subjects under different viewing situations: (a) and (b) viewing a sequence of photographic images; (c) viewing a video presentation; (d) viewing a computer monitor; and (e), (f), and (g) free viewing while walking.

In Figure 12.5a and b, we have presented some of the preceding results in a more quantitative format. Figure 12.5a corresponds to Figure 12.4d and is a plot of average cumulative retinal illuminance as a function of radial distance from the fovea. Figure 12.5b corresponds in a similar way to Figure 12.4e.

## 12.4 DISCUSSION

AMD is a localized disease primarily affecting the central retina, and light is a possible contributing factor. The purpose of this study was to investigate the feasibility of using eye-tracking technology to measure the distribution of light on the retina, averaged over time and under relatively natural viewing conditions. To this end, we assumed a one-to-one relationship between the spatial distribution of light (relative) in the object—the scene being perceived by the subject—and the corresponding distribution of light (relative) in the retinal image. Ideally, the latter can be recorded with an imaging photometer, but these instruments are too large and heavy for use with a head-mounted eye tracker. Instead, we relied on the eye tracker's miniature 8 bit scene camera. This immediately introduces a problem associated with its small dynamic range. The camera assigns a pixel value of 255 to the brightest feature in the scene and 0 to the darkest. The brightness of the scene might range over several orders of magnitude, making it impossible for the scene camera to provide an accurate luminance record. In addition, features of intermediate brightness are assigned pixel values according to the camera's gamma correction, a nonlinear function that results in a compression of brightness at the upper end of the brightness scale and an expansion of brightness at the lower end. In an attempt to remove this nonlinearity, we adjusted each pixel value by means of a function that was the inverse of the gamma correction. However, we recognized that this procedure would not allow us to accurately recover the "true" pixel values from the gamma-corrected ones close to 255.

Consequently, the range of cumulative retinal illuminance depicted in Figures 12.3 through 12.5 will be compressed, probably greatly so, and the rather modest enhancements in light flux that we sometimes observed in the macula may belie a considerably more dramatic situation. For example, when a subject glances, even briefly, at a very bright object the effect on cumulative retinal illuminance will be disproportionately large, yet the scene camera will be unable to give the object its due weight. Thus, although the retinal light distributions that we present should be interpreted as semiquantitative, we may still reach the conclusion that there are activities, such as computer usage, universally producing maximum cumulative illuminance at the center of the macula. Similarly, there may be individuals whose gaze is naturally attracted to bright objects and features in the visual field just as there are others, for example, with a glare disability, with an aversion to bright objects. In cases of higher cumulative retinal illuminance at the center of the macula, the risk of photooxidative damage in this region may be increased along with a higher propensity for AMD.

Such photooxidative damage is believed to be mitigated by the presence in the central macula of the macular carotenoids lutein, zeaxanthin, and *meso*-zeaxanthin, which not only partially screen the vulnerable tissues from blue light but also act in their capacity as antioxidants to quench reactive oxygen species and free radicals (Beatty, Koh et al. 2000). Protection by the macular carotenoids against

# Light Distribution on the Retina

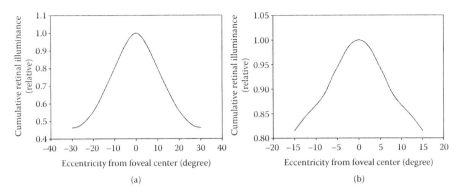

**FIGURE 12.5** Radial plots of the average relative cumulative light distribution on the retina as a function of eccentricity from the fovea: (a) derived from Figure 12.4d—viewing a computer monitor and (b) derived from Figure 12.4e—free viewing while walking.

blue-light-induced damage from 150 μm diameter exposures has been observed in monkey retinas, just as lack of protection has been noted in monkeys raised on a carotenoid-free diet (Barker, Snodderly et al. 2011). (In the latter case, protection was restored after reintroducing carotenoids into the diet.) The density distribution of the macular carotenoids in the retina is such that beyond about 5° from the fovea the protection they afford is likely to be negligibly small (Snodderly, Mares et al. 2004). The broader light distributions seen in Figure 12.5 suggest that the macular carotenoids have not evolved for the purpose of protecting the entire central retina. On the other hand, there is a striking similarity between the spatial distributions of macular carotenoids and cone photopigments (Bone, Brener et al. 2007), suggesting that the cones are the specific target for protection by the carotenoids.

## ACKNOWLEDGMENTS

This chapter was supported by the National Institutes of Health grant SC3GM083671 and by Four Leaf Japan, Ltd., Oska, Japan. Figures 12.1 through 12.5 reproduced from Bone et al. (2012). "Light distributions on the retina: relevance to macular pigment photoprotection." *Acta Biochimica Polonica* **59**: 167–169 by kind permission of *Acta Biochimica Polonica*.

## REFERENCES

Barker, F., D. Snodderly, et al. (2011). "Nutritional manipulation of primate retinas, V: Effects of lutein, zeaxanthin, and n-3 fatty acids on retinal sensitivity to blue-light-induced damage." *Investigative Ophthalmology and Visual Science* **52**: 3934–3942.

Beatty, S., H. H. Koh, et al. (2000). "The role of oxidative stress in the pathogenesis of age-related macular degeneration." *Survey of Ophthalmology* **45**: 115–134.

Bone, R. A., B. Brener, et al. (2007). "Macular pigment, photopigments and melanin: Distributions in young subjects determined by four-wavelength reflectometry." *Vision Research* **47**: 3259–3268.

Bone, R. A., J. T. Landrum, et al. (1992). "Optical density spectra of the macular pigment in vivo and in vitro." *Vision Research* **32**: 105–110.

Chang, M., S. Bressler, et al. (2008). "Racial differences and other risk factors for incidence and progression of age-related macular degeneration: Salisbury Eye Evaluation (SEE) project." *Investigative Ophthalmology and Visual Science* **49**: 2395–2402.

Chucair, A., N. Rotstein, et al. (2007). "Lutein and zeaxanthin protect photoreceptors from apoptosis induce by oxidative stress: Relation with docosahexaenoic acid." *Investigative Ophthalmology and Visual Science* **48**: 5168–5177.

Cruickshanks, K. J., R. Klein, et al. (1993). "Sunlight and age-related macular degeneration." *Archives of Ophthalmology* **111**: 514–518.

Darzins, P., P. Mitchell, et al. (1997). "Sun exposure and age-related macular degeneration: An Australian case-control study." *Ophthalmology* **104**: 770–776.

Delcourt, C., I. Carrière, et al. (2001). "Light exposure and the risk of age-related macular degeneration. The Pathologies Oculaires Liées à l'Age (POLA) Study." *Archives of Ophthalmology* **119**: 1463–1468.

Eye Disease Case-Control Study Group. (1992). "Risk factors for age-related macular degeneration." *Archives of Ophthalmology* **110**: 1701–1708.

Fletcher, A., G. Bentham, et al. (2008). "Sunlight exposure, antioxidants, and age-related macular degeneration." *Archives of Ophthalmology* **126**: 1396–1403.

Friedman, D. S., B. J. O'Colmain, et al. (2004). "Prevalence of age-related macular degeneration in the United States." *Archives of Ophthalmology* **122**: 564–572.

Hafezi, F., J. Steinbach, et al. (1997). "The absence of *c-fos* prevents light-induced apoptotic cell death of photoreceptors in retinal degeneration in vivo." *Nature Medicine* **3**: 346–349.

Klein, B., P. Kerri, et al. (2012). "The relationship of cataract and cataract extraction to age-related macular degeneration: The Beaver Dam Eye Study." *Ophthalmology* **119**: 1628–1633.

Kooijman, A. C. (1983). "Light distribution on the retina of a wide-angle theoretical eye." *Journal of the Optical Society of America* **73**: 1544–1550.

Mann, A. (1993). "Age-related macular degeneration: a review of the effects of light, oxidation and nutrition." *The South African Optometrist* **52**: 113–117.

McCarty, C. A., B. N. Mukesh, et al. (2001). "Risk factors for age-related maculopathy. The visual impairment project." *Archives of Ophthalmology* **119**: 1455–1462.

Mitchell, P., W. Smith, et al. (1998). "Iris color, skin sun sensitivity, and age-related maculopathy: The Blue Mountains Eye Study." *Ophthalmology* **105**: 1359–1363.

Pflibsen, K., O. Pomerantzeff, et al. (1988). "Retinal illuminance using a wide-angle model of the eye." *Journal of the Optical Society of America A* **5**: 145–150.

Schalch, W. (1992). "Carotenoids in the retina—a review of their possible role in preventing or limiting damage caused by light and oxygen." *Free radicals and aging*. I. Emerit and B. Chance. Basel, Birkhauser: 280–298.

Snodderly, D. M., J. A. Mares, et al. (2004). "Macular pigment measurements by heterochromatic flicker photometry in older subjects: The carotenoids and age-related eye disease study." *Investigative Ophthalmology and Visual Science* **45**: 531–538.

Taylor, H., S. West, et al. (1992). "The long-term effects of visible light on the eye." *Archives of Ophthalmology* **110**: 99–104.

Van Norren, D. and J. Vos (1974). "Spectral transmission of the human ocular media." *Vision Research* **14**: 1237–1244.

Vojniković, B., S. Njirić, et al. (2007, Jan). "Ultraviolet sun radiation and incidence of age-related macular degeneration on Croatian island of Rab." *Collegium Antropologicum* **31**(Suppl 1): 43, 44.

Young, R. W. (1988). "Solar radiation and age-related macular degeneration." *Survey of Ophthalmology* **32**: 252–269.

# Index

## A

AAMD, *see* Advanced AMD (AAMD)
ABC, *see* ATP-binding cassette
ABCA1
    immunolocalization of, 107–108
    inhibitor, 108
    transporter gene, 176
*ABCA4* gene, 116
*ABCA4* variants, 116–117
Absorption properties, 109
Actilease technology, 138
Acuity hypothesis, 153–155
Adaptive cellular response, 113–115
Adhesion complex function, 114–115
Adhesion complex proteins, 114
Adipose tissue, 207
Advanced AMD (AAMD), 94
Age-related eye disease study 2 (AREDS2), 171
Age-Related Eye Disease Study (AREDS), 64, 65, 77
Age-Related Eye Disease Study I (AREDS I), 1–2
Age-Related Eye Disease Study II (AREDS II), 2
Age-related macular degeneration (AMD), 1–2, 14, 41, 63–64, 75, 94, 150–153, 156, 171
    advanced stages of, 65
    associated genes encoding factors, 100–102
    *BCO2* with, 112
    description, 223–225, 232–233
    with DNA sequence variation, 98–99
    genes associated with, 46–47
    incidence of, 44–45
    influence of MXs On, *see* Macular xanthophylls (MXs)
    light distribution on retina
        cumulative retinal illuminance, 229–232
        data analysis, 227–229
        eye tracker, 226
        subjects, 226–227
    macular carotenoids, 68
        interventional studies, 78–80
        proof of principle, 77
        trials awaiting completion, 81
    macular xanthophylls protective activity against
        *cis* isomers, 211–212
        high incorporation yield, 208
        indirect antioxidant action of, 210–211
        lutein, 209–210, 214–215
        POS membrane, 212–213
    MP on, 69
    neovascular, 42
    pathogenesis of, 49–52
    prevalence of, 42–44
    protective role of carotenoids in, 77
    risk factors for, 23–24, 45
        age, 24–28
        light, 30–31
        modifiable, 48–49
        nonmodifiable, 45–46
        smoking, 28–29
    risk models, 49
    VLCPUFAs, 216
Age-related macular disease (AMD), 189, 197
Age-related maculopathy (ARM), 31
Age, risk factor for AMD, 24–28
Agilent ChemStation software, 193
AIN-93G-diet-based feeding, 112
Aldehydes, detoxification of, 116
*ALDH3A2* gene, 116
Alginate matrix, starch *vs.*, 138–139
Alginate technology, 138
Alginic acid, 138
Aliphatic aldehydes, 116
AMD, *see* Age-related macular degeneration; Age-related macular disease
AMD-related vision loss, 65
Ames test, 113
Angiogenesis, AMD, 51–52
Antibody staining, 11
Anti-inflammatory effect, 29
Anti-inflammatory response in model systems, 113
Antioxidant-rich diets, 69
Antioxidants, 81
    evidence for relationships with, 64–65
    potential of MXs, 113
    role of MPs, 78
APOA1, *see* Apolipoprotein A1
ApoE, *see* Apolipoprotein E
*APOE* gene, 31, 106
*APOfε2* transgenic animals, 107
Apolipoprotein A1 (APOA1), 109
Apolipoprotein E (ApoE), 106–107, 207
Apoptotic cell death, 108
*Archiv für Ophthalmologie*, 4
AREDS, *see* Age-Related Eye Disease Study
AREDS2, *see* Age-related eye disease study 2
AREDS I, *see* Age-Related Eye Disease Study I
AREDS II, *see* Age-Related Eye Disease Study II

ARM, *see* Age-related maculopathy
*ARMS2* gene, 46, 68
ARPE-19 cells, 176
Astaxanthin, 207
Astrocytes, 106–107
ATP-binding cassette (ABC), 107–109
Autofluorescence, 28
Avian model, 108

## B

Barbados Eye Study, 43
Barrier function, 114–115
BCDO1, *see* Beta-carotene cleavage oxygenase 1 (BCO1)
BCDO2, *see* Beta-carotene cleavage oxygenase2 (BCO2)
BCMO1, *see* Beta-carotene cleavage oxygenase 1 (BCO1)
BCMO2, *see* Beta-carotene cleavage oxygenase2 (BCO2)
BCMO1-binding motifs, 111
BCMO1 gene, 31, 68, 111, 112
BCMO1 variants, 111
BCO1, *see* Beta-carotene cleavage oxygenase 1
BCO2, *see* Beta-carotene cleavage oxygenase 2
BCO2 gene, 112
Beaver Dam Eye Study, 44
Beijing Eye Study, 44
β-cryptoxanthin, 188
β-ionone ring, 188
Beta-carotene carotenoids, 180
Beta-carotene cleavage oxygenase 1 (BCO1), 177
Beta-carotene cleavage oxygenase2 (BCO2), 177
BHT, *see* Butylated hydroxytoluene
*Biochemistry of the Eye, The* (Krause), 8
Biologic plausibility, 95
Blue light exposure in RPE cells, 108
Blue-light hazard, 149–150
Blue light vulnerability, rhesus monkeys, 130–134
BM, *see* Bruch's membrane
Body mass index (BMI), risk factor for AMD, 30
Brain, function within, 161–164
Brightness induction, 158
Bruch's membrane (BM), 50–51, 104
Butylated hydroxytoluene (BHT), 191

## C

CA, *see* Chromatic aberration
Canthaxanthin, 207
Carbenoxolone, 114
CAREDS, *see* Carotenoids in Age-Related Eye Disease Study
CARMA study, *see* Carotenoids in Age-Related Maculopathy study
Carotenoid-binding protein, 172, 173, 178
Carotenoid-protein binding, 109
Carotenoids, 111–112, 177–178, 195
  in AMD, 77
  antioxidant properties of, 208–209
  cleavage enzymes, 175
  with growth factors, 115
  in human blood, 188
Carotenoids in Age-Related Eye Disease Study (CAREDS), 66, 68, 197
Carotenoids in Age-Related Maculopathy (CARMA) study, 78
CD36, *see* Cluster determinant 36
CD36 gene, 31, 105, 177
CDVA, *see* Corrected distance visual acuity
Cell survival, regulation of, 115
Central dip, 84
CFH, *see* Complement Factor H
*CFH* genes, 68
CFH Y402H genotype, 66
Chinese hamster ovary cells, 113
Chiral stationary phases, 190
Choriocapillaris, 50–51, 179
Choroidal neovascularization (CNV) development, 50, 51
Chromatic aberration (CA), 153–154
Chylomicrons
  fraction, 179
  remnants, cellular uptake of, 112–113
  triglyceride-rich components of, 106
*Cis* isomers of macular xanthophylls, 211–212
*Cis-trans* isomerase, 109, 174
Cluster determinant 36 (CD36), 105
CNV, *see* Choroidal neovascularization
Collaborative Optical Macular Pigment Assessment Study (COMPASS), 83
Collagen fibers, abnormal distribution of, 104
Color perception, 6–8
COMPASS, *see* Collaborative Optical Macular Pigment Assessment Study
Complement Component 3 (C3), 45
Complement factor H (CFH), 45, 46
Cone rod dystrophy 3, 116
Contrast engine, 159
Contrast enhancement, eye function, 159
Conventional chromatographic methods, 190
Corrected distance visual acuity (CDVA), 78
COX-2, *see* Cyclooxygenase-2
C-terminal StARD Domain, 174
Curve fitting, 11
Cyclooxygenase-2 (COX-2), 105

# Index

## D

Detergent-resistant fraction (DRM), 212–213
Dietary antioxidants, 81
Dietary carotenoids, 95, 113, 179
Dietary intake
  of Lutein and Zeaxanthin, 194–195
  plasma carotenoid concentrations and, 194–195
Dietary lutein, 65–68
Dietary zeaxanthin, 65–68
Diet, risk factor for AMD, 29
Dioxygenase (DO) cleavage mechanism, 177
Dipalmitoylphosphatidylcholine (DPPC) membrane, 211
Dipolar carotenoid, 207
Dipolar xanthophylls, partition coefficient of, 213
Discomfort glare, 156–157
Dizygotic twins, 96
Dose–response relationship, 28
DPPC membrane, see Dipalmitoylphosphatidylcholine membrane
DRM, see Detergent-resistant fraction
*Drosophila* mutant, 104, 176
Duplex vision, 160

## E

Early AMD, prevalence of, 42
Egg yolks, 195
Elderly, vision loss in, 94
Ensembl Project, 99
Entoptical phenomena, 6–8
Epidemiological studies, 64
Esterified *vs.* nonesterified lutein, 134–138
EUREYE Study, see European Eye Study
EurEye Study, 67
European Eye (EUREYE) Study, 42
Exocytosis, 224–225
Eye Disease Case-Control Study, 95
Eyes
  functions within, 148–149
    Acuity hypothesis, 153–155
    Glare hypothesis, 155–157
    macular pigment and mesopic acuity, 159–161
    Protection hypothesis, 149–153
    Visibility hypothesis, 158–159
  macular pigment, see also Macular pigment
    description, 1–2
    in developing eye, 12–13
    entoptical phenomena, color perception and, 6–8
    measurements of, 8–12
    ophthalmoscopic tools development, 4–6
    reports of, 2–4

Eye tracker, 226
  free-viewing experiments, 227
Eye-tracking technology, 224

## F

FANCC–GSTP1 interactions, 110
FANCC, protein-coding regions of, 110
Food, macular carotenoids in, 194
Foveal sparing, 95
Fovea *vs.* parafovea, 131, 133
Free-viewing experiments, 227
Functional Acuity Analyzer™, 83

## G

GA, see Geographic atrophy
Gamma correction, 228
Ganzfeld illumination, retinal illuminance for, 228
GEE analysis, see Generalized Estimating Equations analysis
Gene associated with AMD, 46–47
Gene chips, 108–110
  discovery and replication, 105
Gene–disease analysis, 97
Generalized Estimating Equations (GEE) analysis, 133
Genetics
  AMD, 45, 49
  risk factor for AMD, 31
Genome-wide association (GWA) studies, 98, 111
Geographic atrophy (GA), 42, 52, 94
GJA1 protein, 114
Glare effects, macular pigment, 158
Glare hypothesis, 155–157
Glutathione S-transferase P1 (GSTP1), 68, 173–174
  protein, 109–110
Glutathione S-transferases (GSTs), 173
Gold standard clinical trial, 76
Green leafy vegetables, 195
GSTA1, 173
*GST* gene, 174
GSTM1, 173
GSTP1, see Glutathione S-transferase P1
*GSTP1* gene, 174
GSTs, see Glutathione S-transferases
GWA studies, see Genome-wide association studies

## H

HA, see Vernier hyperacuity
Haidinger's brushes, 6–7
Handan Eye Study, 44

HapMap cohort, 116
Hardy–Weinberg equilibrium (HWE), 97
HDL-based intraretinal lipid transport mechanism, 207
HDLs, see High-density lipoproteins
Hemodynamic model, 51
Henle fiber layer, 171
Hepatic lipase, 66
Heterochromatic flicker photometry (HFP), 8–9, 24
Heterozygosity, SRB1, 68
HFP, see Heterochromatic flicker photometry
High-density lipoproteins (HDLs), 30, 96, 175–176, 207
  peripheral tissue to, 108
High-performance liquid chromatography (HPLC), 28, 135, 190
  reversed-phase, 136, 140
High-pressure liquid chromatography (HPLC), 189, 191
Human cytosolic GSTs, 173
Human macular membranes, 173
Human plasma responses, 141
Human RPE cell cultures, 108
HWE, see Hardy–Weinberg equilibrium
Hydrophobic carotenoids, 180
Hypermetropia, 48

## I

IGF signaling system, see Insulin-like growth factor signaling system
Imaging techniques, macular pigment measurement, 11–12
Immunohistochemical localization, 104
Immunolocalization study, 114
Ingenuity® Knowledge Base, 110
Insulin-like growth factor (IGF) signaling system, 115
Intercellular signaling within RPE, 114
International HapMap Project, 99
Interphotoreceptor retinoid–binding protein (IRBP), 175
Intraocular lenses (IOLs), 31
Intraocular scatter, 156, 157
*In vitro* analysis, 87
*In vivo* MP spectrum, 174
IOLs, see Intraocular lenses
IRBP, see Interphotoreceptor retinoid–binding protein
Irish Longitudinal Study of Ageing (TILDA), 87–88
Isocratic chromatography, 192

## J

Japanese cohorts, 105

## L

Laminae, retinal cell types and, 98
Large-scale genotyping projects, 98
LAST study, see Lutein Antioxidant Supplementation Trial study
Late AMD, prevalence of, 43–44
Lateral chromatic aberration, 153–154
LDLs, see Low-density lipoproteins
Light, risk factor for AMD, 30–31
Linear trapezoidal rule, 140
*LIPC* gene, 31, 66
Lipid-bilayer membrane, 206, 209–212
Lipid metabolism, 106
Lipid peroxidation, 214
Lipid polyene chains, alcohol metabolism and peroxidation of, 116
Lipofuscin retina, accumulation of, 152
Lipoprotein Lipase (LPL), 112–113
Liquid chromatography methods, 9
Liver X receptor (LXR) agonists, 108
Longitudinal chromatic aberration, 153
Low-density lipoproteins (LDLs), 175–176, 207
LPL, see Lipoprotein Lipase
*LPL* gene, 112–113
Lutein
  in adipose tissue of chicken, 112
  baseline concentration, 140
  BCO2, 177
  binding protein for, 16
  carotenoid-binding proteins, 172
  carotenoids, 171
    hydrophilic xanthophyll, 180
  colocalization of, 214–215
  dietary carotenoids, 95
  dietary supplementation with, 99
  esterified *vs.* nonesterified, 134–135
  esters, 105
  fractions, 190
  human diet and serum, 9–10
  human macula, 175
  kinetic parameters comparison, 142
  macular carotenoids, 176
  metabolism, steroid hormones affecting, 30
  multifunctionality of, 164
  in neural retina, 107
  physiological activity of, 161–162
  plasma response, 135–137
  proportions of, 13
  in retina membranes, orientations of, 209–210
  StARD3, 174
  xanthophyll carotenoids, 173
  and zeaxanthin, 95
    isomer fraction, separation of, 191–192
Lutein Antioxidant Supplementation Trial (LAST) study, 78

# Index

Lutein-binding protein, 174
Lutein-rich diets, 69, 108
LUXEA II trial, 160, 161
LXR agonists, *see* Liver X receptor (LXR) agonists
Lycopene carotenoids, 180
Lypolysis, 113

## M

Macaque retina, 15
*Macula lutea, see* Macular pigment
Macular carotenoids, 1, 7, 9, 15, 171, 189
    clinical trials, 76–77
        in normal subjects, 81–84
        observational studies, 84–88
        in subjects with AMD, 77–81
    distribution of, 11–12
    in food analysis, 194
Macular degeneration, age-related, *see* Age-related macular degeneration
Macular ganglion cell layer, 108
Macular membrane stabilization, 109–110
Macular pathology, 80
Macular pigment (MP), 23–24, 159–161, 171, 173, 223
    absorption characteristics of, 153
    age and, 24–28
    age-related decline in, 84
    antioxidant role of, 78
    augmentation and changes, 87
    carotenoids binding proteins, 173–175
        GSTP1, 173–174
        StARD3, 174–175
    description, 1–2
    in developing eye, 12–13
    dichroic properties of, 81
    entoptical phenomena, color perception and, 6–8
    light and, 30–31
    measurements of
        early to mid-twentieth century, 8–9
        late-twentieth century through the present, 9–12
    ophthalmoscopic tools development, 4–6
    properties of, 75
    reports of, 2–4
    smoking and, 28–29
    spatial vision, CA effects, 154–155
    uptake and transport
        pathways, 177–180
        proteins, 175–177
    visual disability and discomfort, 156–157
Macular pigment optical density (MPOD), 68–69, 156, 177, 197
    age and, 24–28
    macular carotenoids
        formulations on, 80
        supplementation on, 75
    in MOST, 80
    responses, 131, 132, 137, 143
    spatial profile, 84
    and visual performance, 83
Macular xanthophylls (MXs), 94
    accumulation of, 206–207
    against AMD, protective activity of
        *cis* isomers, 211–212
        high incorporation yield, 208
        indirect antioxidant action of, 210–211
        lutein, 209–210, 214–215
        POS membrane, 212–213
    binding, transport, and accretion of, 99, 104
    characteristics, 205–206
    cleavage of, 110–111
    distribution of, 204
    in DNA preservation and repair, 113
    hepatic accumulation of, 108
    hypothesis, 204–205
    macular pigment and concentration of, 95
    mechanisms and metabolic fate of, 95
    molecular genetics for examining, AMD
        with DNA sequence variation, 98
        putative influence, 96–98
    putative role of, 96
    retinal localization of, 103
    6-carbon ring structure of, 114
    stabilization, 207–208
Macula sparing, 14
Malay Eye Study, 44
Malondialdehyde–thiobarbituric acid (MDA-TBA), 214
Marigold lutein, 134
Maxwell's spot, 7
MDA-TBA, *see* Malondialdehyde–thiobarbituric acid
Membrane solubility, macular xanthophylls, 205
Mesopic acuity, 159–161
Mesopic contrast sensitivity, 83
Mesopic glare disability, 83
Meso-zeaxanthin (MZ), 80, 95, 171, 189, 191, 192, , 204
Meso-zeaxanthin Ocular Supplementation Trial (MOST), 80
    vision, 83
Metabolic enzymes, 177
Microsoft PowerPoint slideshow, AMD, 227
Microtubules, 106
MLN64, *see* Steroidogenic acute regulatory domain protein 3 (StARD3)
Modifiable risk factors for AMD, 48–49
Monooxygenase (MO) cleavage mechanism, 177
Monopolar carotenoids, 210
Monozygotic twins, 96

MOST, *see Meso*-zeaxanthin Ocular Supplementation Trial
MP, *see* Macular pigment
MPOD, *see* Macular pigment optical density
Müller cells, astrocytes and, 106–107
Multiethnic Study of Atherosclerosis, 44
MX–GSTP1 complex, 109
MXs, *see* Macular xanthophylls
MX–tubulin interactions, 106
MZ, *see* Meso-zeaxanthin

## N

Narrow type distribution, macular carotenoids, 12
Neovascular AMD, 42
Neural efficiency hypothesis, 161–164
Neural retinal cells, RPE and, 115
Neuronal nitric oxide synthsase (nNOS), 153
N-glycosylated, 176
*NinaD* gene, 176
nNOS, *see* Neuronal nitric oxide synthsase
Nonesterified lutein
   esterified *vs.*, 134–138
      plasma responses comparison to, 138
         design of study, 139–140
         results of study, 140–143
         starch *vs.* alginate matrix, 138–139
Nonheme iron monooxygenases, 110–111
Nonmodifiable risk factors for AMD, 45–47
Nonpolar carotenoids, 205, 207, 210
   orientation of, 208–209, 212
N-terminal MENTAL domain, 174
Nuclear hormone receptors, 94

## O

Ophthalmoscope, 4, 8, 11
Oregon National Primate Research Center, 130
Oxidative stress, AMD, 50

## P

Paired t-tests, 140
Parafovea, fovea *vs.*, 131, 133
Pathogenesis of AMD
   angiogenesis, 51–52
   genetics, 49
   hemodynamic changes, 51
   hydrodynamic changes, 50–51
   subclinical inflammation, 50
Pathologies Oculaires Liées à (POLA) Study, 44
Pathways, macular pigment, 179–180
Permutation test, 97
Peroxisome proliferatoractivated receptor γ (PPARG), 108

Phagocytosis, 224–225
Phenotype–genotype relationship, 97
Photophobia, 156–157
Photoreceptor cells, 179
Photoreceptor outer segment (POS) membrane, 204, 212–213, 216
Photosensitizers in retina, 223
Photostress recovery, macular pigment and, 156
Placebo-controlled clinical trial, 111
Plasma carotenoid concentrations, 194–195
   subclinical inflammation on, 197
Plasma concentrations
   lutein and zeaxanthin, 140–141
   of xanthophyll, 132, 196
Plasma kinetics study, 139
Plasma lipoproteins, 175–176
Plasma response
   esterified *vs.* nonesterified lutein, 134–138
   supplementation with xanthophylls, 131
Plasma retinol, 111
Plethora, MP and, 81
Plexiform layers, 115–116
Polar carotenoids, 207
   orientation of, 208–209, 212
POLA Study, *see* Pathologies Oculaires Liées à Study
Population-based studies, prevalence of AMD, 42–44
POS membrane, *see* Photoreceptor outer segment membrane
Postmortem, eyes
   macular pigment in, 5, 6
   odds ratios for AMD, 14
PPARG, *see* Peroxisome proliferatoractivated receptor γ
Prevalence of AMD, 42–44
Primary human cultures, 104
Protection hypothesis, 149–153
Protein–protein interaction, 109

## R

RA, *see* Resolution acuity
Raman spectroscopy, 28
Randomized controlled trial (RCT), 76
RARB, *see* Retinoic acid receptor beta
RARs, *see* Retinoid A receptors
RBP4, *see* Retinol binding protein 4
RCT, *see* Randomized controlled trial
Reactive oxygen species (ROS), 14–15, 48, 50
"Red-eye" artifact, 5
Reflectance spectroscopy, 28
Replication cohort chip, 108–109, 117
Resolution acuity (RA), 154

# Index

Resonance Raman spectroscopy, 11–12
Retina
　function within, 161–164
　light distribution on
　　cumulative retinal illuminance, 229–232
　　data analysis, 227–229
　　eye tracker, 226
　　subjects, 226–227
　lipofuscin accumulation, 152
　macular xanthophylls
　　accumulation of, 206–207
　　characteristics, 205–206
　　distribution of, 204
　　stabilization, 207–208
　sight-threatening disease of, 95
　xanthophylls role in, 13–16
Retina circulation, 178–179
Retinal cell types, 98
Retinal ganglion cells, 104
　electrophysiologic properties of, 114
Retinal pigment epithelium (RPE) cells, 5, 41, 50, 51, 149, 224–225
　free and esterified cholesterol in, 108
Retinal pigment epithelium–choroid (RPE–choroid), 173
Retinal response, macular carotenoids, 87
Retina membranes
　cis isomers in, 211–213
　lutein orientations, 209–210
Retinex algorithms, 159
Retinoic acid receptor beta (RARB), 94
Retinoid A receptors (RARs), 108
Retinoid X receptor (RXR), 108
Retinol binding protein 4 (RBP4), 175
Reverse cholesterol transport, 175
Reversed-phase HPLC, 136, 140
Reykjavik Study, 44
Rhesus monkey, experiments in, 130–134
Rhodopsin, colocalization of, 216
Risk models of AMD, 49
Rod outer segments, components of, 177
ROS, see Reactive oxygen species
Rotterdam Eye Study, 66
Rotterdam Study, 44
RPE cells, see Retinal pigment epithelium cells
RPE–choroid, see Retinal pigment epithelium–choroid
RPE65 gene, 31
RXR, see Retinoid X receptor

## S

Saponification process, 138
*SCARB1* gene, 31
Scavenger receptor class B member 1 (SR-B1), 176–177
Scavenger receptor class B type I (SR-BI), 104–105
Scene camera, 226
　gamma correction, 228
Serum carotenoid concentrations, interpretation of
　dietary intake
　　of Lutein and Zeaxanthin, 194–195
　　plasma carotenoid concentrations and, 194–195
　Lutein and Zeaxanthin in diet, seasonal factors and sources of, 195
　sex, 195–196
　smoking and subclinical inflammation, 196–197
Serum lutein levels, 177
Serum, macular carotenoids, 87
Serum/plasma extraction, 191
Sex, 195–196
　risk factor for AMD, 30
S-glutathionylation, 174
Shaded cells, 99
Shihpay study, 44
Short-wave (SW) energy, 153
SHRs, see Spontaneous hypertensive rats
Single-nucleotide polymorphism (SNP), 31, 45
Sjögren–Larsson syndrome (SLM), 115–116
SLM, see Sjögren–Larsson syndrome
Smoking, 196–197
　and AMD, 48
　risk factor for AMD, 28–29
Smoking-adjusted pooled analysis, 107
SNP, see Single-nucleotide polymorphism
Spatial vision, chromatic aberration role, 154–155
Spontaneous hypertensive rats (SHRs), 105
SR-B1, see Scavenger receptor class B member 1; Scavenger receptor class B type I
Staining, 114
StARD3, see Steroidogenic acute regulatory domain protein 3
StAR domain, 175
Stargardt disease, 116–117
Stereoisomers, 189
Steroid hormones, affecting lutein metabolism, 30
Steroidogenic acute regulatory domain protein 3 (StARD3), 68, 110, 173, 174–175
Stroop effect, 162
Subclinical inflammation, pathogenesis of AMD, 50
Subretinal deposits
　deposition and clearance of, 105
　formation of, 105
Sun light, risk factor for AMD, 30–31
Supplemental dietary antioxidants, 77
SW energy, see Short-wave energy

## T

Tangiers disease, 68
TC7 Caco-2 cells, 111
Thailand National Survey of Visual Impairment, 44
Third National Health and Nutrition Examination Survey, 197
TILDA, see Irish Longitudinal Study of Ageing
Transmembrane orientation, antioxidant properties of carotenoids, 208–209
Transmembrane protein, 68
TUBGCP3, see Tubulin gamma complex associated protein 3
Tubulin, 106, 173
Tubulin gamma complex associated protein 3 (TUBGCP3), 94

## U

Ultracarb ODS column, 193
United States, AMD in, 224
Unpaired t-tests, 140
Unusual macular pigment distribution, 12

## V

Vascular choroid, 5
Vascular endothelial growth factor (VEGF), 51, 104
VEGF, see Vascular endothelial growth factor
Vernier hyperacuity, 154
Very-long-chain polyunsaturated fatty acids (VLCPUFAs), 215–216
Very low-density lipoprotein (VLDL), 106
Visibility hypothesis, 158–159
Vision loss in elderly, 94
Visual perception, 163
Visual psychophysical techniques, 8–9
Visual system
  eye functions, 148–149
    acuity hypothesis, 153–155
    glare hypothesis, 155–157
    macular pigment and mesopic acuity, 159–161
    protection hypothesis, 149–153
    visibility hypothesis, 158–159
  lutein and zeaxanthin, 147–148
  retina and brain function, 161–164

VLCPUFAs, see Very-long-chain polyunsaturated fatty acids
VLDL, see Very low-density lipoprotein

## W

Wavelength-dependent scattering, 158
WHAM, see Wisconsin hypoalpha mutant
White matter, 163
Wisconsin hypoalpha mutant (WHAM), 108, 176

## X

Xanthophyll-binding proteins, 204–206, 209
Xanthophyll carotenoids, 173
Xanthophyll concentrations, calculation of, 193
Xanthophyll-free status, of rhesus monkeys, 130–131
Xanthophyll–membrane interactions
  description, 204–205
  macular xanthophylls protective activity
    cis isomers, 211–212
    high incorporation yield, 208
    indirect antioxidant action of, 210–211
    lutein, 209–210, 214–215
    POS membrane, 212–213
Xanthophylls, 129–132, 188, 189
  role in retina, 13–16

## Y

Yellow filters, 155, 157–161

## Z

Zeaxanthin, 9–11, 147–148, 189, 191, 204
  accumulation of, 105
  fractions, 190
  isomers, 190
  isomers effects, 211–212
  levels, 107
  multifunctionality of, 164
  orientation of, 209
  physiological activity of, 161–162
  plasma concentrations, 141
  protective effect of, 150
  stereoisomers, 189, 190, 192
Zeaxanthin-binding protein, 174